DESIGN and
ANALYSIS
of
ECOLOGICAL
EXPERIMENTS

DESIGN and ANALYSIS of ECOLOGICAL EXPERIMENTS

Editors

Samuel M. Scheiner
Department of Biological Sciences
Northern Illinois University

and

Jessica Gurevitch
Department of Ecology and Evolution
State University of New York–Stony Brook

CHAPMAN & HALL
New York • London

First published in 1993 by
Chapman & Hall
One Penn Plaza
New York, NY 10119

Published in Great Britain by
Chapman & Hall
2-6 Boundary Row
London SE1 8HN

Printed in the United States of America on acid-free paper

Library of Congress Cataloging in Publication Data

Design and analysis of ecological experiments / [edited by] Samuel
    M. Scheiner and Jessica Gurevitch.
        p.   cm.
    Includes bibliographical references (p.   ) and indexes.
    ISBN 0-412-03551-0. — ISBN 0-412-03561-8 (pbk.)
    1. Ecology—Statistical methods.   2. Experimental design.
    I. Scheiner, Samuel M., 1956–   .   II. Gurevitch, Jessica.
    QH541.15.S72D47   1993
    574.5′0724—dc20   96-36187
                                                                CIP

British Library Cataloguing in Publication Data available on request

Dedication

SMS
For Edith, Judy, and Kayla and the memory of my father Sayre

JG
Dedicated with much appreciation to my father, Louis Gurevitch, and in loving memory of my mother, Esther Gurevitch

Contents

Preface   ix
Acknowledgments   xi
Contributors   xiii

1   Introduction: Theories, hypotheses, and statistics   1
     Samuel M. Scheiner
2   Exploratory data analysis and graphic display   14
     Aaron M. Ellison
3   ANOVA: Experiments in controlled environments   46
     Catherine Potvin
4   ANOVA and ANCOVA: Field competition experiments   69
     Deborah E. Goldberg and Samuel M. Scheiner
5   MANOVA: Multiple response variables and multispecies interactions   94
     Samuel M. Scheiner
6   Repeated-measures analysis: Growth and other time-dependent measures   113
     Carl N. von Ende
7   Time-series intervention analysis: Unreplicated large-scale experiments   138
     Paul W. Rasmussen, Dennis M. Heisey, Erik V. Nordheim, and
     Thomas M. Frost
8   Nonlinear curve fitting: Predation and functional response curves   159
     Steven A. Juliano
9   Multiple regression: Herbivory   183
     Thomas E. Philippi
10   Path analysis: Pollination   211
     Randall J. Mitchell
11   Population sampling and bootstrapping in complex designs: 232
     Demographic analysis
     Mark A. McPeek and Susan Kalisz

12 Failure-time analysis: Emergence, flowering, survivorship, and other   253
    waiting times
      Gordon A. Fox
13 The bootstrap and the jackknife: Describing the precision of ecological   290
    indices
      Philip M. Dixon
14 Spatial statistics: Analysis of field experiments   319
      Jay M. Ver Hoef and Noel Cressie
15 Mantel tests: Spatial structure in field experiments   342
      Marie-Josée Fortin and Jessica Gurevitch
16 Model verification: Optimal foraging theory   360
      Joel S. Brown
17 Meta-analysis: Combining the results of independent experiments   378
      Jessica Gurevitch and Larry V. Hedges

    References   399
    Author index   427
    Subject index   435

Preface

The genesis of this book was the result of a conversation while eating a bento (a Japanese box lunch of rice, sushi, pickles, and other delicacies) on a sidewalk in Yokohama in 1990. The conversation had turned to statistics and both of us were commenting on statistical issues and techniques that were either underused or misused by ecologists. For some time, Jessica had been contemplating a book on statistical techniques in experimental design and analysis, written for ecologists. The goal of such a book was both to encourage the correct use of some of the more well known approaches, and to make some potentially very useful but less well known techniques available to ecologists. We both felt strongly that such a book was timely, and would be useful to ecologists working on applied as well as basic problems. Had Sam not intervened, this idea would undoubtedly have met the fate of many other fine ideas that have never made it into the light of day.

It was apparent to both of us that we did not have the skills to write such a book ourselves, nor the time. However, we were able to compile a list of topics and potential contributors whom we knew were knowledgeable about those topics. (An inital outline for the book was composed while sitting for hours, stalled in traffic going to and from Mount Fuji. At the next INTECOL meeting keep your eyes out for "I survived the trip to Mt. Fuji" t-shirts.) We batted (actually e-mailed) these ideas back and forth for nearly a year. The success of a symposium organized by Phil Dixon at the 1991 annual meeting of the Ecological Society of America on the design of ecological experiments encouraged us to continue our endeavors, and we managed to secure commitments from many of our contributors during that week. The enthusiasm for this undertaking expressed by colleagues we spoke with buoyed us, as did the interest and encouragement of Greg

Payne from Chapman and Hall. Therefore, despite warnings about the travails of editing a book, we forged ahead. So—beware of the dangers of conversation over raw fish.

Samuel M. Scheiner
Jessica Gurevitch

Acknowledgments

We wish to thank a number of people who contributed to the production of this volume. We are grateful to those colleagues who painstakingly reviewed chapters: Norma Fowler, Charles Goodnight, Ed Heske, Charles Janson, Martin Lechowicz, Jim McGraw, Tom Meagher, Tom Mitchell-Olds, Art Weis, and Neil Willits. We thank Aline Click for art/photo production and Andrea Meyer for the cover art. We offer a special thanks to Greg Payne for his continued enthusiasm and his efforts in seeing this book through to production. We must also acknowledge INTERNET, without which this book could not have been produced in so timely a fashion. A fellowship from the Arnold Arboretum of Harvard University, responsible for sending J.G. to Japan and thus for the inception of this book, is gratefully acknowledged. S.M.S. was supported for his trip to Japan by funds from the Department of Biological Sciences, the College of Liberal Arts & Sciences and the Graduate School of Northern Illinois University and the City of Yokohama; the Department of Biological Sciences provided support during the editorial process. S.M.S. thanks Judy Scheiner for all of her patience and forbearance during the production of this book. J.G. extends many thanks to Todd Postol for being in her corner all of these years.

With much appreciation, we acknowledge the contribution of Dr. Bryan F. J. Manly, University of Otego, who reviewed the entire manuscript of the book. His careful comments have increased its utility and accuracy. All responsibility for errors and omissions are, of course, our own.

List of Contributors

Joel S. Brown
Department of Biological Sciences
University of Illinois—Chicago
Chicago, Illinois 60680

Noel Cressie
Department of Statistics
Iowa State University
Ames, Iowa 50011

Philip M. Dixon
Savannah River Ecology Laboratory
University of Georgia
Aiken, South Carolina 29802
    and
Biomathematics Program
Department of Statistics
North Carolina State University
Raleigh, North Carolina 27695

Aaron M. Ellison
Department of Biology
Mount Holyoke College
South Hadley, Massachusetts 01075

Marie-Josée Fortin
Centre d'études nordiques
Université Laval
Sainte-Foy, Québec G1K 7P4

Gordon A. Fox
Division of Environmental Studies
University of California
Davis, California 95616

Thomas M. Frost
Center for Limnology
University of Wisconsin
Madison, Wisconsin 53706

Deborah E. Goldberg
Department of Biology
University of Michigan
Ann Arbor, Michigan 48109

Jessica Gurevitch
Department of Ecology and Evolution
State University of New York—Stony
    Brook
Stony Brook, New York 11794

Larry V. Hedges
Department of Education
University of Chicago
Chicago, Illinois 60637

Dennis M. Heisey
Madison Academic Computing Center
University of Wisconsin
Madison, Wisconsin 53706

Steven A. Juliano
Department of Biological Sciences
Illinois State University
Normal, Illinois 61761

Susan Kalisz
W. K. Kellogg Biological Station
Michigan State University
Hickory Corners, Michigan 49060

Mark A. McPeek
Department of Biological Sciences
Dartmouth College
Hanover, New Hampshire 03755

Randall Mitchell
Department of Biology
University of New Mexico
Albuquerque, New Mexico 87131

Erik V. Nordheim
Department of Statistics
University of Wisconsin
Madison, Wisconsin 53706

Thomas E. Philippi
Department of Biology
University of North Carolina
Chapel Hill, North Carolina 27599

Catherine Potvin
Department of Biology
McGill University
Montreal, Québec H3A 1B1

Paul W. Rasmussen
Bureau of Research
Wisconsin Department of Natural
  Resources
Madison, Wisconsin 53706

Samuel M. Scheiner
Department of Biological Sciences
Northern Illinois University
DeKalb, Illinois 60115

Jay M. Ver Hoef
Alaska Department of Fish and Game
1300 College Road
Fairbanks, Alaska 99701

Carl N. von Ende
Department of Biological Sciences
Northern Illinois University
DeKalb, Illinois 60115

1

Introduction: Theories, Hypotheses, and Statistics

Samuel M. Scheiner

> Picture men in an underground cave, with a long entrance reaching up towards the light along the whole width of the cave. . . . Such men would see nothing of themselves or of each other except the shadows thrown by the fire on the wall of the cave. . . . The only truth that such men would conceive would be the shadows.
>
> —Plato, *The Republic*, Book VII

> I hold that philosophy of science is more of a guide to the historian of science than to the scientist.
>
> —Lakatos (1974)

1.1 The Purposes of this Book

Ecology has become more and more an experimental science. Ecologists are increasing their use of experiments to test theories regarding organizing principles in nature (Hairston, 1989). However, ecological experiments, whether carried out in laboratories, greenhouses, or nature, present many statistical difficulties. Basic statistical assumptions are often seriously violated. Highly unbalanced designs are often encountered due to loss of organisms or other biological realities. Obstacles such as the large scale of ecological processes or cost limitations hinder treatment replication. Often it is difficult to appropriately identify replicate units. Correct answers may require complex designs or elaborate and unusual statistical techniques. To address these problems, we have fashioned this book as a toolbox containing the equipment needed to access advanced statistical techniques.

Most ecologists leave graduate school with only rudimentary training in statistics learned on their own or from a basic one-semester or one-year statistics course. Thus they lack familiarity with many advanced but useful methods for addressing ecological issues. Some methods can be found only in statistics journals that are inaccessible to the average ecologist. Other methods are presented in very general or theoretical terms that are difficult to translate into the actual computational steps necessary to analyze real data. Developments in applying these specialized approaches to ecological questions cannot be found in conventional statistical texts, nor are they explained sufficiently in the brief methods sections of ecological papers. These barriers put them out of the hands of the majority of ecologists, leaving them unavailable to all but a few.

One result of this gap between need and training has been a recent spate of notes and papers decrying the misuse of statistics in ecology. The criticisms include problems of improperly identifying the nature of replicates and pseudoreplication (Hurlburt, 1984; Gurevitch and Chester, 1986; Potvin et al., 1990b), improper multiple comparison tests (Day and Quinn, 1989), the use of ANOVA when more powerful techniques are available (Gaines and Rice, 1990), the misuse of stepwise multiple regression (James and McCulloch, 1990), the improper reporting of statistical parameters and tests (Fowler, 1990b), and the overuse of significance tests (Yoccoz, 1991).

This book is designed as a how-to guide for the working ecologist and for graduate students preparing for research and teaching careers in the field. The ecological topics were chosen because of their importance in current research and include competition (Chapters 4, 15, and 17), plant-animal interactions (Chapters 9 and 10), predation (Chapters 5, 7, 8, and 16), and life-history analyses (Chapters 5, 6, 11, and 12). The statistical techniques were chosen because of their particular usefulness or appropriateness for ecological problems, and for their current widespread use or likely use in the future. Some may be familiar to ecologists but are often misused (e.g., ANOVA, Chapters 3 and 4; multiple regression, Chapter 9), other approaches are rarely used when they should be (e.g., MANOVA, Chapter 5; repeated-measures analysis, Chapter 6; nonlinear curve fitting, Chapter 8), while others are newly developed yet are important for ecologists (e.g., spatial analysis, Chapters 14 and 15; meta-analysis, Chapter 17). Some of the statistical approaches presented here are well established within ecology, while others are new or controversial. We have attempted to mix the tried and true with the innovative. Each statistical technique is demonstrated by applying it to a specific ecological problem, however, that does not mean that its use is in any way limited to that problem. Clearly many of the approaches will be widely applicable. With this book we hope to encourage investigators to expand their repertoire of statistical techniques.

This book deals primarily with the use of statistics in an experimental context, reflecting the increasing use of experimental studies in ecology, especially in natural settings. The book emphasizes manipulative experiments, although many of the techniques are also useful in the analysis of observational experiments (Chapters 5, 10, 11, and 17) and nonexperimental observations (Chapters 2, 7, 8, 9, 13, 14, and 15). In doing so, we in no way slight other forms of ecological experiments and research programs. A number of other books are available that deal with the analysis of nonexperimental observations (Ludwig and Reynolds, 1988; Digby and Kempton, 1987; Manly, 1990b, 1991). The list of ecological problems covered here is not meant to be prescriptive, but illustrative of the wide range of problems amenable to experimental analysis. For example, Chapter 7 deals with large-scale phenomena not usually dealt with by manipulative experiments. Our hope in compiling this book is to demonstrate that even relatively advanced statistical techniques are within the grasp of ecologists, thereby improving the ways that ecological experiments are designed and analyzed.

1.2 Theories, Hypotheses, and Statistics

1.2.1 Building Theories

The vast majority of scientists are, consciously or unconsciously, metaphysical and epistomological realists. That is, we believe that there is an objective universe that exists independent of us and our observations. Science is, then, a process of building theories or models of what that universe consists of, how it is put together, and how it functions. Those theories consist of a set of assumptions about the universe, pieces of the grand truth about the universe that we are striving to discover, with our assumptions being the closest that we can ever come to that truth. In building and testing the theories and accumulating our assumptions, we believe that we are making successively closer approximations to the truth, although we can never be certain that we have ever actually arrived at the truth. Those theories are built and tested using two logical processes, deduction and induction, in which statistics plays several roles (Fig. 1.1).

Inductive reasoning begins with a set of nonexperimental observations. Assembling these observations one discovers patterns, or the lack of them, from which are induced theories—the generalizations. Statistical procedures are used here in two ways, to detect patterns (e.g., various ordination techniques; Ludwig and Reynolds, 1988; James and McCulloch, 1990) and to indicate when those patterns are likely to be due to chance (e.g., multiple regression, Chapters 9; resampling

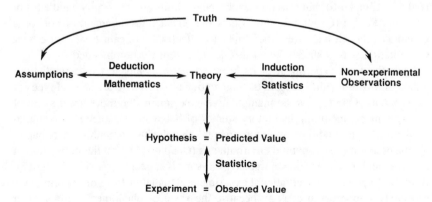

Figure 1.1. The roles of statistics in theory building and theory testing.

tests, Chapter 13). For example, the observation of a tropical to polar gradient in species richness led to various theories about the mechanisms controlling species diversity (Pianka, 1966). Ellison (Chapter 2) demonstrates the use of graphical techniques in pattern identification.

Deductive reasoning begins with a set of assumptions about the universe. From those assumptions generalizations are developed by using rules of deductive logic, including mathematics. For example, Juliano (Chapter 8) presents three types of predator response curves derived from different assumptions about predator behavior. Of course, in any particular instance of theory building in the natural sciences, both inductive and deductive processes are likely to be used. The above description is clearly incomplete; for example, it does not address the issue of how one decides which observations, deductions, and generalizations are worth pursuing. For a general review of the philosophy of scientific theories, see Suppe (1977).

1.2.2 Testing Theories

Next comes a process of testing the theories by means of experiments, using statistics to design and analyze them. From a theory one derives a set of hypotheses. The hypotheses are predictions that, if a set of assumptions is true, then certain parameters will have specified values. Those predictions are then tested with a series of experiments. A test of any theory, whether arrived at inductively or deductively, is a test of the truth of its joint set of assumptions and generalizations—although some assumptions are more in need of testing than others. Brown (Chapter 16) demonstrates this process with theories of optimal foraging.

In the design of experiments, it is critical to consider the statistical analyses that will be used to evaluate the results. Statistical techniques can go only so far in salvaging a poorly designed experiment. These considerations include not only the number and types of treatments and replicates (e.g., repeated-measures analysis, Chapter 6), but also how experimental units are physically handled. For example, Potvin (Chapter 3) shows how the physical arrangement of pots on a greenhouse bench can affect the ability to detect a significant treatment effect. Consultation with a statistician at this stage is highly recommended.

I use the term experiment in its broadest sense as any test of a prediction, although this usage differs from the conventional one. There are several types of experiments. One type is the manipulative experiment, the more usual sense of an experiment, including laboratory studies of behavior (Chapter 8), manipulations of growth conditions in environmental chambers or greenhouses (Chapter 3), the creation of seminatural environments (Chapter 16), and the imposition of experimental treatments on natural environments (Chapters 4, 7, 14, 15, and 17). A second type is the observational experiment. This latter form of experiment is especially important in ecology because the scale of phenomena addressed or ethical considerations often preclude manipulative experiments. For example, the theory may predict that prairies will be more species rich than forests. One would test this hypothesis by measuring (observing) species numbers in each of those habitats. This procedure is a valid test of the theory, as long as the observations used for the test differ from the original, nonexperimental observations used to induce the theory. Intermediate between manipulative and observational experiments are natural experiments in which one predicts the outcome of some naturally occurring perturbation (Diamond, 1986; time-series analysis, Chapter 7). [Some (e.g., Hairston, 1989) would lump natural experiments in with observational experiments.]

In testing theories, statistical procedures are used to indicate the probability that the observed parameter value does or does not equal the value predicted from the assumptions. Often the prediction is not about a specified parameter value, but simply that two experimental treatments have equal parameter values (e.g., growth will be the same in the presence or absence of putative competitors, Chapter 4), or that two parameters will be correlated (e.g., leaf toughness and herbivory rate, Chapter 9). Based on the results of the experiment, the hypothesis is either falsified (the prediction is not met) or not falsified. If the hypothesis is falsified one then alters the set of assumptions and revises the theory. This procedure is called strong inference (Platt, 1964). In the ideal situation, the experimentalist starts with more than one mutually exclusive alternative hypothesis and designs an experiment that will discriminate among them. For example, Mitchell (Chapter 10) illustrates how path analysis can be used to test whether a set of observations fits some preconceived set of relationships among variables and to choose among competing hypotheses.

The term "assumptions" as used here encompasses three concepts. First are explanatory assumptions, those that one is making about the universe (e.g., the assumption that conspecifics will compete more than distantly related species). These assumptions make up the theory that an experiment is attempting to test. Second are simplifying assumptions, those made for analytic convenience (e.g., the assumptions that mating is completely random). These assumptions have either been tested by previous experiments or have been shown to be robust (will not seriously bias results if false) based on independent experiments or analyses. Third are statistical assumptions, those that underlie the statistical procedure used (e.g., the assumption in parametric statistics that the tested variable has a normal distribution). The statistical assumptions are the basis for the stated probability (the P value) that the deviation of the observed parameter value from the predicted value is due only to random sampling effects. Meeting these assumptions is critical if the statistical procedure is to provide a correct answer. Unfortunately, these statistical assumptions are often not known, not understood, or ignored by the experimentalist, resulting in wrong conclusions (e.g., Chapters 7 and 8). Throughout this book, critical assumptions underlying statistical procedures are presented along with warnings about the consequences of a failure to meet those assumptions. Knowing the consequences of violating such assumptions can be very useful as sometimes a reasonably robust answer can be obtained despite violations of some assumptions.

There has been much controversy engendered over the use of strong inference in ecology. Part of this controversy has been due to arguments over the relative value of inductive theory building vs deductive theory testing (Salt, 1983). On one side are claims that ecology has been rife with theory building with not enough testing (Simberloff, 1983); on the other side are claims that the hypotheticodeductive method is inappropriate for ecological questions (Roughgarden, 1983; Toft and Shea, 1983; Peters, 1991). I echo those who call for pluralism on these matters in ecology and evolution (Salt, 1983).

It is important to recognize that theories can be confirmed as well as falsified. For example, although referred to as the Theory of Evolution, today few scientists doubt that living organisms evolved from other organisms. The view that theories can be confirmed as well as falsified differs from a strict Popperian viewpoint in which hypothesis can only be falsified (Popper, 1959). However, a distinction must be made between how we fail to falsify individual hypotheses, a statistical issue, and how we eventually confirm a theory, a larger process that involves philosophical and sociological aspects of science (Lakatos, 1974). The falsification doctrine is an important criterion demarcating the boundary between what is science and what is nonscience or pseudoscience. It is a necessary, but not a sufficient, condition for a complete philosophy of science. According to Lakatos (1974), Popper's strictly deductive falsificationism must be connected to a positive inductive principle wherein the process of science leads us closer to the Truth. [Even Popper states that a theory may be corroborated (1959, Chapter 10).] Eventually enough observations are amassed to allow one to assert that a theory is almost certainly true. That is, science is fallible but ultimately positive, the alternative being that science is skeptical, negative and ultimately just a pure game (Lakatos, 1974).

Thus, my approach to theory building and testing is that we rely on a plurality of inductive and deductive processes and that both manipulative and observational experiments are important. This position differs from that of Suppe (1977) who contends that the difference between "primitive" and "mature" science is the move away from inductive processes and toward deductive ones. But Suppe and most other philosophers of science, including Popper, use physics as their standard. Physics is not a good analogue to ecology and evolution because physics deals with universals and the latter fields deal in large part with historic phenomena. That is, the outcome of ecological processes is often contingent on prior events. This contingency property has long been recognized in evolutionary biology where the historical nature of much of the data has made it obvious (Gould, 1986). An example of the importance of induction and contingency in ecology can be seen in the commonly asked question, "What is the relative importance of competition and predation in structuring communities?" Such a question can be answered only inductively, that is, by surveying many particular instances and then drawing a generalization. The answer cannot be derived deductively from some set of first principles because it will depend, in part, on the chance evolutionary and ecological events that led to the particular set of communities that exists today. I do not mean to imply that theory can make no prediction about this issue. The theory might predict, for instance, that in aquatic communities predation is more important. But, the theory cannot predict exact values. This contrasts with physics, which deals with noncontingent universals and does attempt to predict exact values for such phenomena as the ratio of hydrogen to helium in the universe. This difference regarding contingencies is probably a general distinction between the physical and biological sciences.

1.2.3 The "Null" Hypothesis

The distinction between inductive and deductive processes relates to another controversial issue in ecology, that of the use of "null" models (Conner and Simberloff, 1979; Strong et al., 1979; Diamond and Gilpin, 1982). Null models were introduced to the field by Conner and Simberloff (1979). They are based on subjecting a data set to some type of randomization procedures with the assumption that the resulting pattern is what would occur in the absence of the hypothesized ecological process. In brief, the argument centers around the use of a "pure" random pattern as the standard against which any putative ecological pattern should be judged. This "pure" random pattern is referred to variously as a "null model" and a "null hypothesis."

Arguments over null models are due, in part, to confusion between theory building and theory testing processes. The use of statistics to detect the existence of a pattern is a separate operation from its use to test the theory induced from that pattern. During the process of theory building, when one is attempting to determine if a pattern exists, the use of randomness as a standard of comparison, or null model, is reasonable. However, when testing a theory, a null model has no logical primacy (Roughgarden, 1983). In such a case, the predicted pattern is the proper "null" hypothesis, since the rejection of the "null" hypothesis should result in the alteration or rejection of the theory being tested.

Part of the confusion is due to the use of "null hypothesis" as a synonym for "null model." This use of "null hypothesis" should not be confused with its more common usage in statistics as the hypothesis that is being falsified by the specified statistical test. An experiment should always be designed to refute a theory; the probability of type I and II errors is given in terms of that refutation. Hence the hypothesis to be falsified is always context dependent. For example, when comparing two samples the usual "null" hypothesis is that the means are equal. But it is just as valid to start with a "null" hypothesis of some nonzero difference. The most vociferous arguments over the "null" hypothesis came about in a situation where the same observational data were being used to test a theory as to build it (Diamond and Gilpin, 1982; Simberloff, 1983). By appropriately separating these activities, such an argument becomes moot.

Quinn and Dunham (1983) argue that because ecological phenomena are complex, the "null" hypothesis cannot be identified and critical tests are often not possible. An even more extreme position is taken by Peters (1991) who argues that most or all ecological theories can never be tested. However, these objections disappear if we allow both inductive and deductive tests. I contend that a recognition of plurality in methods of testing does allow for critical tests. In particular, the use of observational experiments permits inductive tests of theories that are not amenable to testing by manipulative experiments. Diamond (1986) refers to this as an epidemiological methodology. For example, a proposed association can be tested by observing whether it varies predictably over populations with different relations to a putative causal link. The complexity of ecological phenomena ensures that any claims for generality will almost always require a mix of manipulative experiments in a few systems and observational experiments in a wide number of systems. Hairston (1989) contends that observational data are always plagued by a posteriori reasoning. I disagree; properly designed observational experiments are possible. For any ecological question that approaches generality, there are often numerous previously unmeasured systems against which one's predictions may be tested. See Chapter 17 for a discussion of the issue of generalizing results in ecology.

1.2.4 Type I and Type II Errors

One way that statistics is often misused involves the falsification decision. Two types of errors can arise over the decision to reject a hypothesis being tested. A type I error is the declaration of the hypothesis to be false when it is actually true; a type II error is the failure to falsify the hypothesis when it is actually false. Statistical procedures are designed to indicate the likelihood or probability of those errors. Most familiar are the probability of type I errors, the α and P values reported in Results sections. Somewhere along the line, a value of $P < 0.05$ became a magic number: reject the hypothesis if the probability of observing a parameter of a certain value by chance alone is less than 5%, do not reject (or accept) if the probability is greater. While some sort of objective criterion is necessary so that one simply does not alter one's expectations to meet the results (throwing darts against the wall and then drawing a target around them), three additional matters must be considered.

First, one must consider the power of the test—the β value or probability of a type II error. (Technically, the power equals $1 - \beta$.) Often in ecology, because of the scale of the phenomena under study, large numbers of replicates of experimental treatments are often not used and sometimes not possible. In that situation, it may be acceptable to use a higher α level as a cutoff for a decision of statistical significance, such as $\alpha = 0.1$, because the power of the test is low.

See Toft and Shea (1983), Young and Young (1991), and Shrader-Frechette and McCoy (1992) for discussions of this issue especially with regard to the relative importance of each type of error in decision making. A very different approach to this problem is offered by the Bayesian school of statistics (Bayes, 1763). Its proponents claim that the typical notions of error rates are wrong, and they use a different procedure for decision making [see Edwards et al. (1983) and Berger (1985) for discussions of Bayesian statistics].

Second, it is important to keep in mind the ultimate goal of any scientific investigation, deciding which of the assumptions about the universe are correct. Ultimately these are yes or no decisions. What the statistical test should indicate is "Yes, the assumption is almost certainly true," "No, the assumption is almost certainly false," or "Maybe, another experiment needs to be performed." Thus, the P values are only guidelines and results that lie in the region $0.011 < P < 0.099$ (to pick two arbitrary values) should be judged with caution.

Third, it is important to distinguish between theory building and theory testing in how one chooses to reach a decision. When looking for patterns and building theories, it is desirable to be less conservative and to use a higher α value so that promising directions of research are not overlooked. On the other hand, when testing theories a more conservative stance is warranted. A more conservative procedure, in statistical parlance, is a procedure that is less likely to reject the null hypothesis.

Sometimes it is difficult to determine what the actual type I error rate is. This rate is supposed to be how often, if the experiment were repeated many times, the hypothesis would be rejected incorrectly when it was true. A problem arises if there is more than one estimation of the same parameter in the course of a study. For example, say one wishes to determine whether two plant populations differ. One might chose some number of individuals in each population, measure 10 different traits of each individual, and compare the estimated parameters. Or, say one measures three different species to test the hypothesis in question. Under what conditions is it necessary to adjust the α value? What is of concern here is whether a single hypothesis is being independently tested more than once. If that is the case, then some sort of correction may be warranted, such as the Bonnferoni procedure (Simes, 1986; Rice, 1990). Another solution, especially if the responses are not independent, is to use multivariate statistics, reducing the problem back to a single statistical test (Chapters 5 and 6).

There has been extensive debate over this issue with regard to the comparison of multiple means (Jones, 1984). One side is concerned that the declaration that two samples are different can depend on what other samples are included in the analysis. Advocates of this position say that only individual pairs of means should be compared and no multiple sample tests such as ANOVA or Tukey's test should be employed (Jones and Matloff, 1986; Saville, 1990). The other side is concerned that such a procedure will cause an inflation of the type I error rate (Day and Quinn, 1989). Both sides have valid positions and each investigator must decide

(1) what is(are) the question(s) being addressed and (2) how conservative one wishes to be. For example, consider an experiment that examines the effects of three pesticides on corn earworm populations in six agricultural fields. If the question being addressed is, "How do corn earworms respond to pesticides?," then the samples are meant to represent a random sample of a larger set, all populations of the species. In that instance a multiple comparison procedure, such as ANOVA, would be appropriate. If instead, the question being addressed is, "Which of these three pesticides is best for controlling corn earworm?," then each sample is of individual interest, and individual comparisons are warranted. Whichever position one takes, in published papers it is important to state clearly what question is being addressed, what statistical procedure is being used to answer it, what are the magnitudes of the type I and type II error rates, and what are the criteria for choosing the procedure and rates.

It is important to keep in mind that statistical significance is not equivalent to biological significance. A large enough experiment can detect very small effects, which in a world where stochastic factors are present will just get swamped by environmental variation. For example, an experiment testing pairwise competitive interactions could be devised that would be able to detect a 0.001% advantage by one species. Such an effect would be meaningless in a situation in which the actual encounter rate is low and large disturbances perturb the community every few years. On the other hand, one must be cautious about failing to recognize the importance of weak processes that can become consequential over very large scales of space and time. For example, macroparasites (mites, ticks, etc.) in the aggregate can have large effects on the population dynamics of their host over long time periods, yet those effects could be difficult to detect because of low parasitism rates on any given individual. One technique for detecting such weak processes is meta-analysis (Chapter 17), a method of combining the results from different experiments.

1.2.5 Using and Reporting Statistics

Ultimately statistics are just one tool among many that we, as scientists, use to gain knowledge about the Universe. To be used well these tools must be used consciously, not by rote. One message that should be conveyed by this book is that there is often no single "right" way to use statistics. On the other hand, there are many wrong ways. Several chapters present alternative procedures for analyzing similar problems (e.g., Chapters 3, 14, and 15). Which procedure is best depends on several factors including how well the data meet the differing statistical assumptions of the various procedures. Most importantly, choosing the statistical procedure depends on what question is being asked. Procedures differ in the types of question that they are best equipped to address. For example, von Ende (Chapter 6) discusses several different types of repeated-measures analysis that are alternatively useful for asking such questions as whether treatments differ in their overall means or whether they differ in the shapes of their response curves. This book should provide a sense of the wide variety of statistical procedures available for addressing many common ecological issues. It should encourage a greater use of uncommon techniques and a decrease in the misuse of common ones.

Equally important to using statistics correctly is reporting such usage completely. Probably the most common statistical sin committed is not describing procedures adequately in publications (Fowler, 1990b). It is important to explicitly report what was done. A scientific publication should permit a reader to draw an independent conclusion about the assumptions being tested. While the ecological, or other biological, assumptions behind an experiment are usually explicitly laid out, statistical assumptions are often not addressed, or worse, the exact statistical procedure is not specified. Thus, when a conclusion is reached that the assumptions have been falsified, the reader does not know if it is the ecological assumptions or the statistical assumptions. This need for accurate reporting is even more important when nonstandard statistical procedures are used. No one would hope to get published an experiment that manipulated light and nutrient conditions without specifically describing each treatment. Yet, commonly the only statement about statistical procedures will be "ANOVA was used to test for treatment effects." Ellison (Chapter 2) discusses some of the simple types of information that need to be reported (and often are not). For example, often only P values are reported without also including information on sample sizes and error variances. If the analysis is done with a computer, it is as important to identify the software package used. Different packages use different algorithms that can reach different conclusions even with the same data. Or, at some future date a "bug" may be found in a software package invalidating certain analyses done with it. I hope that with an increase in the overall level of statistical sophistication of writers, reviewers, and editors will come an increase in the quality of how statistical procedures are reported.

1.3 Using this Book

The material presented here presumes a basic knowledge of statistics, typical of a one-semester introductory course. Any basic statistics book (e.g., Siegel, 1956; Sokal and Rohlf, 1981; Zar, 1984; Snedecor and Cochran, 1989) can be used as a supplement to the procedures presented here and as a source of probability tables. As this is largely a how-to book, the underlying statistical theory is not detailed, although important assumptions and other considerations are presented. Instead, references to the primary statistics literature are presented throughout the text. Although sometimes opaque, readers are encouraged to delve into this literature, especially with regard to any technique which is central to their research program. The effort taken to really understand the assumptions and mathematical properties of various statistics procedures will more than pay for itself in the precision of the conclusions based on them.

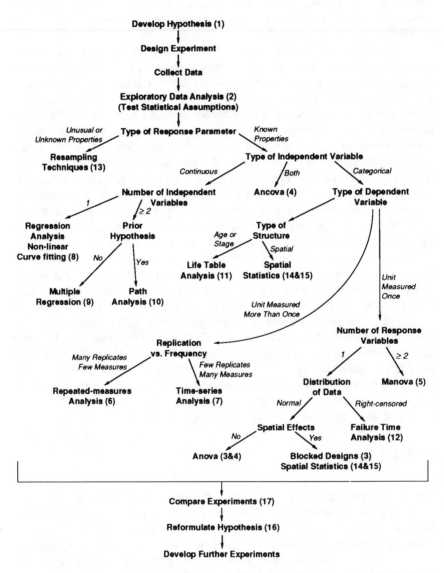

Figure 1.2. A flow chart for applying the topics and techniques presented in this book to the design and analysis of ecological experiments. The techniques are organized with regard to data type.

Each chapter is designed as a roadmap to addressing a particular ecological issue. Unlike most conventional statistics books, each statistical technique presented here is motivated by an ecological question establishing the context of the technique. The chapters are written by ecologists who have grappled with the issues they discuss. As statistical techniques are interrelated, each chapter provides cross-references to other relevant chapters, but each can be used independently. Also, a given statistical technique may be pertinent to a number of ecological problems or types of data; Fig. 1.2 provides a preliminary guide to the matching of data types and techniques. Each chapter presents a step-by-step outline for the application of each technique, permitting one to begin using the statistical procedures presented here with no or minimal additional reference material. Some techniques, however, are difficult and sophisticated, even for someone with more experience. Readers are encouraged to use the techniques presented here in consultation with a statistician especially at the critical stage of experimental design. Because most statisticians are not familiar with ecological problems and the optimal way to design and analyze ecological experiments, they would not necessarily arrive at these methods themselves, nor make them understandable to ecologists. Thus, this book can be used as a bridge between ecologists and their statistics colleagues.

We also recognize that this is the age of the computer. Many problems can only be solved by computer, either because of the enormous mass of the data or because of the extensive number of mathematical steps in the calculations. Each chapter, where appropriate, has an appendix containing the computer code necessary for applying that statistical procedure. We chose the SAS (SAS Institute Inc., 1998, 1989a,b, 1990, 1992) statistical package because of its widespread availability on both mainframe and personal computers and its extensive use by ecologists. In addition, the syntax of SAS is easy to understand and thus to translate to other statistics packages. For example, SYSTAT (Wilkinson, 1988), a common personal computer package, has a syntax very similar to that of SAS. In some instances procedures are not available in SAS; in those cases another statistics package is suggested. In no instances are system commands given and readers unfamiliar with mainframe or personal computers are advised to check manuals and their local computer consultants, mavens, and hackers. A good source for information on statistical packages is the Technological Tools feature in the Bulletin of the Ecological Society of America (e.g., Ellison, 1992) and software reviews in the *Quarterly Review of Biology* and computer magazines.

Acknowledgments

I thank Jessica Gurevitch, André Hudson, Steve Juliano, Marty Lechowicz, and Mark VanderMuelen for suggestions that greatly improved this chapter and Carol Garner for the figures.

2

Exploratory Data Analysis and Graphic Display
Aaron M. Ellison

2.1 Introduction

You have designed your experiment, collected the data, and are now confronted with a tangled mass of information that must be analyzed, presented, and published. Turning this heap of raw spaghetti into an elegant *fettucine alfredo* will be immensely easier if you can visualize the message buried in your data. Data graphics, the visual "display [of] measured quantities by means of the combined use of points, lines, a coordinate system, numbers, symbols, words, shading, and color" (Tufte, 1983:9) provide the means for this visualization.

Graphics serve two general functions in the context of data analysis. First, graphics are a tool one can use to explore patterns in data prior to formal statistical analysis (Exploratory Data Analysis, or EDA sensu Tukey, 1977). Second, graphics communicate large amounts of information clearly, concisely, and rapidly, and illuminate complex relationships within datasets. Graphic EDA yields rough sketches to help guide you to appropriate, often counterintuitive formal statistical analyses. In contrast to EDA, presentation graphics are final illustrations suitable for publication. Presentation graphics of high quality can leave a lasting impression on readers or audiences, while vague, sloppy, or overdone graphics easily can obscure valuable information and engender confusion. Ecological researchers should view EDA and sound presentation graphic techniques as an essential component of data analysis, presentation, and publication.

This chapter provides an introduction to graphic EDA, and some guidelines for clear presentation graphics. More detailed discussions of these and related topics can be found in texts by Tukey (1977), Tufte (1983, 1990), and Cleveland (1985). These techniques are illustrated for univariate, bivariate, and classified quantitative (ANOVA) data sets that exemplify types of data sets encountered commonly in ecological research. Sample data sets are described briefly in Section

2.3; formal analyses of three of the illustrated data sets can be found in Chapters 13 (univariate data set), 8 (predator-prey data set), and 3 (ANOVA data set). You may find some of the graphics types presented unfamiliar or puzzling, but consider them seriously as alternatives for the more common bar charts, histograms, pie charts, etc. The majority of these graphs cannot be produced by SAS or SAS/Graph, the statistical package used in many chapters in this volume. Therefore, for each example, I describe in detail how it was constructed. All figures in this chapter were produced with an IBM PS/2-70 computer and a PostScript laser printer. I produced Figs. 2.8 and 2.9 with S-plus *DOS version 2.0* (Becker et al., 1988; Chambers and Hastie, 1992), and the remainder (save Fig. 2.1) using SYGRAPH *version 5.01* (the graphics module of SYSTAT: Wilkinson, 1990). Examples of SYGRAPH code used to construct the figures are given in Appendix 2.1.

2.1.1 Guiding Principles

The question or hypothesis guiding the experimental design should guide the decision as to which graphics are appropriate for exploring or illustrating the dataset. Sketching a mock graph, without data points, *prior* to beginning the experiment usually will clarify experimental design and alternative outcomes. This procedure also clarifies a priori hypotheses that will prevent inappropriately considering a posteriori hypotheses (suggested by EDA) as a priori ones. Often the simplest graph, without frills, is the best. However, graphs do not have to be simple-minded, conveying only a single type of information, and they need not be assimilated in a single glance. Tufte (1983) and Cleveland (1985) provide numerous examples of graphs that require detailed inspection before they reveal their messages. Besides the aesthetic and cognitive interest they provoke, complex graphs that are data and information rich can save publication costs and time in presentations. Regardless of the complexity of your illustrations, you should adhere to the following four guidelines in EDA and production graphics:

1. Underlying patterns of interest should be illuminated, while not compromising the integrity of the data.

2. The data structure should be maintained, so that readers can reconstruct the data from the figure.

3. Figures should have a high data:ink ratio and no chartjunk—"graphical paraphernalia routinely added to every display" (Tufte, 1983:107), including excessive shading, grid lines, ticks, special effects, and unnecessary three dimensionality.

4. Figures should not distort, exaggerate, or censor the data.

With the increasing availability of hardware and software able to digitize information directly out of published sources, adherence to these guidelines has become increasingly important. Gurevitch (Chapter 17; Gurevitch et al., 1992), for example, relied extensively on information gleaned by digitizing data from many different published figures to explore common ecological effects across many experiments via meta-analysis. Readers will be better able to compare published data sets that are represented clearly and accurately.

2.2 Graphic Approaches

2.2.1 Exploratory Data Analysis (EDA)

Tukey (1977) established many of the principles of EDA, and his book is an indispensable guide to EDA techniques. You should view EDA as a first pass through your dataset prior to formal statistical analysis. EDA is particularly appropriate when there is a large amount of variability in the data (low signal-to-noise ratio) and when treatment effects are not immediately apparent. You can then proceed to explore, through formal analysis, the patterns illuminated with graphic EDA.

Since EDA is designed to illuminate underlying pattern in noisy data, it is imperative that the underlying data structure not be obscured or hidden completely in the process. Also, as EDA is the predecessor to formal analysis, it should not be time consuming. Personal computer-based packages such as SYSTAT, S-plus, and Sigma-plot permit rapid, *interactive* graphic construction with little of the effort needed for formal analysis. Finally, EDA should lead you to appropriate formal analyses and models. A common use of EDA is to determine if the raw data satisfy the assumptions of the statistical tests suggested by the experimental design (see Sections 2.3.1 and 2.3.4). Violation of assumptions revealed by EDA may lead to use of different statistical models from those you had intended to employ a priori. For example, Antonovics and Fowler (1985) found unanticipated effects of planting position in their studies of plant competitive interactions in hexagonal planting arrays. These results led to a new appreciation for neighborhood interactions in plant assemblages (e.g., Czárán and Bartha, 1992).

2.2.2 Production Graphics

Graphics are an essential medium of communication in scientific literature and at seminars and meetings. In a small amount of space or time, it is imperative to get out the message and fix it clearly and memorably in the audience's mind. Numerous authors have investigated and analyzed how individuals perceive different types of graphs, and what make 'good' and 'bad' graphs from a cognitive perspective (reviewed concisely by Wilkinson, 1990; and in depth by Cleveland, 1985). It is not my intention to review this material; rather, through example, I hope to change the way we as ecologists display our data to maximize the amount of information communicated while minimizing distraction.

Cleveland (1985) presented a hierarchy of graphic elements used to construct data graphics that satisfy the guidelines suggested in Section 2.1.1 (Fig. 2.1). Although there is no simple way to distinguish 'good' graphics from 'bad' graphics, we can derive general principles from Cleveland's ranking. First, color, shading, and other chartjunk effects do not as a rule enhance the information content of graphs. They may look snazzy in a seminar, but they lack substance,

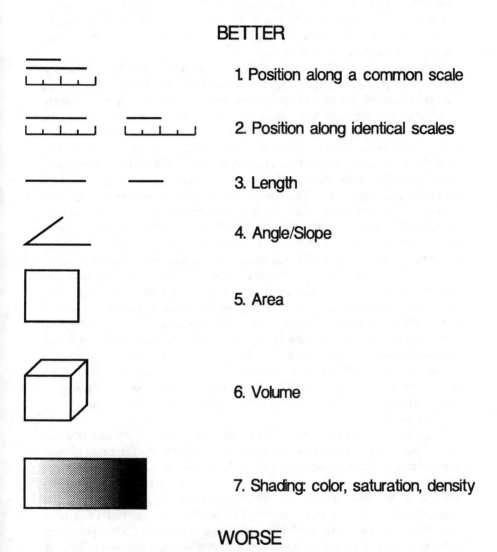

BETTER

1. Position along a common scale

2. Position along identical scales

3. Length

4. Angle/Slope

5. Area

6. Volume

7. Shading: color, saturation, density

WORSE

Figure 2.1. Ordering of graphic features according to their relative accuracy in representing quantitative variation (after Cleveland, 1985).

and use a lot of ink. Second, three-dimensional graphs that are mere extensions of two-dimensional ones (e.g., ribbon charts, three-dimensional histograms, or pie charts) not only do not increase the information content available, but often obscure the message (a dramatic, if unfortunate set of examples can be found in Benditt, 1992). These graphics, common in business presentations and increasingly rife at scientific meetings, violate all of the suggested guidelines. Finally, more dimensions often are used than are needed; e.g., "areas" and lines where a point would do. Simken and Hastie (1987) discuss exceptions to Cleveland's graphic hierarchy. In general, when designing graphics, adhere to the Shaker maxim: form follows function.

High-quality graphical elements can be assembled into effective graphic displays of data (Cleveland, 1985). First, emphasize the data. Lines drawn through data points should not hide the points themselves. Second, data points should never lie on axes themselves, as the axes can obscure data points. If, for example, there are many points that would fall along a 0 line, then extend that axis beyond 0 (Fig. 2.6). Third, reference lines, if needed (which they rarely are) should be deemphasized relative to the data. This can be accomplished with different line types (variable thicknesses; dotted, dashed, or solid, etc.) or shading. Fourth, overlapping data symbols or data sets should be clearly distinguishable. You can increase data visibility and minimize overlap by varying symbol size or position, separating datasets to be compared onto multiple plots, or changing from arithmetic to logarithmic scales. Exemplars include the jitter plot, which avoids overlap of identical values (Fig. 2.3B), and spreading of responses to categories across an axis (Fig. 2.12D). Fifth, the plot must be easily readable following reduction for publication or when projected as a slide to a seminar audience. Finally, Cleveland recommends using a full rectangular plot frame, not the more common bottom axis/left axis only combination seen in many papers. This, together with tick marks *outside* the plot frame (1) emphasize the data and (2) help the reader accurately place individual data points. Tufte (1983) disagrees, as the extra axes are an excessive use of ink and convey no information. Examples in this chapter illustrate most of these possibilities. In the final analysis, many of these rules reflect not only insight into cognitive perception, but also aesthetic judgments by you, the author.

From the above discussion, we could ask, isn't all this too much trouble? Should we dispense with graphs altogether in favor of tables? Because of their conciseness, graphics are almost always preferable in oral presentations. Graphs illustrate more clearly relationships among variables, and can display rapidly multivariate information. However, where exact values are important (as in final publications), tables are more precise. The need for precise tables has been obviated by the increasing availability of digitizing software. When presenting data graphically, however, you must present unbiased and uncensored data. A discussion of what data should be provided, in either graphs or tables, follows in Section 2.4.

2.3 Examples

2.3.1 Univariate Data: Frequency (Density) Distributions

Distributions of height, biomass, or other size metrics are often the primary descriptor of populations or communities. As an example of size distributions, I use a data set containing the number of leaf nodes of 75 *Ailanthus altissima* plants. The experimental design and formal analysis of these data are given in Chapter 13.

With univariate data, two questions are paramount: (1) how are the data distributed (including summary statistics such as the mean, variance, median) and (2) are the data normally distributed or can they be transformed to make them amenable to parametric analyses? Investigators often explore these questions via histograms or normality plots.

A histogram is an example of a *density* plot; that is, what is illustrated in each bar is the frequency, or density, of the values occurring in the dataset between the lower bound and the upper bound of each bar. Histograms are commonly confused with bar charts (see Section 2.3.4). The latter are used to illustrate some summary measure (often the mean, sum, or percent) of all the values within a given treatment category. Histograms of the *Ailanthus* data are shown in Fig. 2.2.

For three reasons a histogram is not the best method for answering the above two questions. First, the raw data are hidden. In this example, there are 75 plants, which have been divided into 12 biomass groups, or *bins* (Fig. 2.2A). It is impossible to know, for example, if the third bar (range 12–14 nodes) contains 10 observations of 12 nodes, 10 observations of 14 nodes, or any other of the possible combinations of 12–14 nodes into 10 observations. Second, the division into 12 bins is arbitrary; it was the default of the graphics program. One could just as easily use 24 or 6 bins, both of which change the apparent shape of the distribution (Figs. 2.2B,C) without conveying additional information. Third, summary statistics cannot be computed from the data illustrated in the histogram. Thus, a histogram does not enable one to answer key questions about univariate data. In addition, histograms fall low on Cleveland's hierarchy of graphic primitives. Bars in a histogram use vertical lines, horizontal lines, and shading in concert to present information embodied in the single point indicated by the top of the bar.

Tukey (1977) introduced the stem-and-leaf diagram as the simplest alternative to the histogram (Fig. 2.3A). The main advantage of the stem-and-leaf diagram is that the raw data are presented in toto. Summary statistics can be derived easily from or incorporated into the figure. Nevertheless, stem-and-leaf diagrams suffer visually from one of the same drawbacks as histograms: the number of bins is arbitrary. Two other alternatives to histograms are jitter plots (Fig. 2.3B) and dit plots (Fig. 2.3C). These two figures preserve the underlying data structure (all

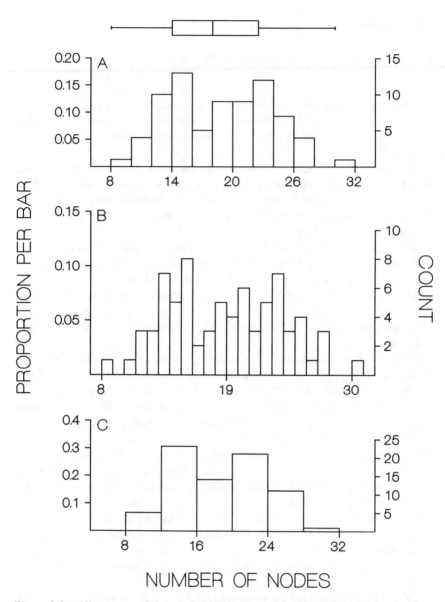

Figure 2.2. Histograms of the number of nodes per plant of 75 surviving *Ailanthus altissima* individuals grown in a 5 × 20 plant rectangular array. Each bar represents the frequency or count (right axis) of observations within the bounds indicated by the ticks on the *x*-axis, and the proportion of the total sample (left axis) represented by each bar. The three plots illustrate the variation in histogram presentation obtained by changing the bin width: (A) default (bin width=4); (B) bin width=2; (C) bin width = 8. At the top of the figure, a box plot (see Fig. 2.4 for construction details) illustrates summary statistics and a better indication of the true data distribution.

values are presented), do not use arbitrary bins, and can be constructed quickly without additional preparation (e.g., sorting) of the data set. Both plots permit rapid assessment of density patterns and are simple to understand.

Stem-and-leaf plots and the density diagrams presented in Fig. 2.3 can be used as simple alternatives to histograms. However, these plots do not convey clearly some of the information that ecologists may want to communicate, and it is difficult to compare the information in two or more of these plots. I suggest the box-and-whisker plot (Tukey, 1977), often called simply a box plot, as a presentation alternative to the univariate histogram (Figs. 2.2 and 2.4A). An advantage of the box plot is that it provides more summary statistical information than a histogram—it includes medians, quartiles, ranges, and outliers (extreme variates)—in much less space and with much less ink. Box plot construction is not dependent on arbitrary bins, so these plots do not exaggerate or distort the data distribution. By notching the box plot (Fig. 2.12E), one can easily add confidence intervals so that plots of several distributions can be compared easily.

Wilkinson (1990; Haber and Wilkinson 1982) developed the fuzzygram (Fig. 2.4B), another alternative to the histogram. Fuzzygrams are histograms with probability distributions superimposed on each bar. Consequently, fuzzygrams present not only the data, but also some estimation of how realistically they represent the actual population distribution. Such a presentation is particularly useful in concert with results derived from sensitivity analyses (Ellison and Bedford, 1991) or resampling methods (Efron, 1982; Dixon, Chapter 13). Haber and Wilkinson (1982) discuss, from a cognitive perspective, the merits of fuzzygrams and other density plots relative to traditional histograms. Histograms (Fig. 2.2), stem-and-leaf plots (Fig. 2.3A), dit plots (Fig. 2.3C), and fuzzygrams (Fig. 2.4B) can indicate possible bimodality in the data. Bimodal data, observed commonly in plant ecology, are obscured by box plots and jittered density diagrams.

Probability plots are common features of most statistical packages, and provide a visual estimate of whether or not the data fit a given distribution. The most common probability plot is the normal probability plot (Fig. 2.5A). Here, the observed values are plotted against their expected values if the data came from a normal distribution; if the data are derived from an approximately normal distribution, the points will fall along a relatively straight diagonal line. There are also numerical statistical tests for normality (e.g., Sokal and Rohlf, 1981; Zar, 1984). If, for biological reasons, the investigator believes the data come from a population with a known distribution different from a normal one, it is similarly possible to construct probability plots for other distribution functions (Fig. 2.5B).

2.3.2 Bivariate Data: Examining Relationships Between Variables

Ecological experiments often explore relationships between two or more continuous variables. Two general questions related to bivariate data can be addressed

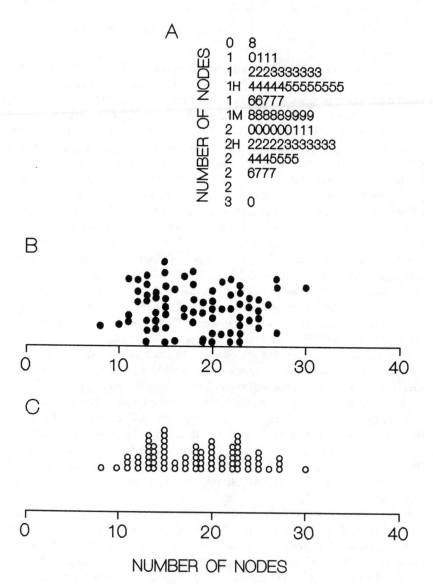

A

	0	8
NUMBER OF NODES	1	0111
	1	2223333333
	1H	4444455555555
	1	66777
	1M	888889999
	2	000000111
	2H	222223333333
	2	4445555
	2	6777
	2	
	3	0

B

C

NUMBER OF NODES

Figure 2.3. Alternative density plots that convey more information than a histogram. (A) A stem-and-leaf plot. In this plot, each line is a *stem*, and the each datum on a stem is a *leaf*. The label for the stem is the first digit (*starting part*) of the number, followed by the value of the leaf. On the first line, the starting part is 0 and the only leaf is 8, indicating a value of 08 nodes. On the second line, the starting part is 1, and there are four leaves, indicating four data points: 10, 11, 11, and 11 nodes. The location of the sample median (M) and upper and lower quartiles (H) are also marked on this plot. (B) A jittered density plot. Each point is placed along the horizontal scale at the exact location of its value. To keep points with equal value from overlapping, they are located at random heights above the *x*-axis. (C) A dit plot. Each point indicates an individual observation, stacked up the *y*-axis at its location along the *x*-axis. In essence, a dit plot is a stem-and-leaf plot with symbols substituted for leaves.

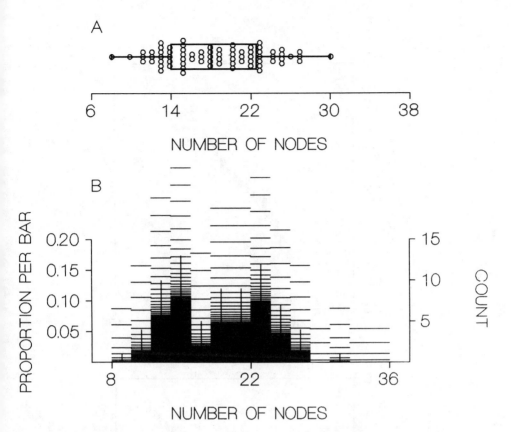

Figure 2.4. Information-rich production alternatives to histograms. (A) A box-and-whisker plot. The vertical line in the center of the box plot indicates the *sample median*. The left and right vertical sides of the box indicate respectively the location of the 25th and 75th percentile of the data (*lower and upper quartiles*, or *hinges*). The absolute value of the distance between the hinges (obtained by subtracting the value of the lower quartile from the value of the upper quartile) is the *hspread*. The whiskers of the box extend *to the last point* occurring between each hinge and its *inner fence*, a distance 1.5 *hspreads* from the hinge. Two kinds of outliers can be distinguished on a box plot. Points occurring between 1.5 hspreads and 3 hspreads (the outer fence) are indicated by an asterisk (see Fig. 2.12E). Points occurring greater beyond the outer fence are indicated by an open circle. The various summary statistics are clearly seen in relation to the raw data, which are overlain on this box plot as a symmetric dit plot. The distance encompassed by the whiskers includes ≈90% of the data (Norusis, 1990). (B) A fuzzygram (Wilkinson, 1990). This plot is a standard histogram (counts and proportions of each bin indicated by the height of the vertical line), with a probability distribution superimposed on each bar. The shading of the bars is based on a gray-scale distribution according to the probability that the *i*th observation will occur in that region: $P_i = P(p_i > \pi_i)$, where $p_i = n_i/n$ is the sample estimate of π_i (the expected proportion of a sample of *n* values from a continuous distribution to fall in the *i*th bin of the histogram. The more likely that $p_i > \pi_i$, the lighter the bar. Consequently, for large sample sizes, the bars will appear in sharp focus, while for small counts, the bars will be fuzzy. See Haber and Wilkinson (1982) for a discussion of the cognitive perception of fuzzygrams.

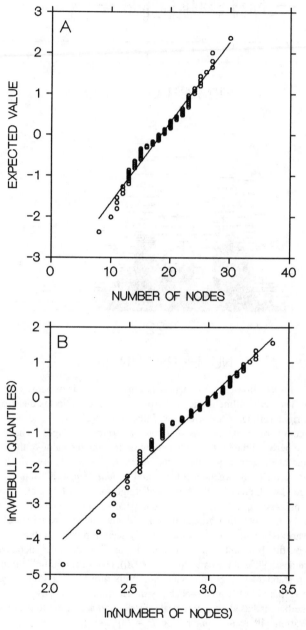

Figure 2.5. Probability plots of the *Ailanthus* data. (A) A normal probability plot. (B) A probability plot with the predicted values coming from a Weibull distribution: $f(y) = 1 - \exp[(-y/s)^t]$, where s is a spread parameter and t is a shape parameter. In this probability plot, the slope of the line is an estimate of $1/t$, and the intercept is an estimate of $\ln(s)$. See Gnanadesikan (1977) for a general discussion of probability plots.

with graphical EDA: (1) what is the general relationship between the two variables and (2) what point(s) is (are) outliers—those points that affect disproportionately the apparent relationship between the two variables? The answers to these questions lead, in formal analyses, to investigations of the strength and significance of the relationship (Chapters 6, 8, 9, and 10). Scatterplots and generalized smoothing routines are illustrated here for exploring and presenting bivariate data. Extensions of these techniques to multivariate data are presented in Section 2.3.3.

Bivariate data sets can be grouped into two types: those where there is a priori knowledge as to which variable ought to be considered independent, leading one to consider formal regression models (Chapters 8 and 9), and those where such a priori knowledge is lacking, leading one to examine correlation coefficients, and subsequent a posteriori analyses. The functional response of *Notonecta glauca*, a predatory aquatic hemipteran, presented experimentally with varying numbers of the isopod *Asellus aquaticus* is used to illustrate the first type of data set; these data are described in detail in Chapter 8. For the latter type of data, I use a dataset consisting of the height, diameter at breast height (dbh), and distance to nearest neighbor of 41 trees in a 625 m^2 plot within an ≈75-year-old mixed hardwood stand in South Hadley, Massachusetts (A. M. Ellison, unpublished data). Data sets of this type are commonly used to construct forestry yield tables (e.g., Tritton and Hornbeck, 1982), and have been used to infer competitive interactions among trees (e.g., Weller, 1987) and forest successional dynamics (e.g., Horn et al., 1989).

For both exploration and presentation, scatterplots are the most straightforward way of displaying bivariate data (Fig. 2.6A). However, scatterplots are merely a display, they do not necessarily reveal pattern. Figure 2.6A illustrates clearly this idea. Three functional response curves (Holling, 1966; Juliano, Chapter 8) could be fit to these data, and it is not clear from the scatterplot itself which one would best fit the data. EDA is particularly useful for dealing with these data, which show high variability and no obvious best relationship between the two variables.

Recent computer-intensive innovations in smoothing techniques (reviewed by Efron and Tibshirani, 1991) have expanded the palette of smoothers developed by Tukey (1977). Basically, to construct a smoothed curve through the data, a best-fit line is constructed through a subset of the data, local to each point along the x-axis. This process is repeated for each point, and a smooth line is constructed by connecting up the intersections of each local regression line. The result of this process, using lowess (robust *lo*cally *w*eighted regr*ess*ion: Cleveland, 1979; Efron and Tibshirani, 1991), is shown for the predator-prey data in Fig. 2.6B. In this case, 50% of the data were used to construct each segment of the smoothed curve. That is, to construct the first segment, the response data from $0 \leq N_0 \leq 50$ were used; to construct the second segment, the response data from $1 \leq N_0 \leq 51$ were used, etc. The apparent type III functional response observed in the smoothed curve is supported by the formal analysis of these data (Chapter 8). The lack of underlying assumptions about the distribution and variance of the data and the ability to elucidate pattern from within very noisy data are two advantages of

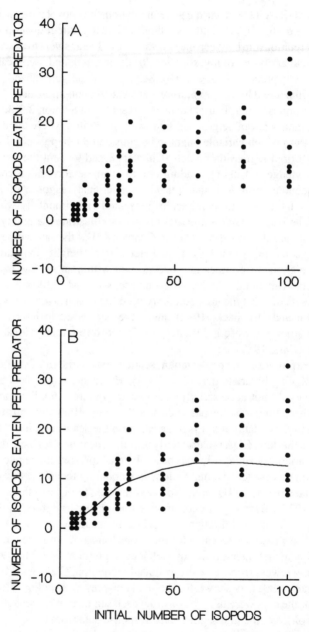

Figure 2.6. Scatterplots of the functional response of *Notonecta* to varying levels of *Asellus*. (A) A simple scatterplot showing the raw data. (B) A scatterplot with a lowess smooth fitted to the data. Note the apparent type III functional response revealed by the smoother (see Chapter 8).

smoothing over traditional regression techniques. Disadvantages of smoothing are that relative weighting of data used for each segment needs to be specified in advance, usually with little or no rational basis for the decision. Moreover, statistical comparison of different smoothed curves is virtually impossible. Finally, with the exception of Tukey's (1977) 3R smoother based on medians, virtually all smoothed curves require sophisticated software (e.g., SYSTAT, Stata, Statistica, Minitab, and S-Plus; see Ellison, 1992 for a review) and fast computers for accurate construction.

Smoothers are used appropriately only when there is clear a priori knowledge of an independent variable and a corresponding dependent variable or variables. When this is not the case, other exploratory techniques are more appropriate for examining relationships between variables. In addition, smoothing does not provide information about potential outliers in the data set. For examining correlations between variables, and to search a posteriori for outliers, influence plots and convex hulls are useful exploratory tools.

A scatterplot of the relationship between tree height and stem diameter (A. M. Ellison, unpublished data) is illustrated in Fig. 2.7A. The raw data are shown, and there appears to be an apparent outlier (a 30-m-tall tree with a dbh > 70 cm). In an influence plot of these data (Fig. 2.7B), the size of each point becomes directly proportional to the magnitude of the change its removal would have on the Pearson correlation coefficient (r) between the two variables. By overlaying a bivariate 50% confidence ellipse, it becomes obvious that outlying points have greater influence on r than do points within the ellipse.

In an influence plot of the logarithmically transformed data (Fig. 2.7C) the apparent outliers have all but disappeared (the large outlier in Fig. 2.7B now has an influence on r of only .01), and the data are better distributed for formal analysis. Fig. 2.7D supports this notion. The outer ellipse is a 95% confidence ellipse centered on the sample (dbh and height) means, with the ellipses' major and minor axes equal in length to the unbiased *sample* standard deviations of height and dbh, respectively. The orientation of the ellipse is determined by the sample covariance. All of the points, save the apparent outlier, fall within this confidence ellipse. For comparison, the inner ellipse is a 95% confidence ellipse with axes computed from the standard errors of the means of each variable and centered on the sample centroid—a graphic illustration of the real difference between the standard deviation and the standard error (see Section 2.4).

Convex hulls and subsequent peeled convex hulls (Barnett, 1976) are useful exploratory tools when the distribution underlying the data is not normal or not known. Convex hulls illustrate order in bivariate or multivariate data, and are used to distinguish distinct groups, outliers, and general shapes of multivariate distributions (a detailed discussion is given in Barnett, 1976). Peeled convex hulls are essentially bivariate smoothers. Figure 2.8 illustrates a convex hull and a subsequent peel around the same data set illustrated in Fig. 2.7. The initial hull (Fig. 2.8A) describes the boundaries of the data—it encompasses the full range

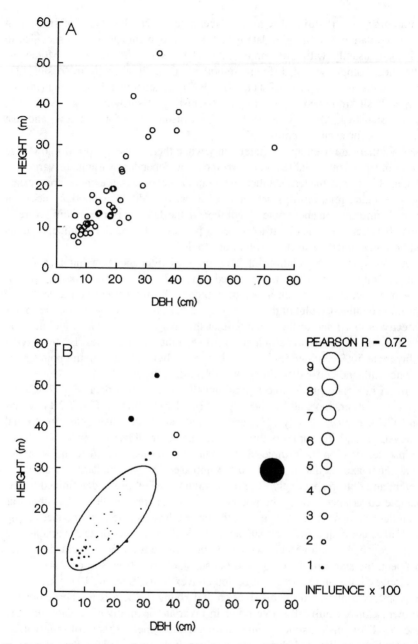

Figure 2.7. Scatterplots of tree diameter vs tree height for 41 trees in a mixed hardwood stand. (A) The raw data. (B) An influence plot, where the size of each point is directly proportional to the magnitude of its influence on *r*. Shading of the points indicates the direction of the influence (open circles have a positive influence on *r*, solid circles a negative influence). In this case, the putative outlier is shown as a large solid point

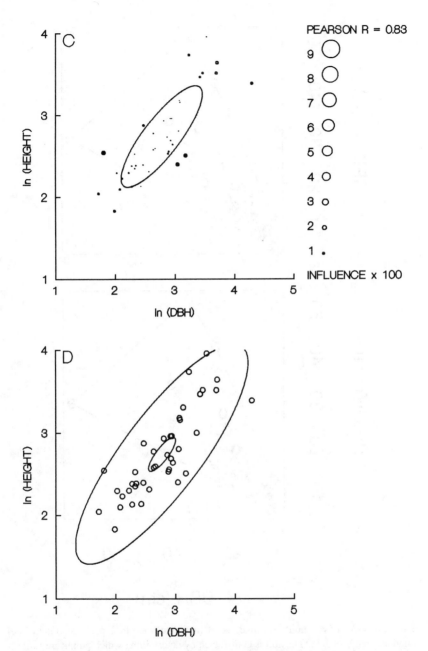

Figure 2.7.—Continued (influence × 100 = 11). Removal of this point alone, therefore, would increase the value of *r* from 0.72 to 0.83. A 50% bivariate confidence ellipse is overlain on the figure. (C) An influence plot of the data following log transformation. (D) Two different 95% confidence ellipses, the outer constructed based on the variables' standard deviations, and the inner constructed based on the standard errors of the means of the variables.

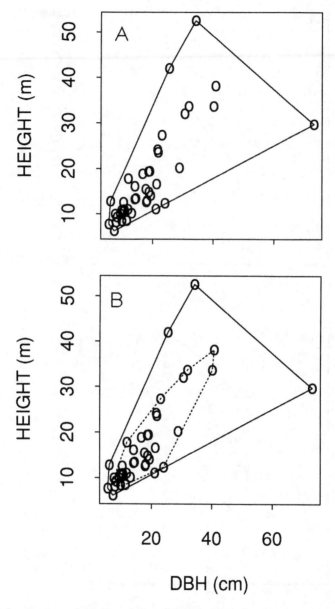

Figure 2.8. A convex hull (A and B, solid line) and a depth 2 peel (B, dotted line) around the tree size data. The hull is constructed by determining which points are furthest from the centroid of the data, and joining those points to form a polygon enveloping the other points. To peel the hull, all the points that lie on the initial convex hull are deleted, and a new convex hull is constructed for the remaining points.

of variation in the data set. The peeled hull, referred to as "peeled to depth 2" (Fig. 2.8B) includes all but the most extreme values of the dataset (compare the points outside the peeled hull of Fig. 2.8B to the points with strong influence on *r* in Fig. 2.7B). This process can be repeated ad infinitum, but normally does not proceed beyond depth 3. This is analogous to Tukey's (1977) running median (3R) smoother, extended in two dimensions. Like smoothers, convex hulls are constructed most easily with pencil and paper, or fast, interactive computer software (S-Plus). Convex hulls are useful for highlighting pattern within noisy data, and make no assumptions about the underlying distribution of the data.

Bivariate plots suitable for EDA are also suitable for final presentation. In preparing these plots for publication, however, there are several conventions often observed in the literature that should be dropped in favor of clarity of presentation. First, it is common in scatterplots to always start each axis at the origin (0,0). In fact, closely adhering to the actual range of the data when scaling axes is far more useful and informative than always including 0, especially if the extreme value of either variable is $<< 0$ or $>> 0$. Restricting the values on the axes to just beyond the extreme values of the data improves clarity and highlights pattern. Axis breaks do not always help, and changing the relative scaling after an axis break usually hinders accurate perception of the data, and can stymie future digitizers.

2.3.3 Extensions of Bivariate Techniques to Multivariate Data Sets

For data sets that include a number of continuous variables, it may not be clear which, if any, pair(s) of variables should be subjected to bivariate correlation or regression analysis, or whether you need to resort to multivariate techniques (Chapter 9). Three-dimensional plots (e.g., Fig. 2.11A) are often used to examine and illustrate higher dimensional data. While aesthetically pleasing, and easy to produce with current graphic software, accurate interpretation and digitizing of these graphs depend on the perspective and orientation of the plot.

The scatterplot matrix, whose origins are shrouded in mystery, provides an alternative exploratory and presentation tool for higher dimensional data. A symmetrical scatterplot matrix of the tree data is shown in Fig. 2.9. This is simply a plot of all possible bivariate combinations of the variables in the dataset. Plots above the diagonal have *x*- and *y*-axes transposed relative to those below the diagonal, which frees the investigator from preconceived notions of "dependent" and "independent" variables. One can, of course, apply the bivariate exploratory techniques described above to each of the scatterplots within the matrix. The possible addition of density plots of each variable along the diagonal gives the investigator a simultaneous feel for the distribution of individual variables (Ellison and Bedford, 1991). The final construction provides an information-rich, but rapidly comprehensible picture of the overall dataset.

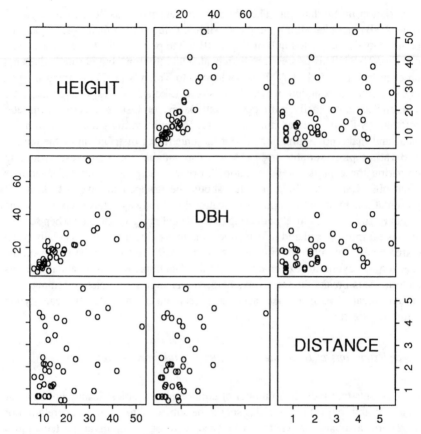

Figure 2.9. A scatterplot matrix of the tree size data. This plot illustrates bivariate relationships between all possible combinations of variables in a multivariate dataset. The variable name in the boxes along the diagonal corresponds to *x*-axis variable of plots below the diagonal, and *y*-axis variables above the diagonal.

2.3.4 Classified Quantitative Data: Alternatives to Bars and Pies

Classified quantitative data are common in many experimental situations. This type of data set consists of responses of a given parameter to discrete treatments. Such experiments may be analyzed by ANOVA (Chapters 3 and 4), and the results expressed in terms of the significance of treatment effects and/or interaction effects. Data from these types of experiments often are not explored prior to formal analysis, although the univariate techniques described in Section 2.3.1 are appropriate for examining the data structure of individual treatment groups. The exception to this generalization are common tests of the critical assumptions of ANOVA: homoscedasticity (variances among treatment groups are equal) and normal distribution of residuals within treatment groups. In particular, failure to test for homoscedasticity is one of the most common statistical errors (Fowler, 1990b), and heteroscedastic (unequal variances) data can complicate or compromise results obtained from ANOVA (Sokal and Rohlf, 1981).

To illustrate EDA and graphical presentation of classified quantitative data, I use data from Potvin (Tables 3.2 and 3.3) that examine effects of genotype (the classifying variable) on fresh mass of *Plantago major*, and the interaction effects of bench position and genotype on stem dry mass of *Helianthus annuus* grown in a latin square design. In each of these data sets, there is only one response variable: plant mass. More complex data sets include responses of several variables to multiple levels of a given treatment. As an example of this latter type of data set, I use data from Ellison et al. (1993). We measured a number of growth and morphological characteristics of *Nepsera aquatica* (an herbaceous species of disturbed areas in tropical wet forests) in response to varying light levels (2, 20, and 40% of full sunlight).

Spread (some measure of variance) vs level (mean, median, etc.) plots (Norusis, 1990) are a rapid, graphic way to examine the within- and between-treatment group variances, and give clues as to appropriate data transformations to bring heteroscedastic data into line. Norusis (1990), modifying the technique of Box et al. (1978), suggests plotting the natural log of the interquartile distance (i.e., the hspread; Fig. 2.4A) vs the natural log of the median for *each* treatment group. An appropriate transformation of the data to remove dependency of the spread on the level is then given as 1 minus the slope of the linear regression line fit to the spread vs level plot. Figure 2.10A illustrates a spread vs level plot for Potvin's *Plantago* data. Note that the raw data are not homoscedastic; the variance increases with the mean. Following Norusis (1990) and Box et al. (1978), the slope of the regression line for this plot is 1.71, suggesting that the data be transformed by raising each observation to the -0.71 power. After such a transformation, the spread vs level plot (Fig. 2.10B) illustrates that the strict dependency of spread on level no longer exists, and the data are somewhat more suitable for ANOVA (the variances are no longer correlated with the mean, although they are still not equalized). Plant size data are often subject to logarithmic transformations to equalize variances within treatment groups. A log transformation of these data does about as well as the negative exponential transform in equalizing these variances (Table 2.1). Box and Cox (1964) and Zar (1984) provide detailed methods on determining the "best" transformation to be used on heteroscedastic data. Such transformations may not make biological sense, but keep in mind that the role of transformations is to bring your data in line with the assumptions and requirements of the statistical model(s) you are testing.

Graphic EDA can also be used to examine interaction effects in data. An example is illustrated in Fig. 2.11 for Potvin's *Helianthus* data. In this experiment, Potvin illustrates how position on a greenhouse bench interacts with genotype to determine plant mass. The top figure illustrates the relative small size of genotype

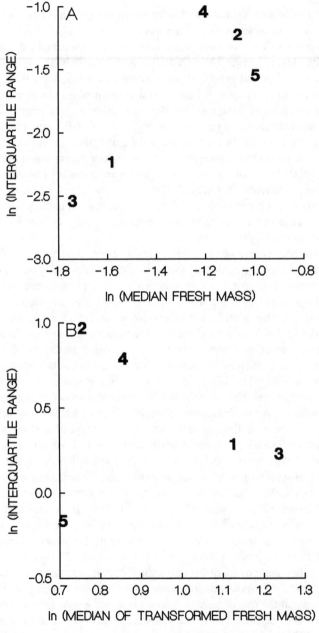

Figure 2.10. Spread vs. level plots of the *Plantago* data. Values plotted on (A) are ln(interquartile distance) on the *y*-axis vs ln(median plant mass) on the *x*-axis of seven replicate individuals of each of five genotypes. Genotype number is indicated on the plot. (B) Spread vs level plot of data following a negative exponential transformation (Norusis, 1990). See text and Table 1.1 for further explanation.

Table 2.1. *Variance (s²)* of *n* = 7 Observations Per Genotype of *Plantago* Fresh Mass[a]

Genotype	Mean	Variance		
		Untransformed	Log transformation	Negative exponential transformation ($y^{-0.71}$)
1	0.198	0.006	0.179	1.245
2	0.309	0.034	0.440	1.798
3	0.109	0.008	0.151	0.710
4	0.298	0.029	0.354	1.302
5	0.412	0.039	0.196	0.392

[a]Variances are shown prior to transformation, following transformation by natural logarithms, and following transformation by the negative exponential suggested by the spread vs level plot (Fig. 2.10).

A and the relatively large size of genotype E. Although a scatterplot matrix might have made this pattern clearer, there is no real point to plotting row × column, or row × genotype, or column × genotype when the point is to illustrate the row × column *interaction* effect on genotype. The lower figure, a contour plot of the top one, illustrates the clear "hot-spot" in the upper left corner of the bench. As interaction effects often involve visualizing data in more than two dimensions, you can use many of the techniques normally applied to multivariate data in the exploration of interactions.

Classified quantitative data are presented poorly in the ecological literature. These problems are illustrated with the data of Ellison et al. on resource allocation and morphological responses to light by *Nepsera* (Fig. 2.12). The most common ways of presenting classified quantitative data are bar charts, separated or stacked (Figs. 2.12A,B), and pie charts (Fig. 2.12C). Separated bar charts (Fig. 2.12A), where a single bar represents the results of a single treatment, suffer from the same problems as histograms. The bars themselves use a lot of ink—horizontal lines, vertical lines, shading of bars of arbitrary width—to convey information about only a single point at the top of the bar (compare Fig. 2.12A with 2.12D). Stacked bar charts (Fig. 2.12B), where treatment groups are divided into subsets and the groups are compared against one another, are virtually unintelligible and never should be used. In this example, the percent allocation to leaves, roots, and stems sum to roughly 100% (allowing for error and missing values). Figure 2.12A (bars side-by-side) at least clearly illustrates the relative allocation to each part. It is not so simple, on the other hand, to determine the relative allocation in Figure 2.12B. Because we use 0 as our reference point, the first guess would be that the allocation to roots in 2% light is ≈70%, and that to stems 100%, when clearly this cannot be true. However, it is difficult to determine visually the beginning point of any of the stacked segments beyond the lowest one. Although measures of variance can be placed clearly on side-by-side bar charts, error bars cannot be placed on stacked bar charts (see Section 2.4). Shadings, hatching,

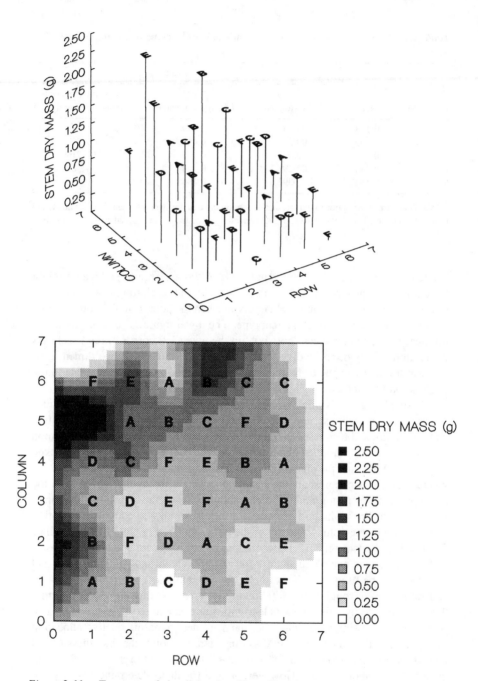

Figure 2.11. Two ways of visualizing the effect of bench position and genotype on stem dry weight of *Helianthus*. The top figure is a three-dimensional scatterplot, with genotype letter (A-F) as the plotting symbol. The addition of sticks connecting each point to its

and other chartjunk used in bar charts also can interfere with accurate perception of the data and decrease the data:ink ratio. Pies share all of the problems of stacked bar charts, and none of the advantages of side-by-side bar charts. I can think of no cases in which a pie chart should be used.

There are several alternatives to bar charts and pie charts. Plots in which the mean value of the response variable is plotted as a single point, along with some measure of error, clearly illustrate the same data as in a bar chart with greater clarity and less data ink (Fig. 2.12D). Sets of box plots better illustrate the underlying data structure and convey more information with less ink and confusion (Fig. 2.12E). These box plots have been "notched" (McGill, et al. 1978) to show 95% confidence intervals. Polar category plots (with or without error bars; the latter are shown in Fig. 2.12F) are the minimalist alternative to bar charts, and are a visually comparable substitute for pie charts. These polar category plots illustrate the response of eight measured variables to the three light environments and clearly convey overall differences between treatment groups.

2.4 A Word about Error Bars

Any reported parameter must include a measure of the reliability of that parameter as well as the sample size. For example, sample means, whether reported graphically or in tables, must be accompanied by the sample size and some estimator of the variance. Error bars on graphs must be correctly identified. Three kinds of error bars are seen commonly in the ecological literature: standard deviations, standard errors, and $n\%$ confidence intervals. Note that strictly speaking, the first is the *sample standard deviation*. The second, more properly referred to as the *standard error of the mean*, is an estimate of the accuracy of the estimate of the mean. We compute it as the standard deviation of a distribution of means of samples of identical sizes from the underlying population (see Zar, 1984:31 for a complete description). Thus, calling error bars simply *standard deviation* bars confounds the two. Measures of error are used to calculate $n\%$ confidence intervals. We can compute easily confidence intervals of normally distributed data from the standard error of the mean (Sokal and Rohlf, 1981). For other distributions, approximations of confidence intervals can be computed using bootstraps, jackknifes, or other resampling techniques (Efron, 1982; Dixon, Chapter 13). *All* of these measures require information about sample size, which must be reported to ensure accurate interpretation of results.

Figure 2.11.—Continued   position on the *x-y* plane permits more accurate perception of the true height along the *z*-axis of each point. The lower figure is a contour plot, with intensity of shading indicating the biomass at a particular row × column location on the bench. These contours were determined by a negative exponential smoothing routine, where the influence of neighboring values decreases exponentially with distance. Shading density increases with biomass.

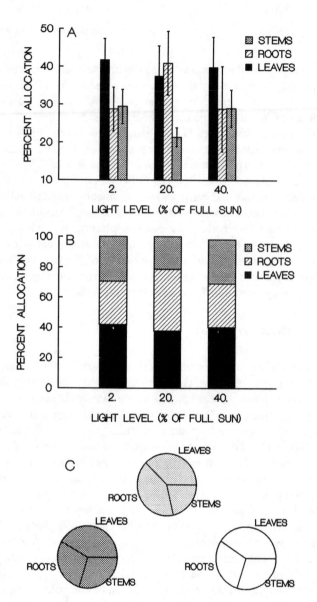

Figure 2.12. Six alternative presentations for presenting classified quantitative data. Data presented are from an experiment examining the effect of three different light levels (2, 20, and 40% of full sun) on growth, resource allocation, and morphology of *Nepsera aquatica*. Each treatment consisted of 20 individually potted plants, harvested after 6 months of growth (Ellison et al., 1993). (A) A side-by-side bar chart illustrating percent allocation to leaves, roots, and stems by plants in each light treatment. Height of the bar indicates mean percent allocation, and error bars indicate 1 standard deviation of the mean. (B) A stacked bar chart illustrating the same data. (C) Pie charts illustrating the relative resource allocation in the three light environments (dark shading: 2% light; intermediate shading: 20% light; no shading: 40% light). Note that it is not possible to place error bars on stacked bar charts or pie charts. (D) Simple category plot of the data illustrated in

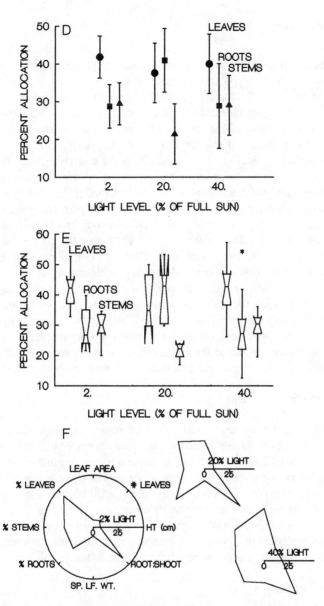

Figure 2.12.—Continued Fig. 2.12A. Each point represents the mean percent allocation to leaves (*circles*), roots (*squares*), and stems (*triangles*); error bars are 1 standard deviation. (E) Notched box plots of the data. Box plot construction as in Fig. 2.4A. Plots are "notched" to illustrate 95% confidence intervals. Where the box reaches full width on either side of the median indicates the limits of the confidence interval. (F) Polar projections of category plots (also known as star plots) of the response of eight measured parameters to the three light treatments. The radius of the circle is equivalent to the y-axis of a rectangular plot; the distance from the center of the circle to each vertex of the polygon is the mean response of each variable to the treatment. Variables are arranged equidistantly around the perimeter of the circle (equivalent to the x-axis of a rectangular plot). One

In general, error bars are useful only when they convey information about confidence intervals. Typically, in the ecological literature, means are plotted along with error bars illustrating one standard error of the mean. For suitably large n, or for samples from a normal distribution, one standard error bar approximates a 68% confidence interval. This conveys little information of interest, since we are accustomed to thinking either in terms of 50, 90, 95, or 99% confidence intervals. Further, most ecological samples are small, or the underlying data distributions are unknown. In those cases, error bars representing one standard error of the mean convey no useful information at all. In keeping with the guidelines for graphical display presented at the beginning of the chapter, I suggest that sample standard deviations or 95% confidence intervals be your error bars of choice. Two-tiered error bars (Cleveland, 1985) that display both quantities are an excellent compromise. Meta-analysis (Gurevitch, Chapter 17) requires sample standard deviations, and if they are reported together with sample size, permit rapid calculation of confidence intervals, standard errors, or most other measures of variation. In the end, the choice of error bar reported lies with you. It is most important that they be identified accurately. Note that if you transformed your data prior to analysis, your calculated standard deviation will be symmetrical relative only with respect to the transformed mean. If you present the results back-transformed (as is common practice), the error bars should be asymmetric.

2.5 Concluding Remarks

Ecologists traditionally have used a limited palette of graphic elements and techniques for exploring and presenting our data. We need to refocus our vision to grasp new or unfamiliar graphic elements and techniques that will permit clear communication of our data. We can now use available computer hardware and software with expanded EDA and presentation capabilities to display our results accurately, concisely, and in aesthetically pleasing ways (Ellison, 1992). We will improve our comprehension and appreciation of data by using many of the graphic techniques presented in this chapter, just as we can increase our appreciation of the diversity of pasta entrées with a trip to a fine Italian restaurant.

Acknowledgments

I thank the late Deborah Rabinowitz for introducing me to EDA and data-rich graphic techniques. Phil Dixon, Steve Juliano, and Catherine Potvin generously

Figure 2.12.—Continued obtains a picture of the overall response of the plant to each light treatment by constructing a polygon whose vertices are equal to the value of the response variable. Different shapes in the different light treatments indicate overall treatment effects. For this type of plot to be effective, all data must be similarly scaled; for this plot, root:shoot ratio (g g^{-1}) was multiplied by 10^2, and specific leaf weight (g cm^{-2}) was multiplied by 10^4. Leaf area is in cm^2, and is a measure of total leaf area per plant.

shared data from their respective chapters. The data on tree size was collected by the 1992 population ecology class at Mount Holyoke College. The work on *Nepsera* was supported by NSF Grant BSR-8605106 to Julie Denslow. Technical support personnel at Systat, Inc. and Statistical Sciences, Inc. helped immensely with final graphics production. Phil Dixon, Elizabeth Farnsworth, Jessica Gurevitch, Catherine Potvin, Sam Scheiner, and one anonymous reviewer provided constructive reviews of early drafts of this chapter that resulted in a much-improved final version. Hardware for graphics production was provided by the BioCIS grant from IBM Corporation. Additional support was provided by NSF Grant BSR-9107195, and the Internet.

Appendix 2.1

SYGRAPH program code to create Figs. 2.2, 2.5, and 2.12.

```
USE '<\PATH\FILENAME.FILETYPE>'              /specifies data file/
SYGRAPH                                      /executes Sygraph/
OUTPUT=PRINTER                               /directs output to printer/
MODE PRINTER=POS1/LPT1                       /defines printer port/

TYPE=STROKE                                  /chooses character set/
CS=1.1                                       /specifies character size/
THICK=1.2                                    /specifies line thickness/

                                                           /Fig. 2.5/
BEGIN                           /To place multiple plots on a single page, the
                                commands must be bracketed by BEGIN and END
                                statements. Everything between these two
                                keywords will appear on  a single page./

ORIGIN=1.125IN,4.75IN                        /location of 1st plot on the page/
PPLOT NODES/NORM,SHORT,                      /normal probability plot/
  SMOOTH=LINEAR,STICK,
  HEIGHT=3IN,WIDTH=3IN,
  XLABEL='NUMBER OF NODES',
  SYMBOL=2,FILL=0,SIZE=.75
WRITE 'A'/HEIGHT=10PT,                                      /plot label/
  WIDTH=10PT,X=0.1IN,Y=2.75IN

ORIGIN=1.125IN,0.75IN                        /location of 2nd plot on the page/
PPLOT NODES/WEIBULL,SHORT,                   /Weibull probability plot/
  SMOOTH=LINEAR,STICK,
  HEIGHT=3IN,WIDTH=3IN,
  XLABEL='ln(NUMBER OF NODES)',
  YLABEL='ln(WEIBULL QUANTILES)',
  SYMBOL=2,FILL=0,SIZE=.75
WRITE 'B'/HEIGHT=10PT,                                      /plot label/
  WIDTH=10PT,X=0.1IN,Y=2.75IN
END                                                        /end of page/
```

```
USE '<\PATH\FILENAME.FILETYPE>'                    /Fig. 2.2/
BEGIN                                     /choose new data file/
ORIGIN=0.125IN,7.85IN              /start page production/
BOX NODES/XLABEL=' ',        /location of box plot on the page/
  AXES=0,SCALE=0,WIDTH=4IN,         /construct box plot; note that
  HEIGHT=2IN,MIN=6,MAX=34                plot widths, and data ranges
                                         match for all plots on this page/

ORIGIN=0.125IN,6.25IN                      /location of Fig. 2.2A/
DENSITY NODES/HIST,                        /construct histogram/
  HEIGHT=3IN,WIDTH=4IN,
  XLABEL=' ',YLABEL=' '
WRITE 'A'/X=0.1IN,Y=1.35IN,
  HEIGHT=10PT,WIDTH=10PT

ORIGIN=0.125IN,3.25IN                      /location of Fig. 2.2B/
DENSITY NODES/HIST,BWIDTH=1,        /construct histogram; note the
  HEIGHT=3IN,WIDTH=4IN,                 bin width is specified with
  XLABEL=' ',YLABEL=' '                      the Bwidth option/
WRITE 'B'/X=0.1IN,Y=2.1IN,                    /plot label!/
  HEIGHT=10PT,WIDTH=10PT

ORIGIN=0.125IN,0.25IN                      /location of Fig. 2.2C/
DENSITY NODES/HIST,BWIDTH=4,    /construct histogram; specify bin width/
  HEIGHT=3IN,WIDTH=4IN,
  XLABEL=' ',YLABEL=' '
WRITE 'C'/X=0.1IN,Y=1.35IN,
  HEIGHT=10PT,WIDTH=10PT

WRITE 'NUMBER OF NODES'/                   /x-axis label; specifies
  Y=-0.75IN,X=.8IN,                        location of label origin,
  HEIGHT=13PT,WIDTH=13PT                   and character size/
WRITE 'PROPORTION PER BAR'/                /left y-axis label/
  Y=2.75IN,X=-.85IN,
  HEIGHT=13PT,WIDTH=13PT,
  ANGLE=90                              /specifies 90° label rotation/
WRITE 'COUNT'/X=4.625IN,                   /right y-axis label/
  Y=4.635IN,HEIGHT=13PT,
  WIDTH=13PT,ANGLE=270
END
```

```
USE '<\PATH\FILE1.FILETYPE>'                    /Fig. 2.12/
BEGIN                                    /select first data file/
                                      /start 1st page production/

ORIGIN=0.75IN,6IN                          /location of Fig. 2.12A/
BAR LEAVES,ROOTS,STEMS*TREAT/                   /construct bar chart/
   FILL=1,4,7,ERROR=SDLVS,                     /choose fill pattern/
   SDROOTS,SDSTEMS,AXES=2,                     /specify error bars/
   YMIN=10,YMAX=50,                         /choose y-axis range/
   XLABEL='LIGHT LEVEL                           /axis labels/
          (% OF FULL SUN)',
   YLABEL='PERCENT ALLOCATION',
   WIDTH=2.75IN,HEIGHT=1.85IN,      /size of plot - consistent for all plots on page/
   LEGEND=2.35IN,1.35IN            /location of legend with respect to plot/
WRITE 'A'/X=0.1IN,Y=1.75IN,
   HEIGHT=8PT,WIDTH=8PT


ORIGIN=0.75IN,3.25IN                        /location of Fig. 2.12B/
BAR LEAVES,ROOTS,STEMS*TREAT/                   /construct bar chart/
   FILL=1,4,7,STACK,                   /use STACK option for stacked bars/
   AXES=2,YMIN=0,YMAX=100,
   XLABEL='LIGHT LEVEL
          (% OF FULL SUN)',
   YLABEL='PERCENT ALLOCATION',
   WIDTH=2.75IN,HEIGHT=1.85IN,
   LEGEND=2.35IN,1.35IN
WRITE 'B'/X=0.1IN,Y=1.75IN,
   HEIGHT=8PT,WIDTH=8PT


CS=3.2                               /new character size to make pie chart
                                 labels conform to other plots on page/
ORIGIN=0.75IN,0.5IN                        /location of 1st pie chart/
PIE AVG2*PART$/SORT,                  /construct 1st pie chart - 2% light/
   WIDTH=0.75IN,HEIGHT=0.75IN,
   FILL=0.35                        /specifies density of pie's shading/
WRITE 'C'/X=0.1IN,Y=1.75IN,
   HEIGHT=8PT,WIDTH=8PT


ORIGIN=1.75IN,1.375IN                       /location of 2nd pie chart/
PIE AVG20*PART$/SORT,                 /construct 2nd pie chart - 20% light/
   WIDTH=0.75IN,HEIGHT=0.75IN,
   FILL=0.1


ORIGIN=2.75IN,0.5IN                        /location of 3rd pie chart/
PIE AVG40*PART$/SORT,                 /construct 3rd pie chart - 40% light/
   WIDTH=0.75IN,HEIGHT=0.75IN,
   FILL=0                                    /no shading/


END                                       /end of 1st page/
```

```
USE '<\PATH\FILE2.FILETYPE>'                    /choose new file/
BEGIN                                           /begin 2nd page production/
ORIGIN=0.75IN,6IN                               /Fig. 2.12D is actually a composite of 3 overlaid
                                                plots. Plots for roots, stems, and leaves are
                                                constructed separately, placed at the same
                                                y-position (height) on the page, but the leaves plot is
                                                moved left, and the stem plot is moved right relative
                                                to the roots plot. Note the change in origin
                                                location to accomplish this overlay./

CPLOT ROOTS*TREAT/SYMBOL=7,                      /plot roots vs light level/
   FILL=1,AXES=2,YMIN=10,YMAX=50,
   ERROR=SDROOTS,WIDTH=2.75IN,
   HEIGHT=1.85IN,SIZE=1.75,
   XLABEL='LIGHT LEVEL
        (% OF FULL SUN)'
   YLABEL='PERCENT ALLOCATION',

ORIGIN=0.625IN,6IN                               /move left 1/8"/
CPLOT LEAVES*TREAT/SYMBOL=2,                      /plot leaves vs light level/
   FILL=1,AXES=0,YMIN=10,
   YMAX=50,ERROR=SDLVS,
   XLABEL=' ',YLABEL=' ',                         /no axes or axis labels to preserve overlay/
   WIDTH=2.75IN,HEIGHT=1.85IN,
   SCALE=0,SIZE=1.75

ORIGIN=0.875IN,6IN                               /move right 1/8" relative to roots plot/
CPLOT STEMS*TREAT/SYMBOL=3,
   FILL=1,AXES=0,YMIN=10,
   YMAX=50,ERROR=SDLVS,
   XLABEL=' ',YLABEL=' ',
   WIDTH=2.75IN,HEIGHT=1.85IN,
   SCALE=0,SIZE=1.75
WRITE 'D'/X=0.1IN,Y=1.75IN,
   HEIGHT=8PT,WIDTH=8PT

WRITE 'LEAVES'/X=1.9375IN,                        /legend needs to be added on explicitly/
   Y=1.8IN,HEIGHT=6PT,WIDTH=6PT
WRITE 'ROOTS'/X=2.0625IN,
   Y=1.435IN,HEIGHT=6PT,WIDTH=6PT
WRITE 'STEMS'/X=2.1876IN,
   Y=1.3IN,HEIGHT=6PT,WIDTH=6PT
```

```
USE '<\PATH\FILE3.FILETYPE>'            /choose new file/
ORIGIN=0.75IN,3.25IN                    /location of Fig. 2.12E/
BOX PERCENT*TREAT/AXES=2,               /To construct this figure, dummy x-values were
  XLABEL='LIGHT LEVEL                   placed in the data file to accomplish the offsetting
    (% OF FULL SUN)',                   done with multiple plots in Fig. 2.12D. This was
  YLABEL='PERCENT ALLOCATION',          done so that the 95% confidence intervals
  WIDTH=2.75IN,HEIGHT=1.85IN,           specified with the Notch option would be
  SCALE=-2,NOTCH,MIN=10                 constructed correctly./
WRITE 'E'/X=0.1IN,Y=1.75IN,
  HEIGHT=8PT,WIDTH=8PT

WRITE 'LEAVES'/X=0.1875IN,              /Again, the legend was placed on manually, as
  Y=1.625IN,HEIGHT=6PT,WIDTH=6PT        were the x-axis labels. Tick marks were
WRITE 'ROOTS'/X=0.375IN,               later erased by hand/
  Y=1.135IN,HEIGHT=6PT,WIDTH=6PT
WRITE 'STEMS'/X=0.5625IN,
  Y=0.9475IN,HEIGHT=6PT,WIDTH=6PT
WRITE '2.'/X=0.375IN,Y=-0.125IN,
  HEIGHT=6PT,WIDTH=6PT
WRITE '20.'/X=1.3125IN,Y=-0.125IN,
  HEIGHT=6PT,WIDTH=6PT
WRITE '40.'/X=2.3125IN,Y=-0.125IN,
  HEIGHT=6PT,WDITH=6PT

USE '<\PATH\FILE4.FILETYPE>' /Choose new file. Data files for polar category plots
                              needs to be arranged so that each star's 'vertex' is
                              an individual case. To construct this plot, four
                              variables are needed in the file: treat$ - the label for
                              each vertex; avg2 - the response of the given
                              parameter (=label) in 2% light; avg20 - the response
                              in 20% light; avg40 - the response in 40% light. As
                              with Fig. 2.12D, Fig. 2.12F is a composite of 3 plots./

CS=1.8                                  /character size set to conform to previous plots/
ORIGIN=0.75IN,0.5IN
WRITE 'F'/X=0.1IN,Y=1.75IN,
  HEIGHT=8PT,WIDTH=8PT

ORIGIN=0.6IN,.55IN
CPLOT AVG2*TREAT$/AXES=2,               /plot of response in 2% light/
  LINE=1,POLAR,NSORT,SCALE=2,           /the polar option gives the projection/
  TICK=2,YMIN=0,YMAX=50,                /the line option connects the vertices/
  SIZE=0,WIDTH=1.25IN,
  XLABEL=' ',YLABEL='2% LIGHT',
  HEIGHT=1.25IN

ORIGIN=2IN,1.25IN                       /plot of response in 20% light/
CPLOT AVG20*TREAT$/AXES=-2,
  LINE=1,POLAR,NSORT,SCALE=-2,
  TICK=2,YMIN=0,YMAX=50,
  SIZE=0,WIDTH=1.25IN,
  HEIGHT=1.25IN,XLABEL=' ',
  YLABEL='20% LIGHT'

ORIGIN=2.75IN,.15IN                     /plot of response in 40% light/
CPLOT AVG40*TREAT$/AXES=-2,
  LINE=1,POLAR,NSORT,SCALE=-2,
  TICK=2,YMIN=0,YMAX=50,
  SIZE=0,WIDTH=1.25IN,
  HEIGHT=1.25IN,XLABEL=' ',
  YLABEL='40% LIGHT'
END                                     /end 2nd page/
```

3

ANOVA: Experiments in Controlled Environments

Catherine Potvin

3.1 Ecological Issue

When the understanding of ecological questions necessitates partitioning the effects of several environmental factors, ecologists often rely on experiments in controlled environments to do so. Controlled environments enable us to vary single factors in order to isolate their effects. For example, growth cabinets make it possible to raise organisms at identical temperatures but different photoperiods or identical light intensities but different temperatures. Even the composition of the atmosphere can be manipulated. The environmental "background," i.e., all the factors that are not altered voluntarily, should be controlled very precisely to ensure that the responses observed when varying a target factor are not confounded with uncontrolled sources of variation. Controlled environments, mainly growth chambers and greenhouses (glasshouses) are a frequent tool in plant ecology while growth cabinets and aquariums are similarly useful in animal ecology. Although the emphasis of this chapter will focus on growth chambers and greenhouses, the same principles prevail for research in other types of experiments in controlled environments and in the field (Chapters 4, 14, and 15).

The first section of this chapter will look at the amount and pattern of environmental heterogeneity that exist in both growth chambers and greenhouses to determine how uniform controlled environments really are. The core of this chapter will address the design of experiments in controlled environments from a classical statistical viewpoint. Experimental designs that can adequately account for environmental heterogeneity as encountered in controlled environments will be examined. Such designs were developed by agronomists and are solidly implemented in the agronomical literature. Growth chamber and greenhouse users have believed for a long time, and some still believe that technology is able to provide perfectly uniform environments. The faith in technology superseded the need for good design. This attitude was further encouraged by engineers who "sold" their product as highly dependable. Consequently, the designs presented in this chapter are not yet commonly used. In the later section of the chapter, the cost of erroneous designs will be explored. Experimentalists will often have more complex designs then those that are presented here but, once the underlying principles are understood, it becomes relatively simple to elaborate the appropriate experimental designs. Good sources for more details on a wide variety of designs include Cochran and Cox (1957) and Winer et al. (1991). This chapter should be viewed as a take-off point to illustrate the considerations that come into play when it is time to design an experiment.

3.1.1 Heterogeneity in Controlled Environments

Heterogeneity may be present in controlled environments at different scales, the environment may vary within a given growth unit as well as between units. Furthermore the environment may also vary through time. Each of these modes of heterogeneity will dictate a specific experimental design. At the start of an experiment, information needs to be obtained on the pattern and magnitude of environmental heterogeneity in the controlled environment. To assess heterogeneity, environmental factors can be monitored through time and mapped. Comparison among different greenhouses was carried out at the Research School of Biological Sciences of the Australian National University in Canberra. Like most greenhouses, these were equipped with a computerized temperature controller with no control over humidity or light. Temperature differences between two greenhouses were at times as large as 5°C despite identical programming of the temperature controllers (Fig. 3.1). The magnitude of the variability existing within a single greenhouse over only a few meters is also surprising. In a greenhouse of the McGill University Phytotron over a bench surface of 4 m^2, air temperature varied by as much as 5°C even though the compartment was equipped with a "state of the art" temperature controller (Fig. 3.2A). When no supplementary lighting was used, light intensity varied by close to 100 μE m^{-2} s^{-1} over the whole surface area (Fig. 3.2B). Mapping of temperature and light shows that the environment varied through time with stronger gradients in the morning. Both factors varied along two axes, north-south and east-west.

It is easy to accept that greenhouses are heterogeneous because the level of environmental control is relatively coarse. But what about growth chambers? Because every environmental factor is artificially controlled, growth conditions could be more uniform than in greenhouses. Environmental mapping of growth chambers has been reported several times (Lee and Rawlings, 1982; Potvin et al., 1990a) and all results suggest that environmental heterogeneity rather than homogeneity is the norm. The most variable environmental factor appears to be irradiance. In a uniformity trial at the McGill University phytotron (Potvin et al., 1990a) photosynthetic photon flux density ranged from 370 to 425 μE m^{-2} s^{-1},

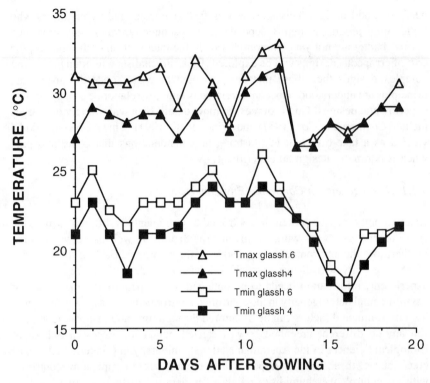

Figure 3.1. Variation in temperature in two different greenhouses (4 and 6) at the Australian National University. Temperature was recorded daily over 20 days. Data were collected by J. Masle, G. Beenister, and C. Wong, Research School of Biological Sciences, Australian National University, Canberra, Australia.

which is similar to the variability reported in the greenhouse. There was a peak of irradiance in the center of two mapped growth chambers probably as a result of the reflection of light from the aluminum-plated walls (Fig. 3.3A.). But the most important aspect of this result is that the pattern of light heterogeneity apparently affects the growth rate of bean and maize (Fig. 3.3B), both species accumulated more biomass when grown in the center of the chamber (Potvin et al., 1990a). If uncontrolled variations in environmental heterogeneity have a direct effect on plant growth, it is essential to account for their effect in data analysis and thus in the experimental design.

3.2 Statistical Issue: Variation in Controlled Environments

Once it is recognized that environmental heterogeneity is substantial in controlled environments, the next step is to examine available experimental designs and their relationship to the patterns of variability. As I compare the various alternatives in

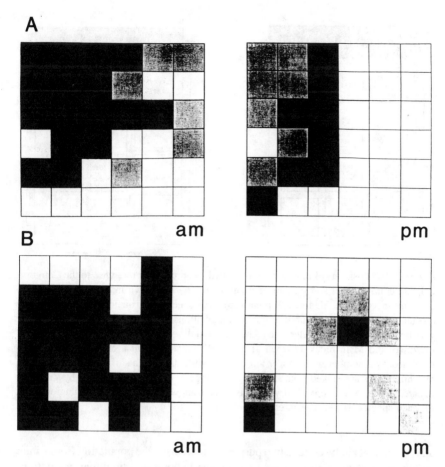

Figure 3.2. Variations in (A) temperature and (B) light over a 2 × 2 m surface in the greenhouse of the McGill University Phytotron. Measurements were taken in the morning and in the afternoon. Differences between two shades of gray are 1°C for temperature and 15 μE m^{-2} s^{-1} for light.

terms of experimental design, it will become increasingly clear that a good knowledge of the pattern of environmental heterogeneity is important in the choice of the most appropriate design. Therefore, I recommend growth chamber or greenhouse users to carry out uniformity trials in their growth environments to document the variability.

3.2.1 Analysis of Variance: A Basic Tool

I will now briefly put aside the peculiar nature of controlled environments and introduce some basic statistical notions. Statistical techniques are used to make inferences about an entire population based on a sample of observed data. Gener-

A

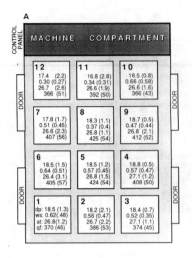

B

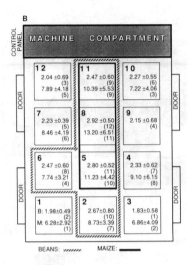

BEANS: ▨▨▨   MAIZE: ▄▄▄

Figure 3.3. Positions of blocks within a growth chamber with reference to the Conviron PGW36 control panel and machine compartment. (A) As shown in block 1, each block gives mean values ± SE for dew point (dp), windspeed (ws), air temperature (at), and quantum flux (qf), as measured in the 12 pot positions, to illustrate the within-chamber variation in physical parameters. (B) Aboveground dry weight (g) for plants of beans and maize (B and M in block 1) when grown in 12 blocks within a chamber. Blocks are identified by a number in the upper left corner, while dry weights of each species within the entire block are ranked from smallest (1) to largest (11, maize; 12, bean) by a number in parentheses. A line separate dry weights that are above and below the distribution mean for each species. (From Potvin et al., 1990a.)

ally, statistical analyses have two purposes: to describe the population from which the observed data originated and to test some hypotheses about that population. The term statistic will be used when a quantity is computed from the observed data while parameter will refer to a quantity that is characteristic of the population. In other words, a statistic estimates a parameter based on one or more samples. Parameter estimation is essential to statistical analyses. Several methods of estimation are available, two of which will be introduced later in this section as they are central to the topics covered by this and other chapters.

Analysis of variance (ANOVA) uses sampled data to test hypotheses regarding a population. The principles outlined below apply as much to agriculture, psychology, engineering, etc., as to ecology. The discussion will assume a basic knowledge of ANOVA, hence I will focus on less known aspects. ANOVAs are based on the specification of linear models used to partition the variance attributable to factor(s) (often treatments). A factor may be represented by any number of levels, i.e., categories into which a factor is divided (Searle, 1971). The parameters of the linear model describing the data can be estimated, among other techniques, by the least-square or the maximum likelihood methods. Least-square estimators, traditionally used in ANOVA, minimize the sum of squares of deviations of the observed data from their expected values (Searle, 1971). In least-squares analysis, if the data set is balanced (i.e., with an equal number of observations per cell), the total sum of squared deviations, can be readily partitioned in sums of squares (SS) contributing to each factor in the design. As a reminder, a deviate is the difference between an observation and the mean. The resulting estimator are minimum-variance, unbiased estimators, which are desirable properties for estimators (Winer et al., 1991). Mean squares (MS), a measure of the average variation existing for each degree of freedom, are then derived by dividing every SS by its degrees of freedom (SS/df). In that sense, MS is equivalent to a statistical variance. An expected value is associated to each calculated MS and Table 3.1 shows that the expected value of a MS is a linear combination of the components of the variance. The statistic, F, on which ANOVA relies for hypothesis testing, is obtained from the ratios of two MSs. Details on the statistical treatment of ANOVA can be found in Searle (1971).

Parameter estimation can also rely on the maximum likelihood method which provides an estimator of a parameter maximizing the probability of obtaining the sample observed as a function of the different values of the parameters underlying the population (Winer et al., 1991). If the data set is balanced, the maximum likelihood equations can be solved explicitly. When some data are missing, however, maximum likelihood estimators are obtained by iterative techniques (Milliken and Johnson, 1984).

Table 3.1. Expected Mean Squares and F-Ratios for a Two-Way ANOVA[a]

Effects	Expected mean squares	F-ratios
A		
A_i	$\sigma^2 + \dfrac{nb}{a-1}\sum\limits_{i=1}^{a}\alpha^2$	MS_A/MS_W
B_j	$\sigma^2 + \dfrac{na}{b-1}\sum\limits_{j=1}^{b}\beta^2$	MS_B/MS_W
AB_{ij}	$\sigma^2 + \dfrac{n}{(a-1)(b-1)}\sum\limits_{i=1}^{a}(\alpha\beta)^2$	MS_{AB}/MS_W
Residual	σ^2	
B		
A_i	$\sigma^2 + n\,\sigma^2_{AB} + nb\,\sigma^2_A$	MS_A/MS_{AB}
B_j	$\sigma^2 + n\,\sigma^2_{AB} + nb\,\sigma^2_B$	MS_B/MS_{AB}
AB_{ij}	$\sigma^2 + n\,\sigma^2_{AB}$	MS_{AB}/MS_W
Residual	σ^2	

[a]The analysis is for (A) fixed effects and (B) random effects models.

The theory underlying the ANOVA technique recognizes an essential difference between two types of factor effects: random and fixed. Factors are random if the levels under study represent a random sample from all conceivable levels of a population of treatment effects. For example, in an experiment designed to study differences in nitrogen acquisition among genotypes of a given species, the genotypes under study may represent a random sample of all possible genotypes, in which case the genotype's effects are random. Conversely, if a factor is represented only by the levels of interest, the factor's effects are fixed. For example, if the aim of an experiment is to assess the responses of plants to an atmosphere with twice the present CO_2 concentration using 350 and 675 $\mu l \ l^{-1}$ CO_2, then the effect of CO_2 is appropriately considered fixed. One can thus view levels of a random factor as being randomly drawn from a larger defined set and levels of a fixed factor as being deliberately chosen by the experimenter. Biologically, the inferential aspect of whether an effect is fixed or random is crucial. If an effect is considered fixed, the conclusions cannot be generalized or extended beyond the levels under study because the levels of the factor have been deliberately chosen. Therefore, to infer to other levels of the treatment factor, the effect of a factor has to be considered random.

The distinction between fixed and random effects in turn leads to three potential types of linear models: fixed, random, and mixed models. In fixed models, all factors but the random error are fixed, while in random models, all factors but the general mean are random. Intermediate situations, combining fixed and random factors, are called mixed models. In a random or a mixed model, the proportion of the variance attributable to a random effect is known as a variance component. When the data set is balanced, i.e., with an equal sample size in each cell of the analysis, computation of the SSs and of the MSs is identical whether a factor is fixed or random (Herr, 1986). However, the expected MSs differ depending on whether the factor has a fixed or a random effect. This is important since the F-ratios are dictated by expected MSs. The simplest case, the two-way model used to examine the effects of two main treatment factors, is illustrated in Table 3.1. Computer codes for the analysis in SAS (SAS Institute Inc., 1989a,b) are given in Appendix 3.1. In fixed models, the expected MS for each effect is the sum of the error variance and of the constant effect of that factor. Therefore, the appropriate MS to be used as the denominator is always the error MS. In random models, the expected MS for each main effect is the sum of the error variance, the interaction variance and the variance due to the effect tested. Thus, F-tests of the main effects use the interaction MS as denominator, while the interaction is still tested over the error MS. In three-way (and higher order) random or mixed factorial models the appropriate denominator is often a combination of MS terms (Winer et al., 1991). I emphasize the central role of the expected MS in determining the appropriate tests of significance, since frequently practitioners use statistical packages that test all factors over the error MS as a default. The above discussion shows that the resulting analysis is valid only for a fixed model whether this is appropriate or not.

Another difference between fixed and random models relates to the estimation methods. If the model is fixed, estimating the parameters of the model using the method of least-squares or that of maximum likelihood is equivalent (Searle, 1971). This is not so if the model contains random effects. Using the least-squares method might produce negative estimates for variance components that are clearly embarrassing as by definition a variance component is positive (Searle, 1971). A possible solution to this problem is to use maximum likelihood estimators in solving random models. Shaw (1987) discusses this alternative at length in the context of quantitative genetics. Although maximum likelihood estimators are attributable to Fisher in the early 1920's, they have gained in popularity only recently possibly because of the ready availability of computing algorithms. Maximum likelihood techniques are available in BMDP3V (BMDP statistical software, 1988) and in the VARCOMP procedure of SAS. Milliken and Johnson (1984) argue, however, that the iteration procedure in SAS is not adequate.

The statistical analysis of mixed models is less straightforward than that of either fixed or random models due to a persisting controversy in the statistical literature (Searle, 1971; Ayres and Thomas, 1990; McClean et al., 1991). Two contending theoretical models exist that differ in their definitions of the variance components themselves (Hartley and Searle, 1969). Assuming a two-way mixed model in which A_i has a fixed effect and B_j has a random effect, the expected MS will dictate the use of either the error MS or the interaction MS as the denominator to test the effect of B_j depending on the model considered. A detailed analysis of the problem has recently been summarized in a biological context by Ayres and Thomas (1990) and suggests that the model using the interaction MS as the appropriate error term is of broader applicability. The confusion in this issue becomes evident if one attempts to analyze the same mixed model using two different statistical packages: SAS and BMDP. The RANDOM. . ./TEST statement of SAS procedure GLM will automatically dictate the use of the error MS as the denominator for the random effect while BMDP8V will use the interaction MS.

An alternative approach to the analysis of mixed models is the use of maximum likelihood estimates (McLean et al., 1991). Mixed models can be divided in two parts, one describing the random and the other the fixed effects. Note that the interaction between a random and a fixed effect is by definition random. Maximum likelihood estimators of the variance components are obtained from the part of the model free of fixed effects and are referred to as restricted maximum likelihood estimators (REML). The estimators for the fixed effect will then be estimated either by least-squares or maximum likelihood. The latest version of SAS (6.07, SAS Institute Inc., 1992) has implemented such a maximum likelihood technique, procedure MIXED. The supplementary manual includes an example (16.1) for the analysis of a split-plot design (Section 3.3.5). One potential advantage of this method is that it will likely prove to be more robust to unbalanced designs (Section 3.2.2) than the standard least-squares techniques. A general disadvantage of the use of maximum likelihood is that computations for complex designs can take a long time and sometimes fail to converge to a solution. Because the documentation on this method is not completely clear, especially for those without extensive statistical training, consultation with a statistician is highly recommended.

3.2.2 Unbalanced Data

In the previous section, I concentrated on the analysis of balanced data, i.e., a data set with an equal number of observations per cell. Despite careful planning, lack of balance is very frequent in ecological experiments. When the design is imbalanced, the total sum of squares can be partitioned in more then one way (Maxwell and Delaney, 1990). The four prevailing methods to compute SSs are referred to as Type I to IV SSs in SAS. For balanced data, these methods yield identical results. However, for unbalanced data, results may differ profoundly. With unbalanced data, Type III SSs are preferred over the other types (Shaw and Mitchell-Olds, 1993). It has been shown, however, that the use of Type III SSs for unbalanced data is prone to type II errors, i.e., accepting the null hypothesis of no treatment effect when it is false. Therefore, the results should be carefully examined, especially if the null hypothesis is not rejected. Shaw and Mitchell-Olds (1993) provide in a comprehensive review of ANOVA for unbalanced data.

There are other alternatives to a lack of balance. The most conservative approach is to artificially impose balance on the data set by randomly removing information from cells having larger sample sizes. The obvious drawback of this method is the loss of information due to the reduction in the size of the data set. Conversely, balance can be impose by estimating input data. This procedure leads to correct estimates of the parameters but to biased tests of significance. Shaw and Mitchell-Olds (1993) therefore recommend the use of computational methods specifically designed for unbalanced data. One such a method is the use of maximum likelihood estimators (Section 3.2.1). This may be a promising avenue as the maximum likelihood technique does not require balanced design (Shaw, 1987). An alternative solution is the analysis of the data using a means model, a technique explicated in Milliken and Johnston (1984).

3.3 Designing an Experiment

Data analysis depends on the design of the experiment itself or, in other words, on how the levels of the factors of interest are assigned to the experimental units. Designing an experiment will also involve the choice of sample size and the physical and temporal layout of the experiment. Knowledge of the design is essential to determine the linear model, which in turn dictates how the ANOVA should be computed. A variety of standard experimental designs are available; each is associated with a mathematical model and analysis. Next I examine four such designs and how each of them can account for specific patterns of environmental heterogeneity. Textbooks on experimental designs present a variety of alternative designs to address specific problems (Cochran and Cox, 1957; Winer et al., 1991). In general the smaller the experimental error, the more efficient the design. An important pitfall to avoid when designing an experiment is the confounding of sources of variation. This concept was made popular by a paper by Hurlbert (1984) under the term of pseudoreplication "the use of inferential statistics to test for treatment effects (factors) with data from experiments where either treatments are not replicated or replicates are not statistically independent".

3.3.1 Completely Randomized Design

The simplest, and most common, design is the completely randomized design. In this design, the levels of the treatment factor of interest are randomly allocated through space. This design is therefore very easy to implement. Note that "randomization" pertains to the allocation of the levels of the factor regardless of whether or not a factor has a random or a fixed effect. The linear model describing the completely randomized design is as follow:

$$X_{ij} = \mu + \tau_i + \varepsilon_{ij} \tag{3.1}$$

where X_{ij} is the response of the jth plant grown under the ith level of the treatment factor τ; μ is the population mean for the response; ε_{ij} the random deviation or "error." The linear model describing the completely randomized design has no term accounting for the variability due to the environment. Strong environmental effects on plant growth will thus result in a large error term and hence in a loss of power in the analysis. Another possible disadvantage in the use of the completely randomized design is that complete randomization does not take into account the pattern of environmental heterogeneity. By chance alone the units receiving the same level of a treatment factor may be grouped together (Dutilleul, 1993). Such aggregation might lead to confounding the effect of the environment with the treatment factor. Take for example the irradiance pattern reported for the growth chambers of the McGill University Phytotron (Fig. 3.3). Suppose that two nutrient levels, high and low, are randomly allocated throughout the growth chambers such that by chance the plants receiving a high nutrient concentration are grouped in the center of the growth chamber, which is also the location where irradiance is the highest. The increase in growth rate that will be observed for those plants might therefore be caused by the confounding effects of high light and high nutrient. Thus, it may be necessary to impose some constraints on the way in which the levels of the treatment factor are allocated to the experimental units.

Despite its appealing simplicity, the completely randomized design should be used with caution; in the face of environmental heterogeneity it may best be avoided (but see Chapter 15).

3.3.2 Blocking

Blocking refers to the grouping of like experimental units when each level of the treatment factor is randomly applied to a different block. Blocking can be used to accommodate environmental heterogeneity. In greenhouses and growth chambers, two types of designs involving blocks are especially useful: the randomized block design and the latin square. In both these designs, experimental units are grouped into blocks within which the environment is relatively constant resulting in similar growth conditions. This enables one to partition the random deviation into a term due to experimental error and one due to the environmental effect. The resulting error term will be smaller and the design will be more powerful than in the completely randomized design. The difference between the randomized block design and the latin square pertains to the pattern of environmental heterogeneity explained. The randomized block design is best used either to account for environmental heterogeneity along a single axis or when the environment is patchy (Fig. 3.4). The latin square will be the most powerful design in the presence of two perpendicular environmental gradients (Fig. 3.5).

There is, however, a cost associated with blocking. In the randomized block design, for example, partitioning of the variance into three terms (treatment, block and error) rather than two (treatment and error) reduces the error degrees of freedom. If blocking is inappropriate and does not account for environmental heterogeneity, a design involving blocks may be less efficient than the completely randomized design. The choice of the most appropriate design and the efficiency of the blocks in accounting for environmental heterogeneity depend on a good understanding of the heterogeneity of the working environment. Blocking will be efficient if the variability between the experimental units in a block is smaller than the variability among blocks. Appropriate blocking will group the experimental units growing in the poorest environment into a block, those growing in the best into a second block, and those in intermediate conditions into yet another block, etc. Each level of the treatment factor is then be allocated to both good and poor growth conditions resulting in a very accurate estimate of the true difference between the levels of the treatment factor (Mead and Curnow, 1983). Potvin et al. (1990a), based on examination of the environmental heterogeneity in growth chambers (Fig. 3.3), suggested that blocking was the best way to account for uncontrolled variability because such an environment is patchy with poorer growth conditions in the corners. Another pitfall in using blocks is that the scale of environmental variation might not match the experimental constraints such as cart or tray size. However, heterogeneity is generally present at more than one scale and consequently most patterns of blocking will be useful (Lee

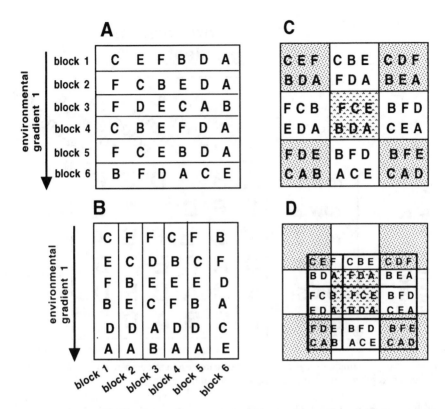

Figure 3.4. Examples of a layout for a randomized block design to compare six levels of a treatment factor (e.g., genotypes of sunflowers): (A) appropriate and (B) erroneous layout in the presence of one environmental gradient or (C) appropriate and (D) erroneous layout when the environment is patchy.

and Rawlings, 1982). Figure 3.4 illustrates optimal and nonoptimal block placement in a randomized block design. A obvious mistake in selecting block placement is to position blocks across different environmental conditions. The above discussion on the use of blocking is made regarding greenhouse and growth chamber experiments, yet spatial heterogeneity is also prevalent in the field. The designs illustrated in this chapter are appropriate and useful for field experiments (see review by Dutilleul, 1993).

3.3.3 Randomized Block Design

Traditionally, the use of blocked designs has been promoted in agriculture where the experimental units were sections of a field from which a single measure of yield was obtained. In that situation there is no replication within the blocks. In ecology, studies often focus on individual responses and produce designs with

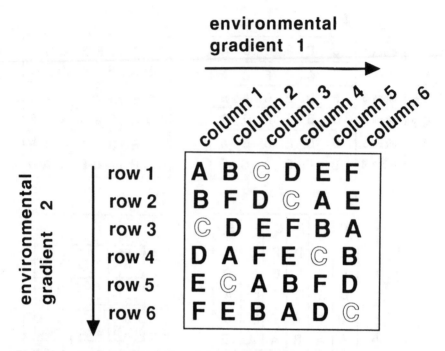

Figure 3.5. Example of layout for a latin square design to compare six genotypes. The design appropriately accounts for two environmental gradients.

within block replication. This section will first consider the classical randomized block design with only one observation per cell (block). In the classic case, each level of the treatment factor of interest will be applied at random to one replicate per block. The number of experimental units within each block will therefore be equal to the levels of the factor under study. To illustrate the analysis of the randomized block design, data comparing the growth of various genotypes of *Plantago major* will be used (Table 3.2).

The experiment involved three fast and two slow growing genotypes. Seeds were germinated in vermiculite and, at the two-cotyledon stage, one replicate plant of each genotype was transferred to an aerated plastic container for hydroponic growth. There were seven containers in the growth chamber. Nutrient solution and growth conditions were identical to those of Poorter and Remkes (1990). Air temperature in the growth chamber at the Research School of Biological Sciences of the Australian National University was held at 20°C. After 12 days of hydroponic growth, the plants were removed from the containers, roots were gently dried with paper, and fresh weight was measured. In this experiment, containers were considered as blocks because they represent the location within the chamber and because, within a container, plants shared the same nutrient solution. The linear model describing the randomized block design is as follows

Table 3.2. (A) Total Plant Fresh Weight (g) of Five Genotypes of P. major *After 12 Days Growth at 20°C and (B) ANOVA of a Randomized Block Design for Log (Fresh Weight)*[a]

A	Genotypes				
Blocks	1	2	3	4	5
1	0.327	0.344	0.206	0.505	0.485
2	0.205	0.114	0.136	0.125	0.272
3	0.141	0.615	0.175	0.305	0.369
4	0.167	0.120	0.127	0.151	0.435
5	0.088	0.172	0.127	0.544	0.203
6	0.249	0.415	0.382	0.299	0.804
7	0.211	0.384	0.178	0.156	0.316

B					
Source of variation	SS	df	MS	F	P
Genotype	0.500	4	0.125	3.81	0.016
Block	0.708	6	0.118	2.60	0.011
Residual	0.786	24	0.033		
Total	1.994	34			

[a]One replicate of each genotype was grown in each of seven blocks.

$$X_{ijk} = \mu + \tau_i + \beta_j + \varepsilon_{ijk} \qquad (3.2)$$

where β_j is the effect of the jth block and the other terms are defined as in Eq. (3.1). The SAS commands are shown in Appendix 3.2. The first step in the analysis of the *Plantago* data was to use a log-transformation to account for heteroscedasticity (see Chapter 2). The randomized block design indicates that a significant portion of the variability was due to the effect of blocks: MS = 0.118, $F = 2.60$, $P = 0.011$ (Table 3.2). Removal of the block effect from the error term enables one to detect a significant difference in the biomass of the different genotypes. The *Plantago* experiment was designed using blocks and it is essential to carry out the analysis accordingly. Blocking a design and then analyzing it as if it were a completely randomized design might produce erroneous results. In the *Plantago* example, ignoring the blocking would artificially inflate the error SS (0.708 + 0.786 = 1.494) and the genotype effect would not be significant resulting in a type II error. The design of an experiment dictates its analysis, not the other way around.

The classic randomized block design, as presented here, can be viewed as a special case of ANOVA with no replication per cell. Consequently, the model does not include an interaction term. In fact, the MS corresponding to the interaction between block and treatment is used as the error term. Although the term β_j is generally considered to be a random effect, there is no requirement that this always be the case (Winer et al., 1991) and the computations of the SSs and the MSs are identical whether it is considered fixed or random. If blocking is done appropriately and the blocks correspond to the various environmental conditions, the β_j term will serve to remove the variability due to environmental heterogeneity from the error term.

A potential drawback to the use of a randomized block design with one observation per cell is the difficulty in dealing with missing data. Mortality or loss of any experimental unit will lead to an imbalance in the design through a missing cell. The main difficulty in an analysis with missing cells is that no information is available for the combinations that are not observed (Shaw and Mitchell-Olds, 1993). To get around that problem, one can estimate the missing value using marginal means (Mead and Curnow, 1983), the mean value of a row or of a column in a two-way data table. In the *Plantago* example (Table 3.2), the marginal means for the columns represent the mean values for each genotype while the marginal means for the rows represent that of blocks. Suppose that mortality in Block 5 lead to the absence of observation for genotype 2, the missing value could be estimated by the marginal mean for that genotype. If the value of 0.172 is omitted, the marginal mean for genotype 2 equals 0.332 and that value would be used to substitute for the missing cell. Underlying this procedure is the assumption that there is no interaction between the block and the treatment factor. This position was challenged by Milliken and Johnson (1984) on the ground that, in reality, interactions are prevalent. In the presence of interactions, it is not possible to make inference about missing treatment combination.

The above example considered a classic randomized block design. It may, however, be advantageous to replicate the observation within each block by level combination. Consider a study where the growth of soybean is compared at three fertilization levels. If soybean plants are grown in individual pots, it is possible to allocate each level of fertilization to more than one pot, say five pots within each block. In the presence of replication, it is possible to estimate the interaction between block and treatment and the appropriate statistical model is

$$X_{ijk} = \mu + \tau_i + \beta_j + \tau\beta_{ij} + \varepsilon_{ijk} \qquad (3.3)$$

where $\tau\beta_{ij}$ is the interactive effect of the ith level of treatment τ and the jth block. Other terms are as defined in Eq. (3.2). The SAS commands for this design are given in Appendix 3.3. This type of blocked design has distinct advantages over the classical design. First, it allows separation of the interaction MS from the error MS potentially resulting in a smaller error term. Most importantly, if there is some mortality or other loss of experimental units, the imbalance does not affect the higher levels of the analysis. A simple way to analyze such data is to compute the mean response for each cell. This rebalances the data and does not inflate the degrees of freedom. The means are then used as observation and the analysis proceeds following Eq. (3.2).

3.3.4 Latin Square Design

The latin square design shares with the randomized block design the blocking of experimental units into homogeneous groups. The main difference is that the blocking in the latin square occurs along two axes. It therefore enables us to control for environmental variability along two gradients rather than one (Fig. 3.5). In controlled environments, mainly in greenhouses, it is often the case that more than one environmental factor varies in this fashion (e.g., Figs. 3.2 and 3.3) and the latin square is often the best design. The linear model describing the latin square is

$$X_{ijk} = \mu + \tau_i + \rho_j + \kappa_k + \varepsilon_{ijk} \tag{3.4}$$

where ρ_j is the effect of the *j*th row and κ_k is the effect of the *k*th column. The SAS commands for this design are shown in Appendix 3.3. Environmental heterogeneity is accounted for by two terms: ρ_j and κ_k. This imposes some constraints on the layout of the latin square. The design requires the same number of rows and columns and, because each level of the treatment factor has to be found in each row and in each column, the total number of replicates equals the number of rows and of columns (Fig. 3.5). If three levels of the treatment factor are compared, the experiment will be laid out in three rows and three columns allowing for three replicates for each level of the treatment factor. In terms of degrees of freedom, the latin square is more powerful when it involves several levels of a factor. The use of the latin square will be illustrated with data on the growth of six different sunflower genotypes (Table 3.3). The plants were grown in liter pots in a greenhouse compartment of the McGill University Phytotron whose environmental conditions are shown in Fig. 3.2. The mapping showed that environmental gradients were present both along the north-south and the east-west axes. To account for the two gradients, the layout followed a latin square as illustrated in Fig. 3.5. Temperature was controlled at 25°/18°C (day/ night) with a relative humidity of 60%. Plants were grown for 4 weeks after which they were harvested and dried at 65°C until a constant dry weight was reached.

Analysis of stem dry weight indicates that environmental heterogeneity associated both with rows and with columns accounts for a large portion of the overall variability (Table 3.3). Removing this heterogeneity from the error term through the use of the latin square enables us to uncover a significant genotypic difference in stem weight. The advantage of blocking along two axes is thus obvious. If this latin square design had been erroneously analyzed as a completely randomized design, the error term pooling of the row, column and error SS would have been 0.731 + 1.257 + 0.685 = 2.673 with 30 degrees of freedom resulting in an *F*

62 / C. Potvin

Table 3.3. (A) Stem Dry Weight (g) of Six Genotypes of Sunflower Grown in a Greenhouse at 22°C[a] and (B) ANOVA of a Latin Square Design for Log (Stem Dry Weight)

A			Column			
Row	1	2	3	4	5	6
1	A 0.8323	B 1.3140	C 0.6236	D 0.9779	E 2.4447	F 0.9214
2	B 0.6129	F 0.3173	D 0.2596	C 1.2969	A 1.1096	E 1.4786
3	C 0.0810	D 0.5751	E 0.3894	F 0.5547	B 1.2165	A 0.5078
4	D 0.5578	A 0.6663	F 0.5957	E 0.6994	C 0.8754	B 1.6749
5	E 0.4748	C 0.2930	A 0.7082	B 0.9279	F 0.7827	D 1.0503
6	F 0.0784	E 0.5280	B 0.5413	A 0.6264	D 0.7397	C 0.5356

B					
Sources of variation	SS	df	MS	F	P
Genotype	0.580	5	0.116	3.39	0.022
Row	0.731	5	0.146	4.27	0.008
Column	1.257	5	0.251	7.38	0.001
Residual	0.685	20	0.034		
Total	3.253	35			

[a]Letters represent the different genotypes associated with each position in the lay-out.

value for the genotype effect of 1.30, $P > 0.25$. As in Section 3.3.3, ignoring the latin square nature of the experiment would lead to a type II error.

3.3.5 Variation Between Growth Compartments: The Split-Plot Design

In the experiments examined thus far, levels of the treatment factor were allocated at the same time within a growth compartment. Very often, the nature of the treatment factor demands that it be applied to a whole growth chamber or greenhouse, for example, temperature or atmospheric CO_2 concentration. Such factors are referred to as between chamber factors since comparison of more than one level requires the use of more than one growth compartment. When between chamber treatment factors are of interest, designing a greenhouse or growth chamber experiment becomes a somewhat difficult task. As noted in the introduction, variation in environmental conditions between different growth compartments imposes unavoidable constraints on experimental design. If the treatment factor is applied to the whole growth chamber, replication between chambers is essential. Therefore, if the treatment factor is represented by two levels, the absolute minimum number of growth environments required in order to allow each level to be repeated twice is four (Potvin and Tardif, 1988). If an experiment involves only a between chamber treatment factor, the design will be a randomized block design as presented in Section 3.3.3, where each growth compartment is a block. Physical and financial constraints usually limit the number of chambers available, hence the degrees of freedom for testing the treatment factor in such an analysis will be extremely small. Because the number of growth chambers available is finite, the block (chamber) effect should be considered fixed and not random.

Very often, experiments will involve both a between chamber factor and a within chamber treatment factor, for example, the temperature responses of different genotypes where temperature is a between chamber factor and genotype is a within chamber factor. Response to light or to atmospheric CO_2 concentration as a function of nutrient limitation likewise involves both a between chamber and a within chamber factor. Such experiments can be appropriately analyzed with a split-plot design. In a split-plot experiment, a single level of one treatment factor is applied to a large plot while all the levels of a second treatment factor are allocated to subplots within that main plot. Two different randomization procedures are used to allocate the treatments. In greenhouse or growth chamber experiments, the levels of the main plot treatment factor are allocated at random, each to a different growth chamber or greenhouse. The growth compartments are then divided into subplots and, in a second randomization procedure, each level of the within chamber factor is assigned to a given subplot.

A theoretical split-plot experiment will be used as an example to compare two atmospheric CO_2 concentrations and six nutrient concentrations on plant biomass increment. The layout of this split-plot design is illustrated in Fig. 3.6 and is representative of many growth chambers studies. The unit of comparison for CO_2

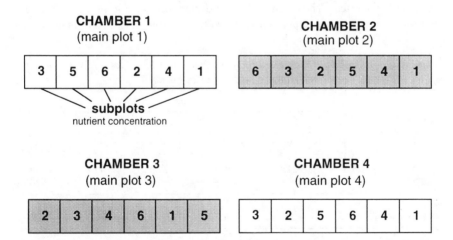

Figure 3.6. Example of layout for a split-plot design involving six nutrient levels as a subplot factor and two CO_2 concentrations as a main plot factor. A replication (block) consists of two growth chambers, each chamber being considered as a main plot. Nutrient levels are represented by number $1-6$ and CO_2 concentration by presence or absence of shading. The chambers are then subdivided into six subplots to which the within-chamber factor is randomly allocated.

concentration is the entire growth chamber while nutrient levels are applied within each chamber. The general linear model to describe the data obtained by such a split-plot design is

$$X_{ijk} = \mu + \alpha_i + \varepsilon_{k(i)} + \nu_j + \alpha\nu_{ij} + \varepsilon'_{k(ij)} \qquad (3.5)$$

where α_i is the effect of the ith level of CO_2 concentration (the between chamber treatment factor), $\varepsilon_{k(i)}$ the main plot error term designates the effect of chamber k within level α_i, ν_j is the effect of the jth nutrient concentration (the within chamber treatment factor), $\alpha\nu_{ij}$ is the effect of the interaction between the ith CO_2 concentration and the jth nutrient level, and $\varepsilon'_{k(ij)}$ is the error term associated with the sub-plot. Other parameters are defined as in Eq. (3.1). Artificial data were generated and analyzed according to this split-plot model (Table 3.4); the SAS commands are given in Appendix 3.5.

The partitioning of the variance in the split-plot ANOVA table is divided into two parts and the analysis distinguishes between two error terms: the whole plot and the subplot error terms. The error term used to test the effect of CO_2 is calculated in terms of the variation between the growth chambers. On the other hand, the effect of nutrient levels is independent from the variation due to growth chambers. Consequently, the error term to test the nutrient effect is determined by the variation between the subplots within the growth chambers (Winer et al., 1991). Note that the interaction term, $\alpha\nu_{ij}$, is independent of the variation due to the whole plot effect. The classic split-plot design illustrated in Fig. 3.6 assumes that only one observation per cell is present. As in the case of the randomized

Table 3.4. A. Artificially Generated Biomass Data. B. A Split Plot ANOVA with CO_2 as the Main Plot Factor and Nutrient as the Subplot Factor

A	Nutrients					
CO_2	1	2	3	4	5	6
350	13.3	19.9	24.5	33.6	38.9	41.4
350	12.4	20.5	22.1	29.4	36.9	42.3
675	15.8	22.4	26.9	38.0	44.8	49.2
675	13.9	21.0	27.5	35.9	46.8	49.0

B					
Source of variation	SS	df	MS	F	P
Main plot					
$\quad CO_2$	130.67	1	130.67	42.959	.0225
Main plot error	6.08	2	1.93	.193	.196
Subplot					
$\quad$ Nutrient	3050.44	5	610.09	386.213	.001
$\quad$ Nutrient $\times CO_2$	35.47	5	7.09	4.491	.021
Subplot error	15.80	10	1.58		

block and the latin square design, split-plot ANOVAs can be extended to allow replication of the experimental units in the subplots.

Split-plot ANOVAs are amenable to more complex designs. Uniformity experiments carried out in the phytotron of North Carolina State University to determine the sources of heterogeneity most important to plant growth, exhaustively used a two-way main effects split-plot design (Lee and Rawlings, 1982). The analysis considered two whole plot treatment factors: chamber and trial, i.e., replication in time. Four different growth chambers were used in four different trials (Table 3.5). The subplot factor, the trucks on which plants were grown, was represented by 24 levels, each corresponding to one of the movable carts on which plants were grown. This design allowed for within subplot replication. Results from the split-plot analysis of soybean growth characteristics (Table 3.5) enables us to quantify the most frequent and, I would argue damaging, error occurring in the statistical analysis of greenhouse or growth chamber experiments.

When the split-plot nature of an experiment involving both within- and between-chambers treatment factor is ignored, the analysis erroneously proceeds using the subplot residual as the error term for all of the main effects. This error pertains to the variability between the subplot within a main plot rather than to the variability associated with the main plot themselves. A comparison of the magnitude of the whole plot and the subplot error MSs indicates that the use of the subplot error underestimates the real error. In the soybean experiment, the discrepancy between the errors was, at times, as large as 400-fold (Table 3.5).

Table 3.5. (A) Mean Squares in Combined Split-Plot Analyses of Variance on Soybean, with a Comparison of Two Main Effects, Growth Chambers and Trials. (B) Proportional Difference in Size of the Two Error Terms[a]

Source of variation	df	Plant height	Leaf area	Petiole length	Fresh weight	Dry weight
A						
Main plot						
Chamber	3	50.05	197.72	39.30	320.63	48.10
Trial	3	48.54	310.83	36.01	464.54	48.10
Main plot error	4	125.72	127.67	50.07	211.63	31.43
Subplot						
Truck	23	9.72	16.90	5.04	30.22	9.14
Chamber × truck	69	7.93	14.59	4.75	24.10	4.08
Trial × truck	69	0.48	1.09	0.27	2.58	0.47
Chamber × trial × truck	292	0.55	1.13	0.26	2.38	0.43
Subplot error	3960	0.29	0.71	0.15	1.62	0.24
B						
Proportional difference		433.5	179.8	333.8	130.6	131.0

[a](From Lee and Rawlings, 1982).

The opposite trend can be found when the comparison involves the error degrees of freedom. In the example, the error degrees of freedom associated with the whole plot is 4 compared to 3960 for the subplot (Table 3.5). Therefore, when the subplot error term is wrongly used to test the chamber and trial main effect, the analysis will proceed with artificially inflated degrees of freedom and a reduced error MS. Similar observations arise from examination of Table 3.4. Needless to say, the erroneous use of the subplot in lieu of the whole plot error term can have a drastic effect on the conclusion of the ANOVA often leading to type I errors.

3.5 Conclusions

This chapter relies on classical statistical methods to address experimental design in controlled environments. The techniques presented are basic and can be incorporated into various other experimental designs or used in conjunction with a wide variety of other analytic techniques. For example, blocked design are referred to in several other chapters (e.g., Chapters 4, 5, 6, 14, 15, and 16). This is because blocking, i.e., the grouping of experimental units subjected to similar environmental conditions, is a powerful way to control for unwanted sources of variation. The potential cost of blocking, a reduction in the degrees of freedom of the error term, should be kept in mind. In situations where the grouping of the experimental units does not correspond to the underlying pattern of environmental heterogeneity, the completely randomized design may be more powerful because of its greater degrees of freedom.

Once the principles of blocking are understood, more complex designs are easy to understand and analyze. An example was provided in Table 3.5 with the analysis of a split-plot model with two main effects. Another common modification of the split-plot design is the use of a randomized block, or a latin square, within the split-plot. In those situations, analysis of the main plot portion of the model remains unchanged while the subplot part is reorganized to account for the special structure of the experiment. Given the pattern of environmental heterogeneity in controlled environments it might often be a most appropriate design. The incomplete block design, where the number of treatment level exceeds the number of experimental units, is also frequently useful (Milliken and Johnson, 1984).

So far environmental heterogeneity has been considered as a function of space alone. An early paper on variability in controlled environment suggested that growth conditions may vary as much through time as in space. Likewise, Potvin and Tardif (1988) have shown that the growth of beans was sensitive to the interaction term of chamber and time. This significant interaction might be linked to several uncontrolled factors such as batch of soil or fertilizer that vary through time, decay of light bulbs, or other aging effect influencing the performance of growth chambers. If a chamber used at two different time affects plant growth differently, it is possible to consider repetition of an experiment in time as true replication. This provides some relief in the design of experiments comparing between chamber factors. The comparison of two levels of a treatment factor may properly be assessed by using two growth chambers in two successive trials. In such a case, each chamber by trial combination (e.g., chamber 1 at time 1) will appropriately be considered as a block and the analysis may proceed following a randomized block design in which block will be a random effect. The importance of time in designing growth chamber experiment has been extensively explored by Potvin and Tardif (1988).

A conclusion that may be unpleasant to hear is that replication is always a necessity even, or especially, when the levels of a factor have to be applied to an entire growth unit. Ignoring the need for replication and using the variability within a growth chamber as the error term for between chamber effect is seriously misleading, as shown above.

Acknowledgments

The *Plantago* experiment was carried out with Drs. J. R. Evans and J. Virgona of the Research School of Biological Sciences, Australian National University, while the sunflower data were collected by Ms. L. Veillette. I am gratefull to all of them. I am indebted to Drs. J. Masle, C. Wong, and Mr. G. Beenister, R.S.B.S., Australian National University, for sharing their data on environmental heterogeneity with me. Dr. Sam Scheiner provided the computer code for the examples.

Appendix 3.1

SAS program code for a two-way ANOVA shown in Table 3.1.

```
PROC GLM;                    /use procedure GLM for unbalanced ANOVAs/
   CLASS A B;                         /indicates categorical variables/
   MODEL X = A B A*B/SS3;    /model statement corresponding to Table 3.1,
                            use type III sums of squares. Warning: results are
                            only approximate if the design is unbalanced./

*** If factors A and B are random effects use the following
      command to obtain the correct F tests ***

   RANDOM B A*B/TEST;        /Declare sites to be a random effect and ask SAS to
                            compute the Satterthwaite correction for an
                            unbalanced design./
```

Appendix 3.2

SAS program code for a randomized block design (Eq. 3.2).

```
PROC GLM;
  CLASS T B;                           /T=treatment levels, B=blocks/
  MODEL X = T B/SS3;                   /all terms are tested over the error/
```

Appendix 3.3

SAS program code for a randomized block design with replication (Eq. 3.3)

```
PROC GLM;
  CLASS T B;
  MODEL X = T B T*B/SS3;               /now an interaction term is present/
  RANDOM B T*B/TEST;                   /the block and block-treatment
                                        interaction terms are random effects/
```

Appendix 3.4

SAS program code for a latin-square design (Eq. 3.4).

```
PROC GLM;
  CLASS T C R;                         /C = column, R = row/
  MODEL X = T R C/SS3;                 /all terms are tested over the error/
```

Appendix 3.5

SAS program code for a split-plot design (Eq. 3.5).

```
PROC GLM;
  CLASS A E N;                         /A = CO2 level, E = chamber (main plot error)
                                        N = nutrient level/

  MODEL X = A E(A) N A*N/SS3;
  RANDOM E(A)/TEST;                    /CO2 level will be tested over the chamber effect.
                                        Nutrient and the CO2-nutrient interaction will be
                                        tested over the error term./
```

4

ANOVA and ANCOVA:
Field Competition Experiments

Deborah E. Goldberg and Samuel M. Scheiner

4.1 Introduction

Competition has occupied a central place in ecological theory since Darwin and
experiments on competition have been conducted on many different organisms
from many different environments (see reviews in Jackson, 1981; Connell, 1983;
Schoener, 1983; Hairston, 1989; Goldberg and Barton, 1992). There are many
different kinds of competition experiments. The main focus of this chapter is how
to choose the appropriate designs and statistical analyses for addressing particular
kinds of questions about competition. Such choices depend on many aspects of
the questions and systems under study. Because the basic statistical methods
appropriate for most of the designs we present, analysis of variance (ANOVA)
and analysis of covariance (ANCOVA), are thoroughly covered in standard
textbooks of experimental design and analysis, we provide somewhat less emphasis on the nuts and bolts of the statistical analyses than do other chapters in this
volume. For an introduction to the basics of ANOVA, see Chapter 3. Although
we focus on competition, many of the points raised are equally applicable to
experiments on other types of species interactions such as predator–prey relationships or mutualisms.

4.2 Ecological Questions about Competition

The simplest question we can ask about competition is whether or not it occurs
in the field. To answer this question, it is necessary to have experimental treatments in which the absolute abundance of potential competitors is manipulated
and to test whether organisms in the treatments with lower abundance of potential
competitors perform better. The difference in performance between such abundance treatments is the magnitude of competition (or magnitude of facilitation if

performance is better with higher abundance). Finding that competition does or does not occur is an important preliminary step in any field investigation of competition but, by itself, is not very informative. Most of the important questions about competition involve comparisons of the magnitude of competition and therefore involve more complex experimental designs and analyses than simply a comparison between two or more abundance treatments (Goldberg and Barton, 1992).

One group of questions requires the comparison of the magnitude of competition among environments (sites or times). For example, field observations might suggest the hypothesis that the distribution of a species is determined by the sum of competition from all other species at the same trophic level. A field experiment to test this hypothesis must compare the magnitude of competitive effects on that focal species in sites where it is abundant with sites where it is absent or rare (e.g., Hairston, 1980; Gurevitch, 1986; McGraw and Chapin, 1989). Similarly, to resolve the current controversy among plant ecologists over whether the importance of competition increases with increasing productivity or stays constant, it is necessary to compare the magnitude of competition on populations or communities among sites that differ in productivity (e.g., Wilson and Tilman, 1991; Campbell and Grime, 1992). Statistical analysis of an experiment to compare magnitudes of competition among sites is described in Section 4.4.1.

A second group of questions requires comparison of the magnitude of competition among taxa, i.e., comparisons of competitive ability. For example, classical competition theory predicts that, for coexisting species, intraspecific competition is greater than interspecific competition. Mechanistic models of competition make predictions about particular traits that are related to competitive ability (e.g., Grime, 1977; Schoener, 1986; Tilman, 1988; Werner and Anholt, 1993) and these require comparisons between taxa that differ in these traits. Similarly, quantifying the magnitude of selection on different traits due to the competitive environment requires a comparison of the magnitude of competition among different phenotypes or genotypes. Statistical analysis of experiments to test hypotheses about rankings of competitive abilities and relationship of competitive abilities to traits of organisms is described in Section 4.4.2.

Finally, some questions require comparisons of both environments and taxa. For example, many of the mechanistic models that generate predictions about traits related to competitive ability also make predictions about the ways in which those traits change between environments or about tradeoffs between competitive ability and response to other processes such as predation or disturbance (see references above). These predictions can be tested by comparing the magnitude of competition between different taxa in different environments (e.g., different resource availabilities, natural enemy densities, disturbance rates). These are necessarily highly complex designs with many independent factors, and the main points of biological interest will often be statistical interactions between factors (e.g., taxon × competition × environment interactions to test whether competitive hierarchies change between environments). However, highly multifactorial designs will often also yield uninterpretable interaction terms as well as requiring enormous sample sizes; the addition of such complexity should be carefully considered in light of the ecological question of interest.

It is important to note that the basic competition experiment described above cannot by itself address the mechanisms underlying competitive interactions. Negative interactions can occur through direct interactions (interference competition), through a variety of shared, limiting resources (exploitation competition), through shared natural enemies (apparent competition), and other complex routes. Understanding these mechanisms is essential for developing generalizations about the role of competition in explaining evolutionary and ecological patterns (Schoener, 1986; Tilman, 1987). However, a broad diversity of approaches (field and laboratory, observational and experimental, manipulation of processes other than competition) is needed to examine specific mechanisms of interaction, making it difficult to provide a general discussion of experimental designs and statistical analyses. Therefore, the focus of this chapter will be on only one part of what is needed for a full understanding of competition: measurement and comparison of the magnitude of competition in the field.

To summarize, we emphasize that many of the important ecological questions about competition entail comparisons beyond simply demonstrating that competition occurs between a particular pair of species at a particular time and place. Therefore the first, and perhaps most important, recommendation of this chapter is that the first step in conducting a competition experiment be careful consideration of the goal of the experiment to identify appropriate comparisons. This may sound trivial and obvious but the literature is replete with statistically perfectly designed and analyzed experiments that do little to explain an observed pattern, to test assumptions or predictions of theory, or to provide a sound basis for management decisions in a particular system.

4.3 Experimental Design

4.3.1 Terminology

Before diving into the details of different types of experimental designs and the questions for which each is appropriate, we present some basic terminology. A *focal taxon* is one whose response to competition we are measuring. An *associate taxon* is one whose effect on the focal we are measuring, i.e., the one whose abundance is being manipulated. (Note that, in some experimental designs, the same species can be both a focal and an associate species.) A *background taxon* is one that is present in all experimental treatments but is not an explicitly designated focal or associate group. Background taxa can include other potential competitors, resources, natural enemies, or mutualists. "Taxon" in these three definitions will often be species, but could also represent genotypes or groups of species (see Section 4.3.6).

A *response parameter* is the aspect of performance of the focal species that is measured. *Individual-level* responses include aspects of behavior, morphology, and physiology, and components of the fitness of individuals (e.g., growth rate, survival probability, or reproductive output). *Population-level* responses include population size or growth rate, where population size can be measured as density, biomass, cover, or other measure of abundance. *Community-level* responses include parameters such as taxonomic or functional group composition, degree of dominance, or diversity.

Competitive abilities can be compared either among focal species (*competitive response*) or among associate species (*competitive effect*) (Goldberg and Werner, 1983) . The distinction is important because different traits can determine ability to suppress other organisms (competitive effect) and ability to tolerate or avoid suppression (competitive response) (Goldberg and Landa, 1991). Competitive effects can be measured at the *natural abundances* of the associate or on a *per-unit amount* basis. Common measures of abundance are density, biomass, and cover (for sessile organisms), but other measures may also be appropriate (e.g., total root length or leaf surface area for plants). The comparison of results using different measures of associate abundance may be informative in itself. For example, the result that species have different per-capita effects but similar per-gram effects would indicate that the major trait influencing per-capita competitive effect is simply biomass per individual.

4.3.2 Basic Experimental Designs

Competition experiments fall into three general categories defined by the way in which density is controlled: substitutive designs, additive designs, and response surface designs (Fig. 4.1; Silvertown, 1987). For all categories, it is essential that densities are experimentally manipulated. If natural variation in abundance is used to test competition, environmental differences potentially correlated with the natural density gradient of the associate may also directly affect the focal individuals, thus confounding competitive effects with other environmental factors, whether biotic or abiotic.

In substitutive experiments (replacement series), total density is kept constant and frequency of each species is varied (Fig. 4.1A). Numerous critiques of substitutive experiments and constraints on its use have been published over the last decade or so and its use is not generally recommended for experiments in natural communities (see Connolly, 1986, and references therein). A substitutive experiment tests only for the relative intensity of intra- and interspecific competition. So, even if all of the rather restrictive assumptions of the design are met, it is never an appropriate design to test for the occurrence of competition or for comparisons of the absolute magnitude of competition. For questions that require only the relative magnitude of competition, however, such as whether or not

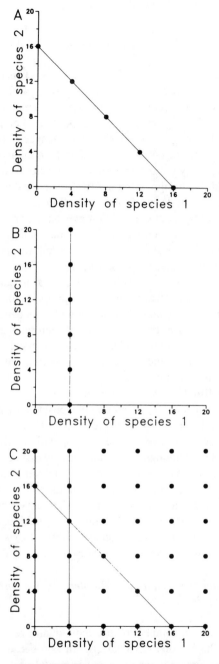

Figure 4.1. Three categories of design for competition experiments where densities of two species (or genotypes or groups of species) are manipulated: (A) substitutive experiments, (B) additive experiments, and (C) response surface experiments. Each point on the graphs represents a single experimental treatment. In (C), the subsections of the phase plane that constitute the entire experiment in (A) and (B) are shown as solid lines.

niche separation occurs, substitutive designs can be useful if the assumptions are met (see Connolly, 1986).

In additive experiments, density of the focal species is kept constant and density of the associate species is varied experimentally (Fig. 4.1B). Although usually described with only two species but many density treatments, this definition actually fits many field experiments on competition. For example, "removal experiments" typically compare response of a focal species between a treatment in which the associate is present at its natural abundance and a treatment where the associate is completely removed. If the focal species is at a constant density, this is a form of additive design with only two associate densities: presence and absence. Keeping the focal density constant in all treatments is important as this keeps the level of intraspecific competition constant. For some situations, this may mean either removing or introducing some individuals of the focal species into experimental plots. If focal individuals are added, it is critical that all focals in all treatments be manipulated the same way, i.e., all introduced or all naturally occurring individuals. Section 4.3.5 describes several important considerations in choosing associate density treatments. The major limitation on additive designs is that they confound density and frequency effects. As density of the associate increases, it also increases in frequency, i.e., it represents a greater proportion of the total mixture of the two species because the focal species is kept at a constant density. If the magnitude of competition is frequency as well as density dependent, this can be a serious problem.

The third design listed above, the response surface experiment (or addition series, sensu Silvertown 1987), gets around this problem by manipulating densities of both focal and associate species (Fig. 4.1C). This type of experiment can provide the data needed to develop realistic population dynamic models (Law and Watkinson, 1987). Although this is the ideal two-species design, the large number of density combinations needed makes such an experiment impractical for many field situations, especially when several species combinations must be investigated. Even in the laboratory, relatively few experiments have explored large combinations of densities and we know of no field experiments that have used it. Therefore, we will not address analysis of these experiments further; the interested reader should see Ayala et al. (1973) and Law and Watkinson (1987).

4.3.3 Level of Response Variables and Problems of Temporal and Spatial Scale

Most field experiments on competition use individual-level response parameters such as foraging behavior or individual growth rate because of constraints on both the spatial and temporal scales of experimental manipulations. Only for small, short-lived organisms is it feasible to measure growth rates of entire populations. This constraint means that conclusions about the population conse- quences of competition for distribution or abundance can be inferred only from individual-level measures rather than directly demonstrated. For long-lived organisms, this raises the question of which life history stage(s) of focal individuals is most appropriate to use. Ideally, experiments should be conducted with all ages or stages, in which case models of age- or size-structured population dynamics could be used to estimate population growth rates under various experimental conditions (e.g., Gurevitch, 1986). In practice, this is not always feasible and the subset of stages to be studied should be chosen based on their likely importance in population regulation. Sensitivity analysis of matrix population models of demographic data can be useful in determining the best ages or stages to study (Caswell, 1989; Chapter 11).

Relatively few field experiments have measured community-level responses to ask how species composition or diversity are affected by competition (Goldberg and Barton, 1992). Possible approaches for analysis of community-level response are to use relative population abundance as the response variable in univariate analyses like those described in Section 4.4, to use diversity indices that summarize community properties in such univariate analyses, or to use absolute abundances of all species in multivariate analyses of variance (Chapter 5).

4.3.4 Absolute vs Relative Response Variables: When to Standardize?

Almost any response variable can be expressed as an absolute value or as a percentage of the value of that variable in the absence of associates. Which of these is used in an analysis can have a major impact on the interpretation of comparisons of the magnitude of competition. Figure 4.2 shows an example in which the growth response of two focal species to a density gradient of the same associate species are compared. Using absolute values of focal response, species 2 has lower growth than species 1 in the absence of any associates (lower intercept) and also has a less steep slope. Thus, species 2 would be considered a better response competitor (less absolute decrease in growth for each associate individual added; Fig. 4.2A). However, when expressed as a percentage of growth in the absence of the associate (intercept for both is 100%), species 1 now shows less of a reduction in growth and so would be considered the better response competitor (Fig. 4.2B).

This potential conflict in interpretation between standardized and unstandardized data is most obvious for comparisons of competitive response of different focal species or of the same focal species in different environments (e.g., high and low productivity) because intercepts will usually be different with unstandardized data. However, it may also appear in comparisons of competitive effect among associate species. Normally, we would expect that a single focal species in a single site would have the same intercept (performance in the absence of any associates) in regressions with different associate species, so that standardized and unstandardized data would yield the same result. However, when the zero-density treatment is a result of experimental removal of the associates, two factors

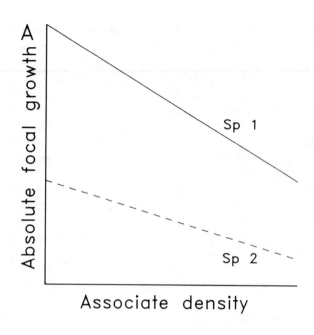

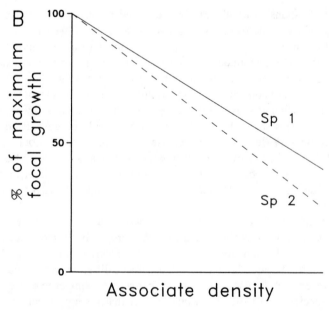

Figure 4.2. A hypothetical example of results of an additive experiment with two different types of response variables for two focal species: (A) absolute growth and (B) percentage of growth in the absence of the associate. Using absolute response, species 2 is the better response competitor, while species 1 is the better response competitor using percentage of maximum growth.

may lead to differences in intercepts among associate species. First, there may be residual effects of the former presence of the associate that could differ between associate taxa. This result is more likely to occur with sessile organisms that could have cumulative effects on a single location. For example, plants may have different residual effects on nutrient availability due to differences in litter quality and subsequent decomposition rates or soil organic matter accumulation. In addition, roots can seldom be removed and the presence of dead and decaying roots could also have significant residual effects. Second, associates can be removed only where they were initially present. If different associate species were initially present or abundant in different microenvironments and these microenvironmental conditions influence the focal species directly, focals may differ in zero-density treatments for different associates, i.e., differ in intercepts. Both of these potential problems point out the importance of having separate zero-density treatments for each of the associate species used in a removal experiment.

There has been remarkably little discussion in the literature about the conditions in which response in the presence of associates should be standardized to response in the absence of associates; the following recommendations should be viewed as preliminary. For questions concerning the consequences of competition for distribution and abundance, relative values are probably most useful. For example, suppose that competition reduces individual growth of a focal species by 10 g in an unproductive site and by 100 g in a productive site, suggesting that competition is more intense in the more productive site. However, if the growth in the absence of competition were 10 and 200 g in the unproductive and productive sites, respectively, the relative responses are 0% and 50%, suggesting competition is more intense in the unproductive site and, in fact, intense enough there to exclude the focal species. [See Campbell and Grime (1992) for an example of contrasting results for standardized and unstandardized data in a competition experiment.]

On the other hand, if the mechanisms of competition are under investigation, absolute values may be more useful. For example, it may be of interest to relate reductions in resource availability due to consumption by an associate species to changes in resource consumption and growth in the focals. Using standardized data would make it difficult to understand the processes underlying these relationships. Absolute responses are also necessary when competitive effects and responses are being quantified to use as input into models of population interactions because most population dynamic models use absolute, not standardized, parameters. For example, in the Lotka-Volterra competition equations, equilibrium density in the absence of competition (K) is an explicit parameter and both K and the per-capita competition coefficients (α_{ij}) have an impact on the dynamics and equilibrium outcome of the interaction.

4.3.5 Manipulation of Competition Intensity: How Many Density Treatments?

The range of density variation in an additive experiment could be simply presence vs absence of the associate or several densities could be experimentally imposed.

Presence/absence comparisons are appropriate where the total magnitude of competitive effect under natural conditions is all that is required. However, unless competitive effects are completely linear, presence/absence comparisons (or comparisons of any two densities) do not allow accurate calculation of per-capita effects (Fig. 4.3). Where the density dependence of per-capita competitive effects has been tested, it is rarely constant (Harper, 1977; Schoener, 1986; Goldberg, 1987; Miller and Werner, 1987), and so interpolating from presence/absence comparisons to per-capita effects across a range of densities is probably not generally safe. Comparisons of competitive effect between associate species and,

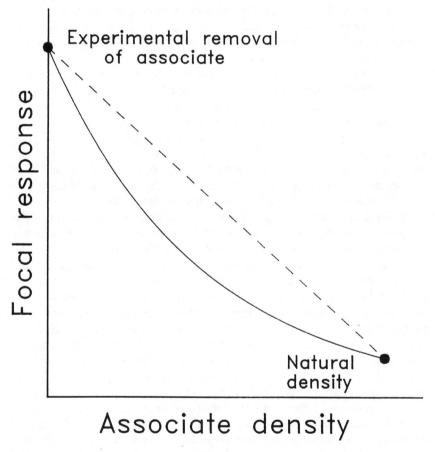

Figure 4.3. An example of inaccurate estimation of per-capita competitive effects from an experiment with two treatments: the associate present at its natural density and the associate completely removed. The true per-capita effect is shown as a solid line. A linear extrapolation from the two treatments (dashed line) underestimates the true per-capita effect at low associate density but overestimates the true effect at high associate density.

especially, relating these competitive effects to traits requires per-capita effects so that effects of total density and per-capita effects (or, more generally, total abundance and per-unit amount) are not confounded.

If multiple densities are chosen, many density treatments with low or no replication will usually be better than few, well-replicated densities. The appropriate analysis will then be a comparison of slopes from regressions of the response variable on density rather than contrasts between means of different treatments (see Section 4.4.2 and Chapter 8). As both the strongest nonlinearities and the greatest variation among replicates tend to be at the lowest densities, concentration of density treatments at the low end rather than an even distribution is recommended. Densities above natural are also useful because (1) naturally occurring densities usually fluctuate considerably and (2) competition might occur only at densities above those normally determined by other factors such as predators. Regardless of the range of densities used in an experiment, it is always important in publications to identify the natural density range so readers can assess the applicability of results to natural conditions.

4.3.6 Pairwise vs Total Competition: Which Associate Groups to Use?

The associate group whose abundance is being manipulated can be a single species or a group of species considered together. Which is more appropriate depends on both the biology of the species under study and on the goal of the experiment. The tendency has been for animal ecologists to measure competition between pairs of species chosen a priori as likely strong competitors while plant ecologists have more often simply quantified total competition from the entire plant community on a focal species. Because plants all require the same few resources, all plants are potential competitors and there is often little a priori basis for choosing subsets. Therefore the most common type of field competition experiment in plant communities is a presence/absence additive design where the associate "species" is the entire plant community except for the focal individuals (Goldberg and Barton, 1992). For many animal species, on the other hand, more restricted guilds of potentially interacting species that share resources can often be defined (e.g., seed-eaters or insect-eaters). The danger here is that while potential competitors are often chosen by taxonomic relatedness (e.g., congeners), distantly related taxa may be just as likely to share resources and compete for them (e.g., seed-eating rodents, ants and birds in deserts; Brown et al., 1986).

When the question of interest concerns comparisons of competitive ability among species and relationships of competitive ability to traits, pairwise experiments with single species as associates and focals are usually most appropriate. The number of species also becomes important because tests of these relationships essentially use each associate or focal species as a "replicate" in a secondary statistical analysis. Use of only a single species with each value of a trait is unreplicated for tests of hypotheses about species characteristics related to competitive ability. For example, if only two associate species are compared and the one with larger leaves has a stronger negative effect, the conclusion that leaf size confers strong competitive effect is unwarranted because of lack of replication within each leaf size class. The taxa may also differ in other correlated traits such as root length or nutrient uptake rate. As leaf size is confounded with these other variables, assigning a causal effect to leaf size alone is not justified. We recommend either replicating species within trait classes or using species with a range of values of a continuous trait.

4.3.7 Direct Effects, Indirect Effects, and Higher-Order Interactions: What to Do with Background Species?

Response of a focal species to manipulation of abundance of an associate species represents the sum of the direct effects of the associate plus all the indirect effects and higher-order interactions involving the associate and all the background species, the *net effect*. If no background species are present, only direct effects are measured (but see below). *Indirect effects* are defined as changes in the effect of an associate on a focal species when background species change the abundance of the associate (linkage indirect effects, sensu Miller and Kerfoot, 1987). *Higher-order interactions* are defined as changes in the effect of an associate on a focal species when a background species changes the per-capita effect of the associate (behavior indirect effects, sensu Miller and Kerfoot, 1987).

If the primary question concerns the effect of the associate species on the distribution and abundance of the focal species, net effect may in fact be what is desired. For example, suppose removal of a dominant has little effect on a focal species because other, unmeasured, species respond to its removal more quickly than does the focal species and thus end up suppressing the focal species just as much as did the designated associate. The conclusion that competition from that associate species is unimportant in controlling the abundance of that focal species is reasonable, even though the associate may have strong direct effects on the focal in a pairwise experiment with no background species. However, if the primary question is which traits of an associate species allow it to become dominant, measuring its net competitive effect will be misleading because traits of the background species will also influence the results.

Therefore our general recommendation is that questions about the relationship of traits to competitive ability or determination of competitive hierarchies are usually best addressed in pairwise experiments that minimize the potential for indirect effects or higher-order interactions. In contrast, questions about the consequences of competition for abundance and distribution are usually best addressed by incorporating any nondirect effects into the response variable, i.e., by leaving background species in all treatments. We point out, however, that strong nondirect effects restrict generalization across sites or times even more than is usual in ecological field experiments because the biotic environment and therefore complex of nondirect effects differ as well as the abiotic environment and history of the sites.

In reality, almost any field experiment of any design has background species and therefore we can never measure only direct effects. It is rarely possible or even desirable to eliminate completely all the biota except for the designated focal and associate species. First, exploitation competition itself is an indirect interaction, mediated through shared food resources. When the resource is a living organism, even if the food is not an explicitly designated associate or focal species, it obviously must be present in the experiment. Similarly, negative interactions may be mediated by shared natural enemies (apparent competition; Holt, 1977; Connell, 1990) or mutualists, and removing "background" species would preclude detecting an important interaction between two "competitors" (but appropriate if quantification of only exploitation competition is desired). Monitoring the dynamics of these intermediary nonmanipulated species, whether resources, natural enemies, or mutualists, can be an important step toward understanding the mechanisms of interactions. Second, microorganisms are part of the biota and may be important links in chains of indirect effects, but are very difficult to remove completely. The important point in terms of designing and interpreting field experiments is that it is critical to recognize which background species are present and to incorporate this knowledge when interpreting results.

4.4 Statistical Analysis and Interpretation of Results

4.4.1 Comparison of Habitats: Effects of Competition on Distribution and Abundance

One of the most common questions about competition is whether it influences the distribution and abundance of a focal species. The recommended approach to this question is to repeat a presence/absence additive design with background species in differing habitats, those where the focal species is present or abundant and those where it is absent or rare. This question is really a population-level one and ideally population-level response parameters should be used (Section 4.3.3). A minimum of two replicate sites in each habitat category (focal abundant vs rare) should be used so that the relevance of habitat differences to focal species distribution can be inferred from the results. With only a single site in each habitat category (admittedly the most common design actually used), the only valid inference that can be made is whether sites differ in the magnitude of competition. That is, site is confounded with habitat.

The choice of associate species can be a single species chosen a priori, a subset of species, or the entire community (see Section 4.3.6). Since the question concerns the net consequences of the presence of some associate species or group of species, it is appropriate to leave all other species present as background species in all treatments, i.e., to quantify the sum of all direct, indirect, and higher-order effects.

The actual experiment would involve establishing an experimental area in each site within which the focal species would be added and the associate species removed as appropriate. It is important to add individuals of the focal species back into its native habitat as well as the habitat from which it is absent to account for transplanting effects. Such a design also ensures that the density of the focal species is the same across all treatments (Section 4.3.2).

For an experiment with a single focal species, a single associate species or group, and several sites of each habitat type, the appropriate ANOVA model is

$$X_{ijkl} = \mu + \tau_i + \theta_j + \tau\theta_{ij} + \Sigma(\theta)_{jk} + \tau\Sigma(\theta)_{ijk} + \varepsilon_{ijkl} \tag{4.1}$$

where μ is the overall mean, τ_i is the deviation due to the ith associate treatment (present at natural abundance or removed), θ_j is the deviation due to the jth habitat (focal species naturally abundant or rare), $\tau\theta_{ij}$ is the deviation due to the interaction of the ith treatment with the jth habitat, $\Sigma(\theta)_{jk}$ is the deviation due to the kth site nested within the jth habitat, $\tau\Sigma(\theta)_{ijk}$ is the deviation due to the interaction of the ith treatment with the kth site nested in the jth habitat, and ε_{ijkl} is the deviation due to the lth replicate within each site-treatment combination. Treatment and habitat are both fixed factors and site is a random factor, so the overall model is a mixed cross-nested design. The degrees of freedom, estimated mean squares E(MS), and F tests for this model are shown in Table 4.1 and SAS commands in Appendix 4.1A. See Searle (1971) for a detailed explanation of the underlying statistical theory. More complex designs can also be used that include blocking factors within sites (Chapter 3).

The important result from this ANOVA is whether the treatment × habitat interaction is significant, i.e., the magnitude of competition differs between habitats where the focal species is abundant and where it is absent or rare. The expected pattern of results is shown in Fig. 4.4, with a larger magnitude of competition where the focal species is absent or rare than where it is abundant. Notice that, in this analysis, the treatment × habitat interaction is tested over the treatment × site interaction and the power of the test is a function of the number of sites. Thus, the most powerful design to use in this case is to maximize the

Table 4.1   ANOVA for a Completely Balanced Cross-Nested Design

Effect	df	E(MS)	F-ratio
Treatment	$t-1$	$\sigma^2_e + n\sigma^2_{\tau\Sigma(\theta)} + hsn\sigma^2_\tau$	$MS_T/MS_{TS(H)}$
Habitat	$h-1$	$\sigma^2_e + tn\sigma^2_{\Sigma(\theta)} + stn\sigma^2_\theta$	$MS_H/MS_{S(H)}$
Treatment × habitat	$(t-1)(h-1)$	$\sigma^2_e + nr^2_{\tau\Sigma(\theta)} + sn\sigma^2_{\tau\theta}$	$MS_{TH}/MS_{TS(H)}$
Site (habitat)	$h(s-1)$	$\sigma^2_e + tn\sigma^2_{\Sigma(\theta)}$	$MS_{S(H)}/MS_e$
Treatment × site (habitat)	$(t-1)h(s-1)$	$\sigma^2_e + n\sigma^2_{\tau\Sigma(\theta)}$	$MS_{TS(H)}/MS_e$
Residual	$ths(n-1)$	σ^2_e	

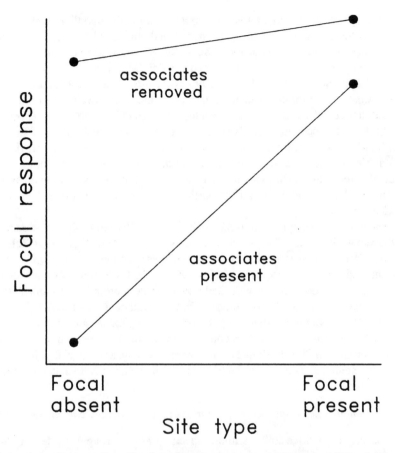

Figure 4.4. Pattern of predicted results from a presence/absence additive design if competition influences the distribution and abundance of the focal species. The key result is that the magnitude of competition (depression due to presence of competitors) is much greater where the focal species is normally absent than where it is normally present (i.e., a habitat × competition interaction). Note that, in this case, the focal species also performs better in the habitat where it is abundant than where it is absent even in the absence of associates and competition is present in both types of habitats so that both habitat and competition would probably be significant main effects.

number of sites within habitat types and minimize the number of replicates within sites. Obviously, such a strategy creates the potential for logistical difficulties if the sites are any distance apart.

When the number of replicates is not the same for each level of all treatment factors, the experimental design is termed "unbalanced" (Searle, 1987). Unbalanced designs are very common in ecological experiments because of insufficient material (e.g., it may be more difficult to get replicates from one site or species or genotype than others) or loss of replicates through mortality. Unbalanced ANOVA designs are more difficult to analyze and, unlike balanced designs, there is no single correct way to analyze them. Readers may wish to consult recent discussions of the analysis of unbalanced designs (Milliken and Johnson, 1984; Fry, 1992; Shaw and Mitchell-Olds, 1993). SAS 6.0 and above will construct an approximate F-test for unbalanced designs using the Satterthwaite approximation for the degrees of freedom (Satterthwaite, 1946; Hocking, 1985; see Appendix 4.1A). If SAS procedure GLM is used for the analysis, Type III sums of squares (SS) must be requested; the F-tests provided by the other types of SSs (I, II, and IV) are incorrect. Because there are several approaches to this problem and no general agreement, the individual researcher must evaluate whether the approach taken by SAS is appropriate for the particular design employed (see discussion in Chapter 3).

A new alternative approach to the analysis of unbalanced designs that contain both fixed and random effects—mixed models—is maximum likelihood (McLean et al., 1991), which is implemented in the latest version of SAS (6.07, SAS Institute Inc., 1992) as procedure MIXED (Appendix 4.1B). One potential advantage of this method is that it will likely prove to be more robust to unbalanced designs than the standard least-squares approach, although this has not yet been proved. A disadvantage is that computations for complex designs can take a long time and sometimes will fail to converge on a solution. Be warned that the documentation of the technique is not completely clear, especially for those without extensive statistical training; consultation with a statistician is highly recommended.

4.4.2 Comparison of Species: the Relationship of Traits to Competitive Ability

Comparing competitive abilities among species (or other groups) and determining the relationship of competitive abilities to traits will usually require per-capita measurements. The recommended approach to compare species is an additive design with multiple densities of the associate, using either a single focal species and several associate species that differ in value of some trait (comparison of competitive effect) or a single associate species and several focal species (comparison of competitive response; Fig. 4.2A). A significant relationship (non-zero slope) between density of the associate and response of the focal species indicates competition (negative slope) or facilitation (positive slope).

The analysis is stepwise (see Zar, 1984, Fig. 18.1). The first step is to test for differences in density effects among focal or associate species, a test for homogeneity of slopes. The ANCOVA model is

$$X_{ijk} = \mu + \Delta_i + \alpha_j + \Delta\alpha_{ij} + \varepsilon_{ijk} \qquad (4.2)$$

where μ is the mean intercept (the value of the response parameter at 0 density), Δ_i is the deviation due to the ith density treatment (the average slope of the line that predicts the effect of associate species density on the response parameter), α_j is the deviation from the mean intercept due to the jth associate or focal species (differences in intercepts among associate or focal species), and $\Delta\alpha_{ij}$ is the deviation due to the interaction of species identity and density (differences in slopes among species; Fig. 4.2A). Δ_i is a continuous variable and α_j is a categorical factor.

In this first analysis, we are concerned only with the interaction term. We ignore the main effects of density and species identity at this juncture because such main effects are meaningful only if the interaction effect is declared not statistically significantly different from 0. The SAS commands for performing this analysis are given in Appendix 4.2. If differences among slopes are found, then a post-hoc Tukey-type multiple comparison test can be performed (Zar, 1984, Section 18.6) to discover which particular species differ from each other. Alternatively, if one has a priori hypotheses concerning differences between particular species or sets of species, then a series of specified comparisons (contrast tests) should be done rather than a single overall ANCOVA.

A potential problem in this analysis arises if only a single zero-density treatment (no competitors) is used for all associates, because it is impossible to assign a treatment level (identity of associate) to replicates at this density. In this case, it is reasonable simply to randomly allocate all of the zero-density replicates among the associate species. (For example, if there are three associate species and 30 zero-density replicates, then 10 replicates would be randomly assigned to each species treatment.) This problem should arise only when all associate individuals are added (e.g., to cages or in greenhouse experiments), because in field removal experiments, there will usually be distinct zero-density treatments for each associate species (see section 4.3.4).

If slopes are homogeneous, one can then proceed to (1) measure the average slope (Δ_i) and (2) test for differences in intercepts (α_j) among species using the following model:

$$X_{ijk} = \mu + \Delta_i + \alpha_j + \varepsilon_{ijk} \tag{4.3}$$

Note that the model is identical to that in Eq. (4.2) except that the interaction term has been deleted. The SAS commands are given in Appendix 4.3. Significant density effects (a nonzero slope) indicate that competition or facilitation is occurring and the average slope measures the general per-capita effect of all associates on the focal species.

In comparisons of focal species, significant species effects (α_j) indicate simply that the focal species differ consistently across all densities of the associate species (e.g., in maximum growth rates or sizes). In comparisons of associate species, significant species effects are more problematic because the focal species remains the same and thus should not differ in response to the associates when no associates are present. Some biological interpretations of significant associate main effects were given in Section 4.3.4. An alternative interpretation of either significant focal or associate effects is that it could be an artifact of using linear regressions on nonlinear data. Figure 4.5 shows an example where two focal species have the same slopes but different intercepts in linear regression analyses. However, inspection of the graphs shows that the intercepts are actually very similar but that species B is strongly nonlinear at low associate densities. If

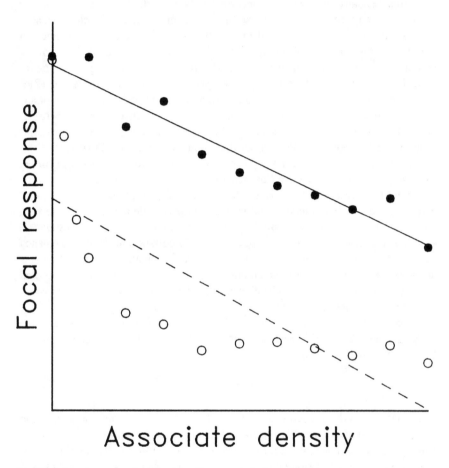

Figure 4.5. An example of misleading results of comparison of slopes (individual competition coefficients) using linear regression models. Using linear regressions, the two species would appear to have different intercepts, but similar slopes, i.e., similar competitive effects if the curves represent two associate species. However, inspection of the data reveals the intercepts are actually very similar and species B (open circles) has a strongly nonlinear effect on the focal species at low densities. Inspection of the residuals of the linear regression for species B would show a nonrandom distribution with associate density.

transformations do not succeed in linearizing the data, a nonlinear regression analysis is warranted (Chapter 8).

Relating per-capita effects or responses to values of a particular trait can be conducted as a second-level statistical analysis by correlating slopes with trait values (e.g., Goldberg and Landa, 1991). With fewer species or with few, discrete values of the trait, slopes can be compared with a priori contrasts among groups of the associate or focal species (Day and Quinn, 1989) or in the context of the ANCOVA where species are nested within species types in a way exactly analogous to sites nested within habitats (see Eq. 4.1).

4.4.3 The "Envelope Effect"

The performance of an organism is likely to be affected by many factors (e.g., genotype, weather, predators) besides the density of competitors. Usually in field experiments such additional factors are either poorly controlled or not controlled at all and effects of competitors may be obscured by the large amount of variation caused by these other factors. This general problem may arise in any type of analysis of field experiments; in regression analyses of competition, it may be manifested as an "envelope effect" (Firbank and Watkinson, 1987; Goldberg, 1987; Fowler, 1990a). Competition may act to keep the response of the focal species below some maximum, while within this competitively determined "envelope" other factors act to further depress the focal species. The result is that a plot of the response variable against density will be a scatter of points forming a triangle in the lower left-hand portion of the graph (Fig. 4.6).

If envelope effects occur, they can cause considerable difficulty in quantifying density effects. First, the effect will be obscured and the slope of the line decreased by the large amount of variation caused by other factors. Second, the estimated per-capita effect will be substantially less than the true effect because a simple linear regression will place the regression line through the middle of the points, rather than along the upper edge of the envelope. Because the variance is smaller at higher competition intensities, the mean will be closer to the upper edge of the envelope at higher competition intensities than at low intensities, thus lowering both the intercept and the slope. Third, the important statistical assumption of homoscedasticity, that the variation around the regression line is the same everywhere, is being violated. Clearly, the variation decreases with increasing density. In this case, there may very well be no transformation of the data that will make the variance homogeneous. Pacala and Silander (1990) developed one solution to this problem. They note that the residuals (the variation around the regression line) of such a distribution are distributed as a gamma function, which has the properties of being unimodal, left-skewed, and bounded at zero (see Chapter 12 for a description of the gamma function). In their experiment, competitive effects followed a hyperbolic function. Thus, rather than the more familiar

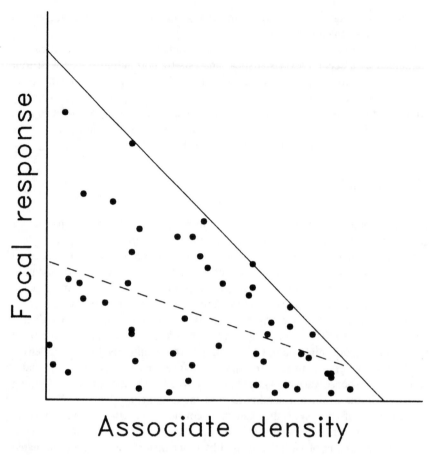

Figure 4.6. Example of the "envelope" effect where the variance in focal species response is much greater at low associate densities than at high associate densities. The solid line at the upper edge of the envelope represents the constraint on maximum focal response imposed by competition. The dashed line represents the competitive effect estimated by a standard least-squares linear regression.

least-squares method, they used a maximum likelihood estimator and iterative Newton method as described in Chapter 8.

Another possible solution to the problem of the "envelope effect" is to monitor other potentially interacting factors and use them as covariates in a multiple regression (Chapter 9). The possible drawbacks of this approach are (1) the logistical difficulty of measuring many additional variables, (2) the potential failure to measure the key factor, and (3) the fact that multiple regression brings with it an additional host of statistical assumptions and difficulties (Chapter 9). A final possible solution is to manipulate or hold constant other variables. Here the drawback is the movement away from as natural a setting as possible, the entire reason for doing a field experiment to begin with.

4.5 Related Techniques

4.5.1 Isotonic Regression

An alternative and intermediate approach to the ANOVA and ANCOVA models presented above is isotonic regression, which can be thought of as a form of ANOVA to be used when there are ordered expectations (Barlow et al., 1972). This technique has been little used up to now in biology. Gaines and Rice (1990) present a cogent description of this technique and offer a microcomputer program for implementing it. The advantage of isotonic regression for detecting density-dependent responses is twofold. First, there are no assumptions as to the shape of the density response except that it is monotonic. For this reason, isotonic regression can be more powerful than standard least-squares regression analysis when the shape of the response is irregular or a step function. Second, fewer density treatments are needed than in ANCOVA because the precise shape of the density effect is not of interest. Instead, replication efforts can be placed elsewhere in more associate or focal species, more sites, or more replicates per site.

However, there are also several limitations to isotonic regression that restrict its usefulness to relatively simple questions about competition. First, per-capita effects cannot be calculated so it is not appropriate if comparisons are needed of species effects independently of their total abundance. Second, the technique is currently developed for analysis of different levels of a single factor. Thus, it could be useful in detecting the occurrence of negative effects of increasing density of a single associate but not in comparing associates or detecting interactions between competition and other factors. However, the method can be used in experiments with one additional factor other than level of competition if the dependent variable is *relative* magnitude of competition. In this special case, competitive effects are included in the response variable rather than as distinct treatments (section 4.3.4). For example, the reduction in response in the presence of competitors at their natural abundance relative to their complete absence could be compared over a series of sites with different expectations of the magnitude of this relative reduction.

4.5.2 Spatial Heterogeneity

This chapter has not dealt with the analysis of spatial heterogeneity except when spatial variation is an explicit part of the ecological question of interest (e.g., variation in magnitude of competition between sites or habitats). Chapter 3 discusses ANOVAs with block designs that explicitly account for variation due to spatial heterogeneity. Chapter 14 discusses an alternative approach to analysis of spatial heterogeneity using variograms. Both of these methods require replication within a treatment, which is impossible using the multiple, unreplicated density design recommended in this chapter when comparisons of per-capita competitive abilities are required. There is currently no satisfactory general method that can both account explicitly for spatial heterogeneity and maximize investigation of potential nonlinearities. Therefore, the individual investigator must decide which is more important in the study and sacrifice number of blocks (replicates per density) or number of distinct densities accordingly. If an unreplicated regression design is chosen and the important environmental variables are known, these can be measured in individual experimental units and used directly as covariates in an ANCOVA. A third possible approach for incorporating spatial heterogeneity effects in designs with unreplicated densities is the permutation method described in Chapter 15, although this method is as yet untried for this application.

4.5.3 Beyond Simple Dependent Variables

The analysis of competitive effects potentially involves examining the response of many different types of response variables. The previous discussion has been in the context of examining a single variable that most likely meets the standard normality assumptions of parametric statistics. Time-based response variables such as survival time or time to emigration, while still single measures, often do not meet standard normality assumptions and require different analytic methods (Chapter 12).

Often the response variable is more complex than a single measure of focal performance. If individual-level measures are being used, for plants one might measure vegetative growth, seed number and seed size. If community-level measures are being used, one might measure the densities of several focal species. If the response variables (traits or species) are likely to be correlated with each other or to interact with each other, then a multivariate analysis of variance is appropriate (Chapter 5). If data are collected over time, for example, as growth curves of individuals or populations, then a repeated-measures analysis should be used (Chapter 6). For long-term experiments, time series analysis may be most appropriate (Chapter 7).

For all of these techniques, the basic design criteria outlined above are still appropriate, although different statistical techniques will be most powerful for different numbers of treatment levels vs replicates within treatments. In all cases, careful thought should be given to the type of analysis to be used at the time of experimental design, rather than waiting until the experiment is complete.

4.6 Some Final Comments on Experimental Design

We have argued that answering questions besides "does competition occur?" is critical to developing and testing general models of competition and that answering these questions will require more complex experimental designs than have often been used in field competition experiments. Nevertheless, there is a real danger that our experimental designs will become so complex that insufficient resources are available to detect any but enormous effects. Decisions on allocation of resources will always depend on the details of any particular study, but a few preliminary steps will aid in the process. We strongly recommend conducting pilot experiments in the field to estimate variances and thus the replication needed to detect effects of a given magnitude (Zar, 1984); results of such power tests may quickly demonstrate that experiments for some questions are simply logistically impossible in that system. If several factors are experimentally manipulated, replication should be focused on the factor(s) that will be used as the error variance(s) (denominator MS) for the most important test(s). As the example in Table 4.1 showed, this is not always at the level of replication of the smallest experimental units or the most detailed treatment combination (e.g., field plots). It is critical to work out the estimated mean squares for any design before beginning an experiment so that replication will be done at the proper level. Brownlee (1965) and Searle (1971) present examples for many common designs; consultation with a statistician at this stage is also highly recommended. The important point is simply that there will always be tradeoffs between the complexity of questions that can be asked from an experiment and the statistical power available for testing particular effects. Which of these is emphasized and which sacrificed is the most important decision underlying any experimental design.

Acknowledgments

We thank Brad Anholt, Jessica Gurevitch, and Earl Werner for valuable discussions of many of the issues raised in this chapter and Yoram Ayal, Drew Barton, Norma Fowler, Jessica Gurevitch, Ariel Novoplansky, and Earl Werner for their helpful comments on earlier drafts. The National Science Foundation provided support for the research on which this chapter was based. This collaboration was made possible by the telecommunications revolution.

Appendix 4.1

SAS program code for analysis of the cross-nested ANOVA shown in Eq. (4.1):

A. Analysis Using Least Squares

```
PROC GLM;
    CLASS T H S;                          /T = associate treatment, H = habitat,
                                                      S = sites within habitat/
    MODEL X = T H T*H S(H)      /use type III sums of squares.  Warning: results are
        T*S(H)/SS3;                        only approximate if the design is unbalanced./
```

*** If the design is unbalanced use the following command to
    obtain the correct F tests for the main effects ***

```
    RANDOM S(H) T*S(H)/TEST;      /Declare sites to be a random effect and ask SAS to
                                           compute the Satterthwaite correction for an
                                                              unbalanced design./
```

*** If the design is balanced use the following commands to
    obtain the correct F tests for the main effects ***

```
    TEST H=T E=T*S(H);                   /test associate treatment over
                                                      the treatment-site interaction/
    TEST H=H E=S(H);                            /test habitat over site/
    TEST H=T*H E=T*S(H);            /test treatment-habitat interaction
                                                over the treatment-site interaction/
```

B. Alternative Analysis Using Maximum Likelihood

```
PROC MIXED;                            /mixed model analysis using reduced
                                           maximum likelihood (the default)/
    CLASS T H S;
    MODEL X = T H T*H;                    /only fixed effects are placed
                                               in the model statement/
    RANDOM S(H) T*S(H);            /random effects are specified separately/
```

Appendix 4.2

SAS program code for analysis of the ANCOVA model in Eq. (4.2):

```
PROC GLM;
    CLASS A;                          /A = associate or focal species treatment/
    MODEL X = D A D*A/SS3;      /D = density, use type III sums of squares/
```

Appendix 4.3

Analysis of the ANCOVA model in Eq. (4.3):

```
PROC GLM;
   CLASS A;                          /A = associate or focal species treatment/
   MODEL X = D A/SS3;                          /use type III sums of squares/
```

5

MANOVA: Multiple Response Variables and Multispecies Interactions

Samuel M. Scheiner

5.1 Ecological Issues

Ecological questions very often concern how multiple components might respond to some change in the environment or might differ among groups. We are often most interested in the interactions among those components and how the interactions might change as the environment changes. Studies in population biology often address issues of constraints or tradeoffs among traits such as seed size vs seed number, sprint speed vs endurance, or development time vs size at maturity. Studies in community ecology often address questions of how a number of coexisting species respond to the removal of a keystone predator or to a change in the environment. In statistical parlance these multiple components are referred to as multiple response variables.

In this chapter I deal with situations in which the independent variables (experimental manipulations) consist of categorical variables, e.g., competitors present or absent or three levels of nutrient availability. When only one response variable has been measured such data are examined by analysis of variance (ANOVA) (Chapters 3 and 4). When more than one response variable has been measured the most appropriate method of analysis is usually multivariate analysis of variance (MANOVA) in which all dependent variables are included in a single analysis. Unfortunately, ecologists often do not use MANOVA when they should. A quick survey of a recent volume of *Ecology* (72, 1991) found that of 62 papers that measured more than one response variable for which a multivariate test would have been appropriate, only 9 actually used MANOVA. The other studies all used multiple univariate analyses, each of the response variables being examined separately. In a number of the latter papers the questions dealt explicitly with interactions among the response variables.

5.2 Statistical Issues

A multivariate analysis is preferred over multiple univariate analyses for two reasons. First, the ecological questions are often multivariate, involving interactions among response variables. Differences that exist among groups may not be a feature of any one response variable alone but, rather, of the entire suite of variables. Second, performing multiple univariate tests can inflate the α value—the probability of a type I error—leading one to conclude that two groups or treatments are different with respect to one or more of the dependent variables, even though the differences are just due to chance. I deal with each of these issues in turn.

5.2.1 Multivariate Responses

We need to be concerned about correlations among response variables in two cases that are opposite sides of the same coin. The more obvious case is when a research question explicitly deals with interactions among response variables. Such interactions are indicated by either (1) shifts in correlations among variables from one treatment to another or (2) correlations across treatments that differ from correlations within treatments. Such correlations among response variables can be revealed only by some form of multivariate analysis. This need for multivariate analysis is especially true when more than two response variables are measured. Simple pairwise correlations cannot reveal the entire pattern of changing relationships among a multiplicity of variables.

A common alternative to multivariate analysis, especially in community ecology, is to analyze a composite variable such as a diversity index (Magurran, 1988). Although such an index has the virtue of simplicity, it can actually hide the features of the data that one is attempting to discover. For example, two samples may have identical diversity values even though individual species may have very different abundances; this would occur if an increase in one species exactly matched a decrease in another species. Such interactions become apparent only by following the behavior of individual response variables within a multivariate analysis.

A not-so-obvious need for multivariate analysis arises when the question is whether two or more groups or treatments differ. For example, we may measure abundances of five species of small mammals on plots with and without predators. Or we may measure 10 aspects of plant morphology in a nutrient and light manipulation experiment. It is possible that tests of individual response variables will fail to reveal differences among groups that a multivariate test would reveal. One form of this effect is when the correlation of variables differs in its sign within and between treatments, termed Simpson's paradox (Simpson, 1951). For example, the number and size of seeds may be positively correlated within treatments because larger plants have more total resources to devote to reproduction. However, across different nutrient levels the correlation may be negative as plants switch from a strategy of many, small seeds at high resource levels to few, large seeds at low resource levels.

Conversely, a multivariate test might reveal that what appears to be a number of independent responses is, actually, a single correlated response. For example, if the removal of a predator led to (univariate) significant increases in two prey species, one might be tempted to conclude that the predator affected two independent species. However, if the prey species abundances were correlated due to other interactions (e.g., mutualism), then there is only a single (albeit complex) response.

5.2.2 The Problem of Multiple Tests

The most important statistical issue involves the use of multiple statistical tests to address a single hypothesis. For example, consider a comparison of two groups that do not actually differ. If one uses a typical nominal type I error rate of $\alpha = 0.05$, and 10 response variables are measured, then the probability of declaring the two groups different for at least one variable is 40%. There are two approaches, plus one hybrid approach, that can be used to solve this problem.

The simplest solution is to go ahead with the standard univariate ANOVAs but use a Bonnferoni correction in setting the α level, the probability level at which an effect is deemed to be statistically significant. For example, if five tests are to be done then the correct $\alpha' = \alpha/5 = 0.01$. This procedure is actually somewhat conservative and a less conservative procedure, a sequential Bonnferoni correction, is now recommended (Rice, 1990).

The alternative approach is to use MANOVA. Only a single analysis is performed and thus no correction of α is needed. The hybrid approach is to first perform the MANOVA, preceding on to the ANOVAs only if the MANOVA yields a significant result. This approach is referred to as a protected ANOVA. The underlying logic is the same as that in which an ANOVA is performed before doing multiple comparison tests of means (Section 1.2.4). The one difficulty is that, unlike multiple comparison procedures like Tukey's test and Student-Newman-Keuls test, there is no generally accepted way to correct for multiple tests even within the "protected" framework. That is, some spurious significant differences are still likely to be declared even when the overall rate of such errors has been lowered. See Harris (1985, Section 4.2) and Stevens (1992, Section 5.5) for discussions of these issues.

Considerable debate exists as to the correctness of each of these approaches. As is often the case in statistics, the best approach depends on the questions being asked. Those who argue for a strictly multivariate approach emphasize the interrelatedness of the dependent variables. Univariate analyses are not able to capture and dissect those relationships except in a very superficial way. On the other hand there are instances in ecology when the behaviors of individual variables are of interest in their own right. For example, theory may make predictions regarding changes in specific traits across treatments. Or predictions as to the probability of local extinction of species require examination of the abundances of each species in a study. Or further theory development may require a detailed examination of individual variable behavior as a preliminary to further refinement of predictions. Huberty and Morris (1989) have a cogent discussion of the relevant considerations for when multivariate tests vs multiple univariate tests are most appropriate.

The above discussion of univariate vs multivariate approaches is couched in terms of determining the effects of individual response variables on the outcome of the MANOVA. For example, of the many traits that were measured, which were actually responsible for the differences among populations? Such information is available directly from MANOVA; univariate ANOVAs are not necessary. In addition, the carrier of the information, the eigenvectors of the MANOVA, also contain information on patterns of variable correlations. Thus, properly interpreted, a MANOVA can provide all necessary information.

5.3 Statistical Solution

5.3.1 A Simple Example

To elucidate the basic principles and procedures for MANOVA, I present data from a study of the herbaceous perennial *Coreopsis lanceolata* L. (Asteraceae). These data have been modified to better illustrate several points. We will pretend that the data are derived from an experiment consisting of plants grown in individual pots in a greenhouse at three different nutrient levels, low, medium and high, with 67, 48, and 80 plants, respectively. Treatments were randomized on the bench with no further blocking. The plants were allowed to flower and set seed. From each plant one flowerhead was collected (for simplicity of analysis) and the number of seeds and mean seed mass were determined. The results are presented in Fig. 5.1A. The data were analyzed using the following model:

$$(X_{1ij} X_{2ij})^T = \mu + \tau_i + \varepsilon_{ij} \qquad (5.1)$$

where $(X_{1ij} X_{2ij})$ is a vector of measurements of traits 1 (mean seed mass) and 2 (number of seeds) of the jth replicate in the ith treatment, T indicates vector transpose, μ is the grand mean of all measured individuals, τ_i is the deviation from the grand mean due to the ith treatment, and ε_{ij} is the deviation of the jth individual from the mean of the ith treatment. These last deviations (ε_{ij}) are assumed to be independently and randomly distributed with mean 0 and variance σ^2_e. These are the same assumptions as for ANOVA except that the error variance is now assumed to be multivariate normal (see also Section 5.4.3). In this instance

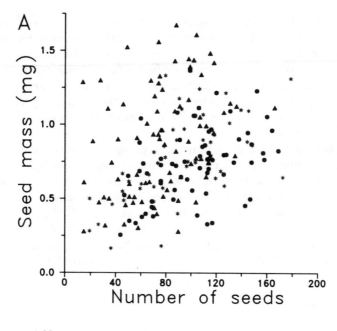

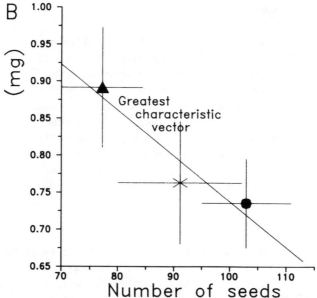

Figure 5.1. (A) Number of seeds and mean seed mass for each individual in the three nutrient treatments: high (●), medium (∗), and low (▲). (B) Bivariate means and 95% confidence intervals for each treatment (note the change in scale from A). The line is the first eigenvector, the greatest characteristic vector.

I am interested in addressing the question of whether differences in nutrient levels affect patterns of reproductive allocation.

The first step is to look for overall differences among the treatments. The basics of MANOVA are just like ANOVA except that, rather than compare the means of groups, one compares their centroids. A centroid is a multivariate mean, the center of a multidimensional distribution. Thus, the programming steps are very simple, although the actual computations are more complex. The steps are (1) choose SAS (SAS Institute Inc., 1989a,b) procedures ANOVA (for balanced data) or GLM (for unbalanced data or covariate analyses), (2) construct the usual MODEL statement, (3) include all relevant response variables on the left-hand side of the equation, and (4) add the MANOVA statement (Appendix 5.1A). (See Chapters 3 and 4 for details on constructing model statements.) As with ANOVA, the test for significant differences is based on examining the variation among groups to see if it is larger than would be expected by chance given the variation within groups, using the standard F statistic. The actual formula is based on the ratio of the among-group sums-of-squares/cross-product (SS&CP) matrix (**H**) divided by the pooled within-group (error) matrix (**E**) (Table 5.1). The diagonals of these matrices are the among-group and error sums of squares of the univariate ANOVAs. SAS takes care of the actual calculations; for details see Harris (1985, Section 4.2).

Table 5.1. *SAS MANOVA Output from the Program Listed in Appendix 5.1A[a]*

GENERAL LINEAR MODELS PROCEDURE
MULTIVARIATE ANALYSIS OF VARIANCE

H = TYPE III SS&CP MATRIX FOR TREATMENT

	NUM	MASS
NUM	24250.912821	−150.8244582
MASS	−150.8244582	1.0084091912

E = ERROR SS&CP MATRIX

	NUM	MASS
NUM	218033.46667	829.14072
MASS	829.14072	18.269723565

MANOVA TEST CRITERIA AND F APPROXIMATIONS FOR
THE HYPOTHESIS OF NO OVERALL TREATMENT EFFECT
H = TYPE III SS&CP MATRIX FOR TREATMENT   E = ERROR SS&CP MATRIX
S=2   M=−0.5   N=94.5

STATISTIC	VALUE	F	NUM DF	DEN DF	PR>F
WILKS' LAMBDA	0.782757	12.442	4	382	0.0001
PILLAI'S TRACE	0.217648	11.723	4	384	0.0001
HOTELLING-LAWLEY TRACE	0.277018	13.158	4	380	0.0001
ROY'S GREATEST ROOT	0.275136	26.413	2	192	0.0001

NOTE: F STATISTIC FOR ROY'S GREATEST ROOT IS AN UPPER BOUND.
NOTE: F STATISTIC FOR WILKS' LAMBDA IS EXACT.

[a]Shown are only those portions of the output directly relevant to the multivariate analysis. **H** is the among-group matrix and **E** is the within-group matrix.

SAS (and other computer packages) supplies four statistics for the test of significant differences among the groups: Wilk's lambda, Pillai's trace, Hotelling-Lawley trace, and Roy's greatest root. The latter is often referred to as the greatest characteristic root or the first discriminant function. All four measures are based on the eigenvectors and eigenvalues of the matrix derived from dividing **H** by **E**. Eigenvectors are linear combinations of all dependent variables [see Harris (1985) or Morrison (1990) for further discussion]. In the analysis of differences among groups, the first eigenvector is arrayed along the axis of maximal among-group variation (one can think of it as a least-squares regression of the group centroids) (Fig. 5.1B). It is the linear combination of variables that results in the greatest amount of among-group to within-group variation and, thus, the greatest F value. Subsequent eigenvectors are each orthogonal (at right angles and uncorrelated) to the previous vector with each arrayed along the axis of remaining maximal variation.

The number of eigenvectors will be equal to either the number of response variables (v) or one less than the number of groups $(k - 1)$, whichever is smaller. [SAS actually gives you as many eigenvectors as dependent variables, but if $v > (k - 1)$ then the last $v - (k - 1)$ eigenvectors will explain 0% of the variation. For factorial or nested MANOVAs the number of eigenvectors is based on the numerator degrees of freedom for the test of each factor rather than $k - 1$.]

The eigenvalues (λ) indicate the amount of variation explained by each eigenvector $[r^2 = \lambda/(1 + \lambda)]$. The first three MANOVA statistics, Wilk's lambda, Pillai's trace, and Hotelling-Lawley trace, are based on either the sum or the product of all eigenvalues. Roy's greatest root is the first eigenvalue. As can be seen from the output in Table 5.1, the statistical conclusions about whether the groups differ are identical for all four measures, as will usually be the case. [Note: while SAS indicates that the F statistic for Roy's greatest root is an upper bound, SYSTAT (Wilkinson, 1988) uses a different calculation, which provides an exact probability value for this parameter which it calls theta.] When they differ, preference should be given to either Pillai's trace or Roy's greatest root. The former has been shown to be the most robust to violations of assumptions while the latter has the greatest power. In addition, Roy's greatest root leads directly and naturally to post hoc procedures. See Harris (1985, Section 4.5) for additional discussion.

5.3.2 Post Hoc Procedures

Once a difference among groups has been established (Table 5.1), two questions are immediately raised: (1) which of the several groups actually differ from each other and (2) which traits and what relationships among them are responsible for the differences?

The statistical procedure for deciding which groups differ will depend on whether the sets of comparisons were decided on prior to analyzing the data. (In statistical parlance such comparisons are called contrasts.) These comparisons can be made either between pairs of groups or between sets of groups. The procedure will also depend on what combination of response variables is used. The biggest difference in either case is based on whether the decisions on the form of the comparisons were made a priori or are a posteriori (after the fact). If they are a posteriori, then you must adjust the α level that will be used to determine significance and accept a decrease in power.

Most typically pairwise contrasts would be tested based on the linear combination of variables obtained from the overall analysis. This procedure will result in $k(k-1)$ different pairwise comparisons. A valid statistical test of these comparisons requires that the α level be adjusted to correct for the large number of comparisons. The simplest correction to use is the Bonnferoni correction mentioned above. However, this procedure will tend to be conservative, especially if the number of groups, and subsequently the number of potential comparisons, is large. For the univariate case, tests such as Scheffé's, Tukey's, and Student-Newman-Keuls have been developed which specifically correct for these multiple comparisons. The closest equivalent in the multivariate case is presented by Harris (1985, Section 4.2). In his procedure the critical value for the F statistic is determined from the formula:

$$F = df_{eff} [\theta/(1-\theta)] \tag{5.2}$$

where df_{eff} is the numerator degrees of freedom from the MANOVA and θ is the critical value for the greatest characteristic root. A table of critical values is available in Harris (1985).

The above procedure can also be used with other combinations of dependent variables that might be suggested by examination of the data. For example, an apparent tradeoff between two traits or species (X_1 and X_2) could be tested by creating a new variable, the difference ($X_1 - X_2$) in values for each replicate, and then performing a univariate ANOVA. This is a valid a posteriori test as long as the critical F value is properly adjusted. Such a procedure greatly strengthens the potential interpretive power of MANOVA and is especially appropriate if it is used to generate hypotheses for future experiments.

For the *Coreopsis* example, I performed two types of follow-up tests, univariate and multivariate. Such follow-up test should be done only if the MANOVA indicates the existence of some significant differences. The univariate tests are not actually necessary, as I show, but are included here for comparative purposes (Table 5.2A). The pairwise contrasts of the individual response variables, using a Bonnferoni corrected $\alpha' = 0.05/3 = 0.017$, show that the high and low treatments differ from each other for both traits and the medium treatment differs from neither other treatment for neither variable. (The same results are obtained using a Tukey's test.) These differences can also be seen by examining the overlap among confidence intervals (Fig. 5.1B). Thus, the differences among groups are

Table 5.2 Pairwise Contrasts of the Three Treatments (High = H, Medium = M, Low = L) Based on Individual Traits and the Combination of Both Traits Using the CONTRAST Statements in Appendix 5.1A

A. Individual Traits

		Number of seeds			Seed mass		
CONTRAST	DF	CONTRAST SS	F VALUE	PR>F	CONTRAST SS	F VALUE	PR>F
H VS M	1	3915.907	3.45	0.0648	0.02139	0.22	0.6359
H VS L	1	24083.173	21.21	0.0001	0.88943	9.35	0.0026
M VS L	1	5768.533	5.08	0.0253	0.49557	5.21	0.0236

B. Multivariate Analysis Showing Roy's Greatest Root

CONTRAST	VALUE	F	NUM DF	DEN DF	PR>F
H vs M	0.02773	2.65	2	191	0.0734
H vs L	0.26597	25.40	2	191	0.0001
M vs L	0.09166	8.75	2	191	0.0002

ambiguous. More powerful are the multivariate follow-up tests (Table 5.2B). The multivariate comparisons show that the high and medium treatments do not differ from each other and they both differ from the low treatment. The difference in results for the comparison of the low and medium treatments shows how univariate tests can be misleading. These conclusions would be the same using either a Bonnferoni correction ($\alpha' = 0.017$) or the critical value from Harris ($F = 8.5052$). Note that the formula from Harris applies only to the greatest characteristic root, although when there are only two groups the F and P values will be exactly the same for all four test statistics.

The comparisons were done using the CONTRAST option in SAS procedure GLM (Appendix 5.1A). The alternative would be to break the data into three sets containing each pair of treatments and redo the MANOVA for each pair. (A two-sample MANOVA is also called a Hotelling's T^2 test, just as a two-sample ANOVA is the equivalent of a *t*-test.) The reason for instead using the CONTRAST statement is the latter's increased power. Note that the denominator degrees of freedom in the pairwise contrasts is 191, just 1 less than that for the full analysis. When contrasts are carried out the pooled error SS&CP matrix is used, just as the pooled error variance is used in a univariate post hoc Tukey's test. Because this error matrix is pooled across all groups, the accuracy of the estimation of its elements is greater, resulting in a more powerful test. Appendix 5.1A lists the contrast statements necessary to make these comparisons. If other types of comparisons are desired (such as the comparison of high and medium vs low), those contrast statements can also be added or substituted. See Neter and Wasserman (1974), Millikin and Johnson (1984), Hand and Taylor (1987), or Snedecor and Cochran (1989) for a discussion of how to construct contrast statements (see also Section 14.3.2).

In some instances all of the contrasts will be specified before the experiment is carried out. For example, one might plan to compare each of several treatments to a single control treatment. In that case the critical F value is determined as

$$F' = \frac{v \, df_{err}}{df_{err} - v + 1} F \qquad (5.3)$$

where F is an F-statistic with degrees of freedom v and $df_{err} - v + 1$, v is the number of response variables, and df_{err} is the error degrees of freedom from the MANOVA.

Determining which combinations of response variables are responsible for the differences among groups can be done by examining the greatest characteristic vector and a related parameter, the standardized canonical variate, which are shown in Table 5.3. Examining Fig 5.1B it is clear that plants grown at low nutrient levels are ripening fewer and heavier seeds than those grown at higher nutrient levels. In this case there are two eigenvectors; the first explains 22% of the total variation (the "squared canonical correlation" reported in Table 5.3B) or virtually all of the explainable variation (the "percent" reported in Table 5.3A). That is not surprising as the three centroids lie on nearly a straight line (Fig. 5.1B).

The canonical analysis has the potential for providing a more complete picture of the statistical significance of the individual eigenvectors and is obtained using the CANONICAL option of the MANOVA statement (second MANOVA statement in Appendix 5.1A). SAS reports a test, given as the likelihood ratio, for the significance of the sum of the ith to nth eigenvectors. That is, if there were five eigenvectors, SAS would sequentially do the following: test all five together, then delete the first and test the remaining four, delete the second and test the remaining three, and so forth. The first test is the equivalent of Wilk's lambda (compare Tables 5.1 and 5.3B). However, Harris (1985, Section 4.5.3) cautions against the reliability of such a procedure. Generally only the greatest characteristic root is of interest, the significance of which is tested separately as Roy's greatest root.

The numbers in the two right hand columns in Table 5.3A under the headings NUM and MASS are the coefficients of the eigenvectors. However, they are not very useful as the magnitudes of the coefficients are dependent on the scale of the data. Better is a comparison of standardized coefficients, obtained by multiplying each of the original coefficients by the standard deviation of each response variable. (This standardization is the equivalent of what is done to go from partial regression coefficients to standardized regression coefficients in multiple regression, Chapter 9.) Unfortunately SAS does not calculate the standardized coefficients for you; for these data the standardized coefficients are -0.075 and 0.063 for number of seeds and seed mass, respectively.

Most informative are the standardized canonical coefficients of the canonical variates (Table 5.3B). (In the SAS output the canonical variates are arranged

Table 5.3. *The Eigenvector (Default) and Canonical Analyses of the MANOVA Option from SAS Procedure GLM Showing the Parameters and Associated Statistics of the Characteristic Vectors and Canonical Variates*

A. Default Eigenvector Analysis from the First MANOVA Statement in Appendix 5.1A

CHARACTERISTIC ROOTS AND VECTORS OF: E INVERSE * H, WHERE
H = TYPE III SS&CP MATRIX FOR TREATMENT   E = ERROR SS&CP MATRIX

CHARACTERISTIC ROOT	PERCENT	CHARACTERISTIC VECTOR NUM	V'EV=1 MASS
0.27513581	99.32	−0.00211225	0.19920128
0.00188219	0.68	0.00103995	0.16269926

B. Alternative Canonical Analysis from the Second MANOVA Statement in Appendix 5.1A

CANONICAL ANALYSIS
H = TYPE III SS&CP MATRIX FOR TREATMENT   E = ERROR SS&CP MATRIX

	CANONICAL CORRELATION	ADJUSTED CANONICAL CORRELATION	APPROX STANDARD ERROR	SQUARED CANONICAL CORRELATION
1	0.464510	0.456687	0.056304	0.215770
2	0.043343		0.071661	0.001879

TEST OF H0: THE CANONICAL CORRELATIONS IN THE
CURRENT ROW AND ALL THAT FOLLOW ARE ZERO

	LIKELIHOOD RATIO	APPROX F	NUM DF	DEN DF	PR>F
1	0.78275689	12.4419	4	382	0.0001
2	0.99812135	0.3614	1	192	0.5484

STANDARDIZED CANONICAL COEFFICIENTS

	CAN1	CAN2
NUM	−1.0343	0.5092
MASS	0.8701	0.7107

BETWEEN CANONICAL STRUCTURE

	CAN1	CAN2
NUM	−0.9949	0.1008
MASS	0.9862	0.1657

vertically while the eigenvectors are arranged horizontally.) Canonical variates are simply eigenvectors scaled to unit variance by multiplying the eigenvector coefficients by the square root of the error degrees of freedom. Again, the most useful form is obtained by standardizing the coefficients by multiplying each by the standard deviation of the variable; in this instance SAS does the standardization for you. For the first variate, the standardized coefficients for NUM and MASS are −1.03 and 0.87, respectively (see Table 5.3B under the heading CAN1). The opposite signs indicate that the two variables are negatively correlated across groups; that is, the pattern of resource allocation is changing across treatments. This contrasts with the positive correlation within groups (mean r = 0.42), which is reflected in the similar signs for the coefficients of the second variate. The greater magnitude of the coefficient for NUM indicates that NUM explains more of the variation among groups than MASS. Interpreting eigenvector or canonical coefficients must be done cautiously as the magnitudes, and even the signs, can change depending on what other response variables are included in the analysis.

SAS also provides information on the canonical structure of the data under three headings, TOTAL, BETWEEN, and WITHIN. The between canonical structure is of greatest interest and is shown in Table 5.3B. These values are the correlations between the individual response variables and the canonical variates; they are sometimes referred to as canonical loadings. In this case, because the three centroids lie nearly on a single line, the magnitudes of the correlations for the first canonical variate are close to ± 1. Whereas the canonical coefficients indicate the unique contribution of a given variable to the differences among groups (i.e., are analogous to partial regression coefficients), the canonical correlations indicate how much of the variation each variable shares with the canonical variate ignoring correlations among response variables. Such correlations can also be calculated for the eigenvector coefficients. The canonical correlations are touted as being more stable to sampling effects than the canonical coefficients, however I feel that the canonical coefficients are more informative. See Bray and Maxwell (1985, pp. 42–45) or Stevens (1986, pp. 412–415) for a discussion of the use of canonical correlations vs canonical coefficients.

5.3.3 A Complex Example

The above example was simple and straightforward to interpret. As the number of response variables increases, interpreting differences among groups becomes more complex and simple graphic examination becomes very difficult. In addition, the right-hand side of the equation, the independent variables, can also become more complex. Any model that can be used in ANOVA or ANCOVA (nested, factorial, etc., see Chapters 3 and 4 for examples) can also be used in MANOVA. Here I present the results from a $2 \times 2 \times 2$ factorial experiment. The data are part of a study on trophic interactions among inhabitants of pitcher-plant leaves examining changes in prey species' community structure as predator species changed (Cochran-Stafira, 1993). In this experiment the presence vs absence of three protozoan predators (*Cyclidium, Colpoda* and *Bodo*) was manipulated in laboratory test tubes cultures. The response variables were the densities of four bacterial prey species (designated A, B, C and D) after 72 hours of growth. There were three replicates per treatment; the design was completely balanced.

The ANOVAs (not shown) indicate that species A had a significant response to all treatment effects and interactions, species B did not respond significantly to any treatments, species C had a significant response to the main effects of *Colpoda* and *Bodo* and the three-way interaction, while species D had significant responses only to the main effects of *Colpoda* and *Cyclidium*. But this description does not capture the extent of the interactions among the species. Examination of the results of the MANOVA (Table 5.4) indicates that all the main effects were significant, the *Colpoda-Cyclidium* interaction was significant, and the three-way interaction was significant. Thus, predator species identity affected prey-species community structure. (Note that in determining the proper numerator and denominator matrices for the tests, the same rules as for ANOVA are followed; in the current case all factors were fixed so the **E** matrix was used for all tests.)

The canonical coefficients show that significant differences were primarily due to changes in the abundance of species A (Fig. 5.2). Additionally, the abundance of species D tended to be positively correlated with that of A while C tended to be negatively correlated with both of them. The correlation of A and D was reversed, however, for the *Colpoda–Bodo* and *Cyclidium–Bodo* interactions. Finally, the full extent of changes in species C and D was more clearly revealed in the MANOVA as indicated, for example, by the relatively large canonical coefficients for the *Colpoda–Cyclidium* interaction, which was not significant for the ANOVAs of either species. This last is an example of how two variables that

Table 5.4. *Multivariate Analysis of the Effects of Protozoan Species on Bacterial Densities*[a]

A. Roy's Greatest Root

SOURCE	VALUE	F	NUM DF	DEN DF	PR>F
COLPODA	75.9066	246.7	4	13	0.0001
CYCLID	8.4839	27.6	4	13	0.0001
COLPODA*CYCLID	9.3905	30.5	4	13	0.0001
BODO	6.5291	21.2	4	13	0.0001
COLPODA*BODO	0.9522	3.1	4	13	0.0539
CYCLID*BODO	0.8017	2.6	4	13	0.0848
COLPODA*CYCLID*BODO	2.5207	8.2	4	13	0.0016

B. Standardized Canonical Coefficients

	A	B	C	D
COLPODA	7.2868	0.2261	−1.8530	2.4058
CYCLID	7.2648	0.5033	−2.5893	2.3589
COLPODA*CYCLID	7.5991	0.3505	−2.3772	2.1048
BODO	4.5085	−0.3285	0.9958	−0.2532
COLPODA*BODO	5.9543	−0.0961	−0.0045	−0.5063
CYCLID*BODO	4.5032	−0.2423	0.9595	−0.9633
COLPODA*CYCLID*BODO	6.9132	0.2715	−2.6776	1.3263

[a]Shown are the significance tests for Roy's greatest root and the standardized canonical coefficients for each main effect and interaction. As each main effect has only two levels (treatments), only one eigenvector exists for each effect.

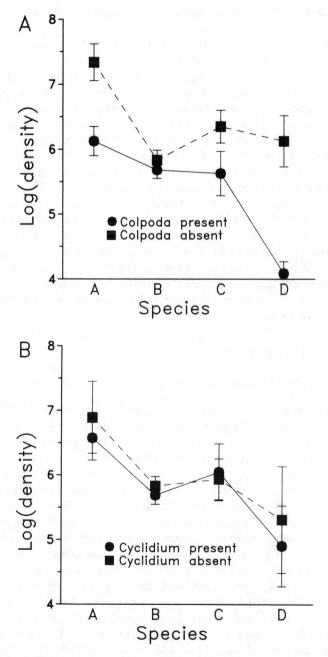

Figure 5.2. (A) Bacterial species densities (mean ± 95% CI) with and without *Colpoda*.
(B) Bacterial species densities (mean ± 95% CI) with and without *Cyclidium*.

are correlated can be jointly important, even though they are not individually significant.

Several different types of ecological interactions could be responsible for these patterns. The bacterial species could be directly competing for resources. The bacterial species could be affecting each other—positively or negatively—indirectly through the production of secondary metabolites. The protozoans could also be producing secondary metabolites creating indirect correlations; for example, as *Colpoda* eats species A and increases in density the protozoan's waste products might benefit species C. At this stage any ecological interpretations of the results must be considered tentative hypotheses to be tested by further experiments (Fig. 1.2). What the MANOVA does is indicate which potential interactions are worth pursuing, information not extractable from the ANOVAs.

In this case, as there are only two treatments (presence vs absence) for each main effect in the MANOVA, post hoc tests of the main effect differences are not necessary. However, the three-way effect is significant and one might wish to determine which of the eight treatments differ from each other. No single best procedure exists. Beyond a simple examination of treatment means, Harris (1985) describes two alternatives: (1) a method of contrasts of contrasts (Section 4.7) or (2) recoding the data as a one-way MANOVA followed by a series of appropriate contrasts (Section 4.8).

5.4 Other Statistical Issues

5.4.1 Experimental Design and Power

The same basic design considerations that hold for ANOVA and ANCOVA, and discussed in detail in Chapters 3 and 4, also hold for MANOVA. The one additional consideration is that of power, the ability to detect a true difference. The power of MANOVA declines with an increase in the number of response variables. Stevens (1992, Sections 4.9, 4.10, 4.12, and 5.15) has a detailed discussion of power considerations in MANOVA including tables for suggested sample sizes for different effect sizes; more extensive tables can be found in Lauter (1978). For example, the *Coreopsis* experiment had a 90% probability of detecting a small to moderate difference in the treatment means. (No actual formulas are available to determine power; one just looks up the values in the appropriate table.) On the other hand, the power for the protozoa–bacteria experiment was low, so that conclusions about the lack of significant two-way interactions involving *Bodo* must be viewed with caution. A priori power analyses are strongly recommended in designing an experiment so that one will have the ability to detect the desired effect. A post hoc power analysis is especially important if one concludes that no differences exist so that the robustness of that conclusion can be assessed.

Because the power of MANOVA declines with increasing number of response variables, the distribution of effort among response variables vs replicates needs to be carefully considered. Very often experimenters will simply measure all possible response variables, examine them in a single huge analysis, and hope that something appears. Beyond problems of power, such an approach will often lead to results that are hard to interpret. A better approach is to start with clearly delineated hypotheses involving specified sets of response variables. As all of the response variables would not be included in any single analysis, the power of the analysis would not be overly diminished. The interpretation of the results would be simplified because they would be compared with prespecified conjectures.

5.4.2 Missing Data and Unbalanced Designs

Missing data can take two forms in MANOVA, either the value of one or more response variables is missing for a subject (replicate) or an entire subject is missing. MANOVA can be performed only when all subjects have been measured for all response variables. If some measurements are missing, the entire subject is deleted from the analysis. As we all are loath to throw away data, one solution is to estimate the missing values using some form of multiple regression analysis of the other response variables. One assumes that the missing values do not deviate in any unusual way from the general distribution of the other values. A graphic analysis (Chapter 2) of the other values for that subject is recommended to determine if that subject was unusual or was an outlier. Other information should also be considered (e.g., the data are missing because the animal was sick and died). Such procedures are generally robust as long as the missing data are sparse ($< 5\%$ of all data, missing values distributed among all response variables, and generally no more than 1 missing value per subject).

When an entire subject is missing, estimating missing values is more problematic because very restrictive assumptions must be made about the entire covariance structure of the data; in particular that it is uniform (as opposed to merely similar) across treatments or groups. However, estimating missing values is not necessary as SAS procedure GLM (and other computer packages) will handle unbalanced designs. In such a case, one must use Type III sums of squares as they are designed to account for unbalanced designs. Generally, unequal sample sizes are not a big problem for one-way MANOVA, but they might bias the results for factorial or nested designs. Unfortunately, little work has been done on this problem for MANOVA, thus statements about the robustness of ANOVA to unbalanced designs (Shaw and Mitchell-Olds, 1993) should be extrapolated to the multivariate case with caution. As a rule, as long as designs are not too unbalanced (the smallest cell not less than 50% the size of the largest cell) results are reliable. As always, conclusions based on probability values close to the critical value ($0.01 < P < 0.10$) should always be tempered.

A relatively new approach to such analyses, maximum likelihood (McLean et al., 1991), has now become available in the latest version of SAS (6.07, SAS

Institute Inc., 1992) as procedure MIXED. One advantage of this approach is that having complete data for all subjects is not necessary so missing values need not be estimated. However, assumptions about homogeneity of subjects (missing values are not unusual) still hold so examination of the data before analysis and consideration of the reasons for the missing data are still important. A disadvantage is that computations for experiments with either many response variables or complex designs can take a very long time for MANOVA and sometimes will fail to converge on a solution. Be warned that the documentation of the technique is not completely clear, especially for those without extensive statistical training; consultation with a statistician is highly recommended.

5.4.3 Assumptions

The assumptions of ANOVA also hold for MANOVA: subjects are independent, all random effects (particularly the within-group or within-cell error effects) are normally distributed, and the variances of those error effects are equal among groups or cells (homoscedasticity). In addition, for MANOVA it is assumed that the error effects are multivariate normal and that the covariances are equal among groups. These latter two assumptions are just multivariate extensions of the last two assumptions from ANOVA. Unfortunately, it is virtually impossible to actually test for multivariate normality because of the very large sample sizes needed. The best that we can do is to test each of the dependent variables separately for univariate normality and then (reasonably) assume that multivariate normality holds. There are tests for equality of variance/covariance matrices, but statisticians are cautious about their use. The reasons are (1) the tests are sensitive to departures from multivariate normality and (2) MANOVA is generally robust to deviations such that a significant departure from equality will be found before it will bias the conclusions. In particular, Pillai's trace has been found to be very robust.

My general advice is that if visual inspection of the data indicates overall agreement with the assumptions, go ahead and do the MANOVA. The obvious case where the assumptions are being violated is when the sign of the correlation between two dependent variables differs among groups. In that case, the most appropriate analysis strategies are to (1) specifically address questions regarding differences in correlations and (2) compare means using univariate analyses or multivariate analyses that do not use that particular pair of dependent variables simultaneously. The other situation in which the assumption of normality will likely be violated is community-level studies that involve species that are not being directly manipulated (e.g., the presence of predators is manipulated while prey species freely move on and off plots). In that instance, several of the species are likely to have low or very variable abundances resulting in a data matrix with many zeros. In such a case no transformation will succeed in normalizing the distribution. One potential solution to the general problem of violation of assumptions is the use of randomization tests (Chapters 11, 13, and 15) which are distribution free. Such randomization tests for the case of a one-way MANOVA are presented by Mielke et al. (1981) and Zimmerman et al. (1985).

5.5 Related Issues and Techniques

I have discussed briefly the very complex subject of multivariate analysis. A number of books exist that explain the basics and provide an entry into the more esoteric aspects of this subject. Nontechnical introductions can be found in Barker and Barker (1984), Bray and Maxwell (1985), and Manly (1986b). More detailed treatments are given in Harris (1985), Stevens (1992), and Morrison (1990). A complete mathematical treatment can be found in Johnson and Wichern (1988). In Chapter 6, von Ende discusses another aspect of this topic, repeated-measures analysis, in which the same subject is measured more than once. He explains a procedure called profile analysis, which can also be used with MANOVA to dissect out effects of particular variables (see also Simms and Burdick, 1988). More complex situations in which the same subject is measured more than once for more than one trait, called doubly multivariate designs, can also be envisaged (see Example 9 of SAS procedure GLM). In some instances the response variables themselves may be further related in some cause and effect schema; in that case the problem should be broken into a series of subanalyses as part of a path analysis (Chapter 10). In Chapter 9, Philippi discusses multiple regression analysis in which the independent variables consist of several continuous effects. If there are also multiple response variables, the result is referred to as a canonical analysis. Such an analysis is actually the most general form of parametric statistical analysis and all other procedures can be thought of as various special cases of this general approach. See Gittens (1985) for a useful description of canonical analysis in an ecological context.

The basic message that I have tried to convey in this chapter is that MANOVA is no more difficult, and often more informative, than ANOVA. Its implementation needs to be encouraged in ecology as the questions that we ask are very often multivariate ones.

Acknowledgments

I thank Liane Cochran-Stafira for kindly letting me use her data and assisting in the interpretation of its analysis. Discussions with Carl von Ende were invaluable for my understanding of MANOVA. Comments by Charles Goodnight, Jessica Gurevitch, Richard King, and Mark VanderMuelen helped improve the text.

Appendix 5.1

SAS program code for multivariate analysis of variance (MANOVA).

A. *Analysis of the* Coreopsis *Experiment*

```
PROC GLM;                              /use GLM for unbalance designs/
   CLASS TREAT;                        /TREAT = treatments/
   MODEL NUM MASS = TREAT/SS3;         /use type III sums of squares
                                        for unbalanced designs/
   CONTRAST 'H VS M' TREAT -1 1 0;     /contrast statements for
   CONTRAST 'H VS L' TREAT -1 0 1;     obtaining the tests of
   CONTRAST 'M VS L' TREAT 0 -1 1;     pairwise differences/
   MANOVA H=SITE/PRINTH                /MANOVA statement for obtaining the multivariate
            PRINTE;                     and the coefficients analysis of the greatest
                                        characteristic roots. PRINTH and PRINTE request
                                        the among-group and within-group
                                        SS&CP matrices/
   MANOVA H=SITE/CANONICAL;            /alternative MANOVA statement for obtaining
                                        the multivariate analysis and the
                                        canonical coefficients/
```

B. *Analysis of the Protozoan–Bacteria Experiment*

```
PROC ANOVA;
   CLASS COLPODA CYCLID BODO;
   MODEL A B C D = COLPODA|            /the "|" specifies a model with
            CYCLID|BODO;                all possible interactions/
   MANOVA H=_ALL_/CANONICAL;           /the "_ALL_" requests multivariate
                                        tests for all model effects/
```

6

Repeated-Measures Analysis: Growth and Other Time-Dependent Measures

Carl N. von Ende

6.1 Introduction

There are many situations in which ecologists make repeated measurements on the same individual, the same experimental unit, or at the same sampling site. Probably the most common is where some characteristic or factor is measured at several different times. For example, one may be interested in how body size, plant size, survivorship, clutch size, population size, nutrient level, or pollutant level change over time for different populations, locations, or experimental treatments. A second kind of study involving repeated measurements is where each individual organism is exposed to different levels of some treatment and its response is measured at each level. For example, in plant ecophysiological studies the same plants often are exposed to a series of different CO_2 levels and their photosynthetic rate measured at each level (Potvin et al., 1990b). In both kinds of studies there is an explicit interest in the pattern or shape of the response, over time, or over the levels of an experimental treatment.

The same questions could be addressed without repeated measurements on individuals. Assume one wanted to study the effect of diet on growth in squirrels, in particular, to examine whether the pattern of growth in squirrels fed acorns was different from that fed hickory nuts over a three month period. Assume the squirrels were fed independently and weighed on the starting date and at two week intervals for a total of six measurements. With six dates and two feeding regimes, there would be 12 Diet × Date combinations. The experiment could be conducted in either of two ways. One could assign, for example, three squirrels to each Diet × Date treatment combination. All squirrels would be started at the same time and fed regularly, but a different subset of squirrels would be weighed on each date for each diet. Thus, each squirrel would be weighed only once and a total of 36 squirrels would be required for the experiment. Alternatively, six squirrels could be assigned to each diet and all squirrels measured on each date, so that each was measured six times. Only 12 squirrels would be required for this second design, a repeated-measures design.

The first experiment could be analyzed as a two-factor ANOVA (Diet, Date) because the squirrels within a Diet treatment level at any one date would be independent of those at other dates, but obviously the experiment would require many more animals than the repeated-measures design. It would be inappropriate, however, to analyze the second design as a two-factor ANOVA because the same animals would be measured repeatedly during the experiment and the weights of each squirrel on the different dates would not be independent of one another. (Independence of replicates is a basic assumption of ANOVA.) Rather a "repeated-measures" analysis, which takes into account the correlation among dates because animals are reweighed, should be used for the analysis of the second experiment. Historically, although many ecologists have collected data in a repeated-measures design, they frequently have either analyzed them incorrectly with an "ordinary" ANOVA, "plugged" the data into a repeated-measures analysis in a statistical package without considering underlying assumptions of some repeated-measures analyses, or thrown out intermediate data and analyzed only the final measurements with ANOVA. As the level of ecologists' statistical knowledge has increased, the first two alternatives have become unacceptable and the last undesirable. Repeated-measures designs have been used in psychology and agriculture for some time (Snedecor and Cochran, 1989; Winer et al., 1991), but have received the explicit attention of ecologists only relatively recently (Gurevitch and Chester, 1986; Potvin et al., 1990b). In this chapter I discuss various parametric methods of analyzing repeated-measures data.

6.2 Statistical Issues—Overview

Repeated-measures designs can be analyzed by parametric methods using either univariate or multivariate approaches. The univariate analyses are ANOVA designs (randomized block, split-plot) that involve blocking (Chapter 3). MANOVA (Chapter 5) is used for the multivariate analyses. Although the univariate approach is computationally simpler and generally considered to be more powerful than MANOVA, it also has more restrictive assumptions. However, see Mead (1988) for a discussion of the philosophical problem of treating Time as a "split-unit" factor in which repeated measurements are taken on the same, not different, experimental units. With the development of sophisticated statistical software over the last decade, the use of MANOVA is no longer limited by computational complexity. Both approaches can handle factorial designs. I will briefly review the univariate and the multivariate approaches, highlight important aspects, and then discuss several examples in more detail.

6.2.1 Univariate Repeated-Measures Analyses

The basic univariate designs for repeated measures are randomized complete block and split-plot designs. In the simplest repeated-measures designs, one is interested in whether different treatment levels applied to the same individuals have a significant effect (the plant photosynthesis example above). These can by analyzed as a randomized-block ANOVA (Chapter 3). Each individual is considered a different block within which the treatment is applied. This is analogous to an agricultural experiment in which different fertilizers are applied to adjacent areas within each of several blocks. In repeated-measures jargon, the treatment is referred to as the "repeated factor," or by psychologists as the "within-subject" factor. The purpose of blocking is to make the analysis more sensitive by removing variance among blocks, or among subjects, from the error term. Conditions are assumed to be more homogeneous within a block than between blocks, so it is within a block, or to the same subject, that the treatment levels are applied. When Time is the within-subject factor, there are observations at different times for each subject.

The split-plot design (Chapter 3) is the natural extension of the randomized block design. A typical agricultural example would be a situation in which one is interested in the effect of irrigation on a fertilization treatment. Imagine several fields or "plots," half of which were irrigated and half that were not. Each "whole plot" would be divided into several subplots, and within each subplot one of the fertilizer levels described in the randomized block example would be applied. In repeated-measures jargon, the Irrigation/No Irrigation treatment is called the "between-subjects" (whole-plot) factor, and again, the Fertilizer treatment the within-subject (split-plot) factor. The squirrel example above in which there are repeated observations on the same squirrels fed different diets can be treated as a split-plot design. Diet is the between-subjects factor, individual squirrels are the whole-plots, or subjects, within each food type, and Date is the within-subject factor. Again, the ecological question is whether the squirrels on the two different diets have the same patterns of weight change over time.

Although randomized block and split-plot designs are different from ordinary ANOVA in that they assume some correlation among treatment levels within a block, many ecologists are unaware that these designs make certain assumptions about the variances and covariances of the levels of the within-subject factor, and the relationship among these variances and covariances. Specifically, these designs assume what is called "circularity" among the levels of the within-subject factor. I will briefly describe this property. One can construct a square variance–covariance matrix for the within-subject factor in these designs. The variances for each level of the within-subject factor will fall along the diagonal of the matrix, and the covariances between all the different levels will occupy the remainder of the matrix. A "circular" variance–covariance matrix has the property that the variance of the *difference* between any two levels of the within-subject factor equals the same constant value (Winer et al., 1991). This means that if one takes two levels of the within-subject factor and subtracts the scores for one level from the scores for another level, the resulting scores must have the same variance for every pair of levels (Maxwell and Delaney, 1990). The assumption of circularity is less restrictive than the assumption of *compound symmetry* in which the variances are all assumed to be equal to one another, and the covariances also equal to one another. For many years it was thought compound symmetry was required for the validity of the F statistic in repeated-measures ANOVA, however, Huynh and Feldt (1970) showed that circularity was sufficient. All matrices that have compound symmetry are circular, but not all circular matrices have compound symmetry.

Another matrix characteristic, *sphericity*, is used in assessing the circularity of a variance–covariance matrix. An orthogonal transformation is one that creates transformed variables that are independent (orthogonal) of each other. If these transformed variates are then scaled such that the sum of squares of the coefficients for each variable is 1, this constitutes a set of *orthonormal* variables (Stevens, 1992). A circular variance–covariance matrix that is transformed to its normalized orthogonal form is *spherical* (Winer et al., 1991). The transformation can be accomplished by first constructing a matrix with the coefficients of orthogonal contrasts as the rows and then normalizing the coefficients (see Winer et al., 1991, p. 244 for an example). This matrix is then used to transform the variance–covariance matrix to an orthonormalized variance–covariance matrix. The circularity of a variance–covariance matrix can be assessed by testing the sphericity of an orthonormalized form of the matrix. If the transformed variance–covariance matrix is spherical, the original variance–covariance matrix is circular. In a spherical variance-covariance matrix the variances of the transformed variables are equal and their covariances are 0.

There are tests for sphericity, the most popular of which is Mauchly's (Crowder and Hand, 1990). The test statistic is W and can range between 0 and 1 (equation in Winer et al., 1991). It uses the orthonormalized form of the variance-covariance matrix. The closer W is to zero, the greater the departure from sphericity (Winer et al., 1991). Huynh and Mandeville (1979) showed W to be sensitive to violations of normality, and for these tendencies to be amplified with increased sample sizes. O'Brien and Kaiser (1985), Stevens (1992), and Winer et al. (1991) recommend against using W.

In the case of the plant experiment above, if the sequence of the CO_2 levels presented to a plant were randomized, and there were no carry-over effects on photosynthetic rate from one CO_2 concentration to the next, the assumption of circularity probably could be met. However, when Time is the within-subject factor, usually data collected on adjacent sampling dates are more highly correlated than are data from separated sampling dates, and the circularity condition is not met. There are numerous instances in the ecological literature in which ecologists have done a repeated-measures ANOVA analysis with Time as the within-subject factor without mentioning the assumption of circularity (or sphericity). The consequence of failing to meet the assumption is that F statistics for the within-subject factors (and their interactions) are inflated so that one is more likely to conclude effects are statistically significant when they are not. The alternatives available when this assumption cannot be met are to adjust the degrees of freedom of the F-test so that it is more conservative, or to use the multivariate approach to analyze the data. These are discussed below.

There is an additional assumption involved in using the split-plot design because of between-subjects factors. It is that the variance–covariance matrices of the *differences* among the levels of the within-subject factor should be homogeneous (equal) for all levels of the between-subjects factor. That is, if one had three levels of a between-subjects factor, and calculated the variance–covariance matrix of *differences* for the within-subject factor for each of those levels, those matrices should be identical. The reason is that the test of the significance of the within-subject factor (and its interaction with the between-subjects factor) is based on the pooled variance–covariance matrix. The matrices should be equal to be pooled. It is this pooled variance–covariance matrix of differences that must meet the criterion of circularity discussed above. The significance tests of the within-subject factor, and its interaction with the between-subjects factor, are sensitive to the circularity assumption as described above for the simple within-subject design. They are robust to the second assumption of homogeneity of variance-covariance matrices as long as sample sizes are equal. However, as sample sizes become less equal, the test becomes less robust (Maxwell and Delaney, 1990)

When within-subject and/or between-subjects effects are found significant, follow-up tests often are desirable. O'Brien and Kaiser (1985), Maxwell and Delaney (1990), and Stevens (1992) discuss post hoc tests for repeated-measures designs. It is important to realize that planned (a priori) contrasts involving the specific error term for each contrast, rather than an overall error term, are not subject to the circularity assumption. See O'Brien and Kaiser (1985) for an example.

Although I have described only the simple experimental designs with a single within- and between-subjects factor, repeated-measures designs can be expanded to include additional within- and between-subjects factors. See Maxwell and Delaney (1990), Winer et al. (1991), and Stevens (1992), and Section 6.4 for examples.

6.2.2 Multivariate Repeated-Measures Analyses

Repeated-measures data also can be analyzed by using MANOVA (Chapter 5). The approach taken is to treat the response variable for each level of the within-subject factor as a different dependent variable. In terms of the plant example above, the photosynthetic rate at each of the CO_2 concentrations would be considered a different response variable. So, if there were five CO_2 levels, the response of each plant would be characterized in terms of five photosynthetic rates, one for each of the CO_2 concentrations. MANOVA is designed to simultaneously analyze the response of several correlated dependent variables. As usually applied to repeated-measures data, it does not require the dependent variables be equally correlated as repeated-measures ANOVA does. It assumes what is called an "unstructured" variance–covariance matrix, which means there is no particular pattern required of the matrix.

To analyze the squirrel example as a MANOVA, the dependent variables would be the weights on each of the six dates, and the treatment would be Diet. The analysis tests whether the mean response vectors for the two diets are different (Chapter 5). Although the problem of circularity inherent in repeated-measures ANOVA is avoided by doing MANOVA, MANOVA also has its limitations. In the repeated-measures design for the squirrel example, I suggested the analysis could be done with six squirrels (n) for each diet. However, the number of dependent variables (k) that can be analyzed in a MANOVA depends on the total number of samples (subjects) (N) and the number of between-subjects treatment levels (groups) (M) (Potvin et al., 1990b). The constraint is that $N - M > k$. Because the squirrel example would have weights on six dates, six squirrels, and two treatment levels, this condition would be satisfied ($12 - 2 = 10 > k = 6$). Although $N = 12$ would satisfy the condition, the test would have low power. It would be best to increase the sample size even more, and/or decrease the number of dates on which the squirrels were weighed. Power increases as the ratio $n{:}k$ increases (Potvin et al., 1990b). This limitation on the number of levels of within-subject factors and associated problems of low power are inherent in using MANOVA (Chapter 5). Stevens (1992) has a detailed discussion of power in MANOVA, and mentions the need to use a larger α if one suspects low power. Maxwell and Delaney (1990) discuss the power of MANOVA in repeated-measures designs and provide tables of estimated sample sizes for different levels of power for one within-subject designs. They recommend doing MANOVA only when $N - M > k + 9$ (Maxwell and Delaney, 1990, p. 676). These restrictions highlight the need to consider the amount of replication and the number of levels of within-subject factors in the initial stages of experimental design rather than after an experiment has been completed. Potvin et al. (1990b) recommend using the minimum number of levels of the within-subject factor required to characterize a response adequately and to increase the amount of replication to yield a more powerful analysis.

The MANOVA analysis combines two sources of differences in the within-subject data when comparing the mean response curves of the levels of the between-subjects treatment: it takes into account differences in the *shapes* of the response curves, and differences in the *levels* of the response curves (Harris, 1985). Consider the squirrel example, but for only three dates (Fig. 6.1). The growth curves could be parallel (Fig. 6.1A) indicating the squirrels had similar patterns of growth on the two diets (shape), or they could diverge (Fig. 6.1B)

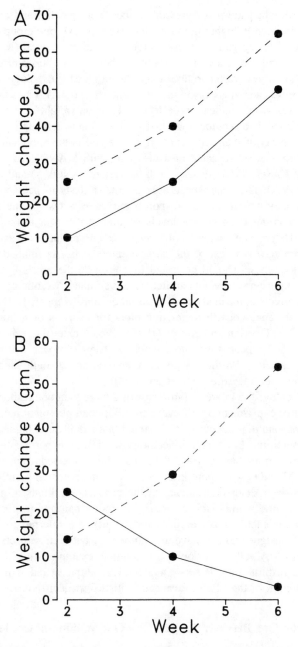

Figure 6.1. Profile analysis: hypothetical weight change over 6 weeks for squirrels fed hickory nuts (dashed lines) and acorns (solid lines). (A) Curves show "parallelism" and "levels" effects. (B) Curves show absence of parallelism, i.e., Diet × Time interaction.

indicating different patterns of growth. If they were parallel and one response curve was consistently higher than the other (Fig. 6.1A), one group of squirrels would be consistently gaining more weight as time progressed (levels). If the curves were parallel and nearly overlapping there would be no effect of Diet. Finally, if the curves were parallel, a significant slope (i.e., slope > 0) would indicate there was a change in weight over time (Fig. 6.1A), whereas horizontal response curves would indicate no effect of Time on weight.

Profile analysis is a methodology that enables one to test the significance of these three aspects of the multivariate response and is the approach most commonly used to analyze repeated-measures data with MANOVA (Harris, 1985; O'Brien and Kaiser, 1985). It uses both univariate (ANOVA) and multivariate tests (MANOVA). The three aspects are examined as tests of specific hypotheses. A comparison of the shapes of response curves is a test of the *parallelism* hypothesis; a comparison of the levels of the curves is a test of the *levels* hypothesis. Determining whether the response curves have an average slope different from zero is a test of the *flatness* hypothesis. As implied above, the parallelism hypothesis should be tested first because the levels hypothesis becomes moot if the curves are not parallel. This is similar to examining interactions before main effects in the analysis of factorial designs (Keppel, 1991). The plots in Fig. 6.1 are analogous to interaction plots for analysis of factorial designs (Keppel, 1991). O'Brien and Kaiser (1985) give an extensive primer on using profile analysis in repeated-measures analyses. Their approach will be followed in the examples below. Profile analysis also can be applied to multivariate analyses other than repeated-measures (Morrison, 1990).

Often an ecologist is interested in measuring more than one response variable for a subject or experimental unit over time. For example, plant height and leaf area could respond in some correlated manner that would be of interest in studies of plant competition. In community ecology studies, an experimental unit (pen, bag, mesocosm) often has an array of species that is subjected to some treatment. Analysis by MANOVA is appropriate because of the potential correlation of the responses of the species cooccurring in the pens. These kinds of examples in which there is one within-subject factor, but more than one response variable is measured for each subject (experimental unit), are referred to as "doubly multivariate" designs and can have any number of between-subjects factors, as in an ordinary MANOVA. A doubly multivariate example is given in the SAS manual (SAS Institute Inc., 1989b, Chapter 24, Example 9) and in the Command Reference section of the SPSS Advanced Statistics manual (Norusis, 1990).

6.3 Example: One Between-Subjects and One Within-Subject Factor

The first example is a split-plot design (Table 6.1). Assume a plant species was grown under two Nutrient regimes (low, high) in pots in a greenhouse, and the number of leaves per plant was monitored for five weeks. The example will be analyzed first by repeated-measures ANOVA (6.3.1), then by MANOVA and profile analysis (6.3.2).

6.3.1 Repeated-Measures ANOVA

Recall in a split-plot design applied to repeated-measures data, subjects are plots and levels of the within-subject factor are subplots. Each subject is assigned to one of the between-subjects treatment levels. In terms of Table 6.1, the within-subject factor is Time, and the between-subjects factor Nutrient level. The model for the data in Table 6.1 is

$$X_{ijkm} = \mu + \nu_i + \psi_{k(i)} + \tau_j + \nu\tau_{ij} + \psi\tau_{jk(i)} + \varepsilon_{m(ijk)}. \tag{6.1}$$

where ν_i is the effect of the Nutrient treatment on plant growth, $\psi_{k(i)}$ is the subject effect nested within the respective Nutrient levels, τ_j is the Time effect, $\nu\tau_{ij}$ is the Nutrient × Time interaction, $\psi\tau_{jki}$ is the Subject × Time interaction, and $\varepsilon_{m(ijk)}$ is the error term. m is a dummy subscript included to indicate that the experimental error is nested within the individual observation. The S × T interaction is included in the model statement to emphasize that the potential exists for this interaction to be present; however, because there is no replication of S × T cells, the interaction cannot be estimated and becomes part of the error term for testing the within-subject factor and its interaction with the between-subjects factor (Winer et al., 1991). The expected means squares for this model assuming Nutrient and Time are fixed factors, and subjects (plants) are random, are shown in Table 6.2.

When repeated-measures designs (i.e., not doubly multivariate) are analyzed in SAS (SAS Institute Inc., 1989a,b) both the repeated-measures ANOVA and profile analysis can be done in one analysis (Appendix 6.1.1). The steps are (1) select SAS procedure ANOVA for balanced designs or procedure GLM for

Table 6.1. Growth of Hypothetical Plant (Number of Leaves per Plant) over 5 Weeks at Low (L) and High (H) Nutrient Levels

Plant	Nutrient level	Week 1	Week 2	Week 3	Week 4	Week 5
1	L	4	5	6	8	10
2	L	3	4	6	6	9
3	L	6	7	9	10	12
4	L	5	7	8	10	12
5	L	5	6	7	8	10
6	H	4	6	9	9	11
7	H	3	5	7	10	12
8	H	6	8	11	10	14
9	H	5	7	9	10	12
10	H	5	8	9	11	11

Table 6.2   Expected Mean Squares for Repeated-Measures ANOVA (One Between- and One Within-Subject Factor)

Source of variation	df	E(MS)
Between-subjects	$np-1$	
Nutrient	$p-1$	$\sigma^2_E + q\sigma^2_S + nq\sigma^2_N$
Subjects within groups	$p(n-1)$	$\sigma^2_E + q\sigma^2_S$
Within-subject	$np(q-1)$	
Time	$q-1$	$\sigma^2_E + \sigma^2_{TS} + np\sigma^2_T$
Nutrient × Time	$(p-1)(q-1)$	$\sigma^2_E + \sigma^2_{TS} + n\sigma^2_{NT}$
Subjects × Time within groups	$p(n-1)(q-1)$	$\sigma^2_E + \sigma^2_{TS}$

unbalanced designs, (2) include the CLASS statement to list the between-subjects factor(s), (3) write the MODEL statement in a MANOVA form, with the dependent variables describing the levels of the within-subject factor on the left side of the equation and the between-subjects treatment on the right, and (4) use the REPEATED statement to list the "label," and the number of levels, of the within-subject factor. The results of this analysis of the data in Table 6.1 are presented in Table 6.3.

Notice that the sources of variation are divided into between-subjects (Nutrient) and within-subject effects. The latter includes the within-subject main effect (Time) and its interaction with the between-subjects factor (Nutrient × Time). One usually uses a repeated-measures design because of interest in the effect of the between-subjects treatment over time, i.e., Nutrient × Time interaction. Hence, it should be the first treatment examined.

The probabilities for the respective F statistics $(P > F)$ in column six (Table

Table 6.3.   Repeated-Measures ANOVA of Split-Plot Design[a]

Between-subjects Source	df	MS	F	P>F
Nutrient	1	16.82	2.60	0.1453
Error	8	6.46		

Within-subject					Adj P>F	
Source	df	MS	F	P>F	G–G	H–F
Time	4	66.85	158.22	0.0001	0.0001	0.0001
Nutrient × Time	4	1.27	3.01	0.0326	0.0666	0.0353
Error (Time)	32	0.42				
		Greenhouse–Geisser $\epsilon = 0.5882$				
		Huynh–Feldt $\epsilon = 0.9531$				

[a] $P > F$ is unadjusted probability, G–G and H–F are Greenhouse-Geisser and Huynh-Feldt adjusted probabilities respectively based on the respective epsilons (see below for details).

6.3) are calculated assuming the data meet the circularity assumption. Based on these probabilities, there is a statistically significant Nutrient × Time interaction ($P < 0.0326$) that indicates plants in the fertilized condition added leaves at a faster rate than the unfertilized plants. The statistically significant Time effect ($P < 0.0001$) indicates the average number of leaves increased over time. The Nutrient main effect was not statistically significant based on this F-test, which is not unusual since the interaction was statistically significant.

The F statistics for within-subject factors (and their interactions) are inflated when the sphericity condition is not met. Several estimators (epsilons, ϵ) have been developed for decreasing the degrees of freedom of the F statistic according to the severity of the violation of the sphericity assumption. The degrees of freedom for the critical F statistic are multiplied by the ϵ. Box (1954) originally proposed the equation for ϵ for a within-subject design (no between-subjects factors) based on the "population" variance-covariance matrix. Geisser and Greenhouse (1958) extended this to the split-plot design. However, since ϵ is not known (because the "population" variance–covariance matrix is unknown), ϵ has to be estimated from the sample variance–covariance matrices. This introduces the added complication of not knowing how inaccuracies in the sampling data will effect the estimate of ϵ, and hence how this will influence the adjustment of the F statistic degrees of freedom. For that reason, Greenhouse and Geisser (1959) suggested a very conservative approach of replacing ϵ with the smallest value it could take, $1/(k-1)$, where k is the number of levels of the within-subject factor. Collier et al. (1967) and others have examined the bias introduced by using the sample variance–covariance matrix for estimating ϵ (Crowder and Hand, 1990). The estimator based on sample data is referred to as $\hat{\epsilon}$. For $\hat{\epsilon}$ greater than 0.75, and n (sample size) less than $2k$, $\hat{\epsilon}$ tends to overcorrect the degrees of freedom and produce a conservative test (Crowder and Hand, 1990; Winer et al., 1991). Huynh and Feldt (1976) proposed a less biased estimate of ϵ based on sample data ($\tilde{\epsilon}$) (described in Winer et al., 1991), although it can be overly liberal in some cases (type I error). SAS provides estimates of both Box's ϵ ($\hat{\epsilon}$) and Huynh-Feldt's adjustment to ϵ ($\tilde{\epsilon}$), as well as the respective probabilities for the adjusted F statistics of the within-subject factors, as part of its repeated-measures ANOVA output (Table 6.3). They refer to $\hat{\epsilon}$ as the Greenhouse-Geisser ϵ as I will). Its value ranges from 0 to 1, whereas $\tilde{\epsilon}$ can be > 1. When the latter is > 1, then 1 is used as the value. The smaller the value of $\hat{\epsilon}$ and $\tilde{\epsilon}$, the greater the departure from sphericity. $\hat{\epsilon}$ is $\leq \tilde{\epsilon}$, so generally $\hat{\epsilon}$ is the more conservative adjustment. They converge as sample size increases (Maxwell and Delaney, 1990).

The epsilons for the analysis of the data in Table 6.1 are at the bottom of Table 6.3. One can see that the adjustment is greater for $\hat{\epsilon}$ than for $\tilde{\epsilon}$. The probabilities for the modified F statistics are in the two right hand columns of the within-subject treatment combinations. When the adjustments are made for the F statistics, the Nutrient × Time interaction is no longer statistically significant according to the

Greenhouse-Geisser adjustment, but is for the Huynh–Feldt adjustment. The probabilities for the Time effect are unchanged. There is not universal agreement among statisticians about the use of these adjustment procedures, especially compared with the alternative of profile analysis. (See section 6.8.2 for a comparison of alternative recommendations.)

6.3.2 Profile Analysis

The MANOVA approach to the analysis of repeated-measures data can consist of just estimating the multivariate test statistic for data from all levels of the within-subject factor considered simultaneously. However, profile analysis is more informative because it analyzes the pattern of response of the within-subject factor. Profile analysis addresses hypotheses about "parallelism," "levels," and "flatness" by transforming within-subject repeated-measures data to a set of "contrasts" or differences, and then doing univariate (t-test, ANOVA) and/or multivariate (Hotelling's T^2, MANOVA) analyses on the contrasts. Potvin et al. (1990b) termed this approach MANOVAR. I will present the details of profile analysis of the nutrient experiment (Table 6.1) first using data from the first 2 weeks, and then from all five dates.

For comparative purposes, I will estimate the overall multivariate statistic for the first 2 weeks of data and then analyze it by profile analysis. The multivariate comparison of the number of leaves in the low and high Nutrient levels for the first 2 weeks (two dependent variables) using Hotelling's T^2 gives $F = 5.69$, $df = 2,7$, $P = 0.034$. This shows that the number of leaves is different between low and high Nutrient levels when Weeks 1 and 2 are considered simultaneously.

To analyze the first 2 weeks of data by profile analysis, first the "parallelism" hypothesis will be addressed by testing whether the "shapes" or slopes of the response curves are the same for both groups (Fig. 6.2). In terms of repeated-measures ANOVA, this is a test of whether there is a significant Nutrient × Time interaction. Comparing the shapes of the response curves is equivalent to asking whether the mean change in number of leaves between Weeks 1 and 2 is the same for both nutrient levels, i.e., how do the mean differences (d) compare for the two groups (Table 6.4)?

The hypothesis tested is:

$$H_o: (\mu_{11} - \mu_{12}) = (\mu_{h1} - \mu_{h2})$$

where the first (ith) subscript refers to low and high Nutrient, and the second (jth) to Week 1 and Week 2. For Weeks 1 and 2, d is 1.2 for the low treatment level and 2.0 for the high Nutrient treatment level. A t-test shows that the two groups are significantly different ($t = 3.54$, $df = 8$, $P = 0.008$) in terms of the means changes. This indicates there is a significant Nutrient × Time interaction and that the number of leaves increased faster in the high Nutrient treatment level.

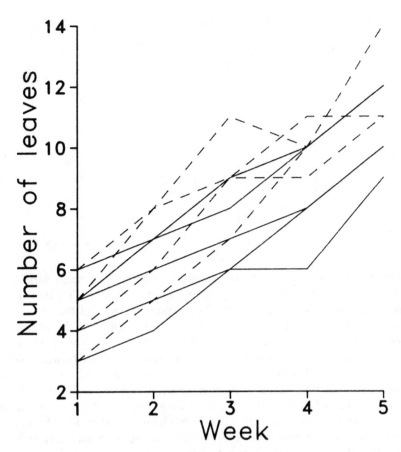

Figure 6.2. Growth of hypothetical plants grown under High Nutrient (dashed lines) and Low Nutrient (solid lines) conditions for 5 weeks.

Since the Nutrient × Time interaction was found significant, the tests of the "levels" and "flatness" hypotheses are less meaningful, but will be explained for demonstration purposes. The test of the "levels" hypothesis is a test of the Nutrient main effect. The Week 1 and Week 2 values are averaged for each subject [(Week 1 + Week 2)/2], and then the means compared between groups with a *t*-test. One is asking whether the "levels" of the response curves are different, i.e., whether, when averaged over Weeks 1 and 2, the high Nutrient plants had more leaves than the low Nutrient plants. In this example there was no uniform effect of the Nutrient treatment in the first 2 weeks ($t = 0.66$, $df = 8$, $P = 0.53$), which was consistent with the significant Nutrient × Time interaction.

Finally, for a test of the "flatness" hypothesis, or the Time effect, the contrast variable (differences between Weeks 2 and 1) is used again, but is averaged over

Table 6.4. Difference in Number of Leaves per Plant for Example in Table 6.1

Plant	Nutrient level	2−1	3−2	4−3	5−4
		Difference between adjacent weeks			
1	L	1	1	2	2
2	L	1	2	0	3
3	L	1	2	1	2
4	L	2	1	2	2
5	L	1	1	1	2
Mean		1.2	1.4	1.2	2.2
6	H	1	3	0	2
7	H	2	2	3	2
8	H	2	3	−1	4
9	H	2	2	1	2
10	H	3	1	2	0
Mean		2.0	2.2	1.0	2.0

Nutrient levels. The test is whether the grand mean of the contrast variable is different from 0, which is essentially a paired t-test (Zar, 1984, Chapter 10), but without regard to the Nutrient treatment grouping. In terms of the original response variables, it is a test of whether there is an increase in the number of leaves between Weeks 1 and 2 (slope > 0) when all plants are averaged together for each week. In our example Time was statistically significant between Weeks 1 and 2 ($t = 7.97$, $df = 9$, $P < 0.0001$), so there was an overall slope different from 0. Summarizing, in the multivariate analysis of a simple design with one between-subjects factor (Nutrient), and a within-subject factor with two levels (Weeks 1,2), one can use Hotelling's T^2 to test for an overall difference between groups, or profile analysis (three separate univariate tests) to separately examine the parallelism, levels, and flatness hypotheses.

Next I will analyze the data in Table 6.1 for all five dates using profile analysis, since this is the more informative approach and is used by SAS (procedure GLM and procedure ANOVA). Profile analysis of the Time effect (flatness) and the Nutrient × Time interaction (parallelism) is based on differences (contrasts) of adjacent weeks. There are four differences because there are data for five weeks. Because these differences can be treated as four dependent variables, MANOVA is appropriate for the analysis. SAS refers to these differences (contrasts) as a *transformation* of the original within-subject data. This is somewhat different from the kinds of transformation most ecologists are familiar with in statistical analyses, e.g., log, square root, arcsine, etc. In the difference transformation the values of two levels of the within-subject factor are "transformed" to one. For a test of the parallelism hypothesis, the set of differences used is Week 2 − Week 1, Week 3 − Week 2, Week 4 − Week 3, and Week 5 − Week 4 (Table 6.3). The desired transformation is identified for the within-subject factor in the REPEATED command (Appendix 6.1A). PROFILE is used for this analysis because it generates the contrast variables based on differences of adjacent levels of the within-subject factor. Other transformations available in SAS are described below.

Table 6.5A shows the results of the MANOVA analysis of the differences. The values of four different multivariate test statistics are shown in the second column, and their equivalent F statistics in the third. The probabilities for the F statistics are in the last column. The probabilities ($P = 0.025$) indicate that the Nutrient $\times$ Time interaction is statistically significant. So the slopes of the growth curves in the high and low Nutrient levels are different by profile analysis, which agrees with the repeated-measures ANOVA analysis (Table 6.3). (A discussion of the different multivariate test statistics is presented in Chapter 5.)

The test for the flatness hypothesis (Time effect) is whether there is a significant increase in the number of leaves over time when averaged over both Nutrient levels. This is a test of whether the grand mean (averaged over Nutrient levels) of the set of the four time differences is zero. Again, the statistically significant Fs in the MANOVA analysis indicate there is an increase in the number of leaves over Time ($P < 0.0001$) (Table 6.5B).

The test for the between-subjects effect(s) in profile analysis (e.g., Nutrient levels) is the same as in repeated-measures ANOVA: both are based on a comparison of the mean within-subject response across the levels of the between-subjects factor (Time) (SAS Institute Inc., 1989b; Potvin et al., 1990b).

If one is interested in identifying the particular time intervals in which the treatment effects are different, individual ANOVAs (F-tests) can be done on each of the contrasts. For example, one can test the significance of the Nutrient $\times$ Time interaction, or the Time effect for each contrast in Table 6.4. The results of such an analysis are shown in Table 6.6. Mean refers to a test of the flatness hypothesis of whether there was a significant Time effect, and Nutrient refers to the test of the parallelism hypothesis of whether there was a significant Nutrient

Table 6.5. MANOVA of Nutrient $\times$ Time Interaction, and Time Effect, of Data in Table 6.1[a]

Statistic	Value	F	Num df	Den df	$P > F$
A. Nutrient $\times$ Time					
Wilks' lambda	0.1447	7.38	4	5	0.0250
Pillai's trace	0.8552	7.38	4	5	0.0250
Hotelling-Lawley trace	5.9074	7.38	4	5	0.0250
Roy's greatest root	5.9074	7.38	4	5	0.0250
B. Time					
Wilks' lambda	0.0084	146.04	4	5	0.0001
Pillai's trace	0.9915	146.04	4	5	0.0001
Hotelling-Lawley trace	116.8333	146.04	4	5	0.0001
Roy's greatest root	116.8333	146.04	4	5	0.0001

[a]Num df and Den df refer to numerator and denominator degrees of freedom respectively.

× Time interaction in the respective ANOVAs for each contrast variable. Notice that although the multivariate Fs (MANOVA) for the Nutrient × Time interaction and the Time effect were both statistically significant when all four contrasts were analyzed together (Table 6.6), the latter was only significant for Week 2 − Week 1 in the individual ANOVAs. This indicates that a significant difference in the change in the number of leaves due to the Nutrient treatment occurred only between Week 1 and Week 2, which can be seen when the means for the contrasts are compared (Table 6.3). The differences between the mean differences for the low and high Nutrient levels for the other dates were not sufficiently large themselves, but they contributed to an overall difference when all were analyzed simultaneously.

Since repeated ANOVAs are done when individual contrasts are analyzed, the experimentwise error rate should be adjusted according to the number of tests run. To maintain an overall α of 0.05, a Bonferroni adjustment of $\alpha = 0.05/4 = 0.0125$ for each contrast could be used for Table 6.6. If the investigator is interested in testing only a single contrast, then α would not be adjusted. Although a statistical package such as SAS automatically does the MANOVA and ANOVAs for all contrasts, one could examine the ANOVA only for the contrast of interest.

It should be emphasized that different transformations (contrasts) can be used for the within-subject factor(s), and it will not affect the outcome of the MANOVA (multivariate F) because of the invariance property of the multivariate test statistic (Morrison, 1990). The particular transformation used depends on the patterns one is looking for in the within-subject effect(s). SAS gives five choices of the transformation: Profile, Contrast, Helmert, Means, and Polynomial. As described

Table 6.6. ANOVAs on Each of the Contrasts of Within-Subject Factor, Time for Data in Table 6.1

Source	df	MS	F	P>F
Contrast Variable: Week 2−Week 1				
Mean	1	28.9	144.50	0.0001
Nutrient	1	2.5	12.50	0.0077
Error	8	0.2		
Contrast Variable: Week 3−Week 2				
Mean	1	32.4	64.80	0.0001
Nutrient	1	1.6	3.20	0.1114
Error	8	0.5		
Contrast Variable: Week 4−Week 3				
Mean	1	12.1	7.56	0.0251
Nutrient	1	0.1	0.06	0.8089
Error	8	1.6		
Contrast Variable: Week 5−Week 4				
Mean	1	44.1	40.09	0.0002
Nutrient	1	0.1	0.09	0.7707
Error	8	1.1		

above, I used the Profile transformation in which differences were constructed from adjacent levels of the Time factor. The Contrast transformation is the default condition, one level of the within-subject factor is considered the control and all other levels are compared to it. Potvin et al. (1990b) used the Helmert transformation: it compares the level of a within-subject factor to the mean of subsequent levels. The Mean transformation compares each level of the within-subject factor to the mean of all the other levels of the within-subject factor.

The Polynomial transformation is particularly useful if one is interested in examining trends in the within-subject data to see if they conform to a particular form. For example, does the number of leaves per plant change in a linear, quadratic, cubic, etc. manner over time? The method relies on a set of polynomials of increasing order that are constructed as a set of independent (orthogonal) contrasts and often is referred to as "trend" analysis (Winer et al., 1991). Orthogonal polynomials enable one to ask whether there is a significant linear (first order), quadratic (second order), cubic (third order), etc., trend in the data (Gurevitch and Chester, 1986; Hand and Taylor, 1987). If k is the number of levels of the within-subject factor, $k - 1$ polynomials can be constructed, although usually one would be interested in the lower order (linear, quadratic) trends only. In the analysis of the within-subject factor(s), MANOVA considers simultaneously all orders of the polynomials that can be obtained from the data. As with the profile transformation, one can examine particular polynomials by individual ANOVAs. In examining orthogonal polynomials, normally one proceeds from higher to lower order polynomials in testing for significance, and stops at the order of polynomial at which significance is found. Since higher order polynomials would seem inappropriate for most ecological data, and examining more tests increases one's chance of committing a type I error, it would be appropriate in many cases to begin with the cubic, or quadratic analysis for testing for significance.

The plant growth data from Table 6.1 were analyzed using orthogonal polynomials (Table 6.7). Only the individual F-tests are shown because the multivariate Fs are the same as in Table 6.5. Only the quadratic and linear trends were examined in this example, although SAS also gives the cubic. α was set at 0.025 because there were two separate analyses (0.05/2). There was a significant quadratic trend for the Nutrient $\times$ Time interaction ($P = 0.0175$). Remember Nutrient in the SAS output indicates the Nutrient $\times$ Time interaction. So the low and high Nutrient treatment levels differed significantly in their quadratic responses of number of leaves over time. Time (Mean) did not have a significant quadratic response, but the linear trend for Time was statistically significant ($P < 0.0001$) indicating there was an overall linear increase in the number of leaves during the experiment.

6.4 Multiple Within-Subject Factors

Situations often arise in ecological experiments and data collection in which repeated observations on the same individual can be classified factorially, or the

Table 6.7. Individual ANOVAs for Different Order Orthogonal Polynomials for Data in Table 6.1

Source	df	MS	F	P>F
Contrast Variable: First Order				
Mean	1	265.690	432.02	0.0001
Nutrient	1	2.890	4.70	0.0620
Error	8	0.615		
Contrast Variable: Second Order				
Mean	1	0.007	0.03	0.8651
Nutrient	1	2.064	8.89	0.0175
Error	8	0.232		

same individual is observed under more than one set of conditions. In terms of the plant growth experiment (Table 6.1), it may be of interest to run the experiment for a second year to see whether the high Nutrient plants also produced more leaves than the low Nutrient plants in the second year (Table 6.8). This would be a design with two within-subject factors, Weeks and Years, and one between-subjects factor, Nutrient.

The within-subject factors are treated as follows: (1) averaging over years, one can test whether there is a significant Week effect; (2) averaging over weeks, one can test whether there is a significant Year effect; (3) by the appropriate contrasts one can test whether there is a significant Week × Year interaction, i.e., averaging over the Nutrient treatment, did the plants grow differently between years?

The between-subjects factor (Nutrient) is cross-classified with each of these, so one must examine the interactions of each of the within-subject main effects (Week, Year), and their interaction (Week × Year), with the Nutrient treatment.

Table 6.8. Growth of Plants in Table 6.1 for Years 1 and 2[a]

Plant	Nutrient level	Year 1					Year 2				
		1	2	3	4	5	1	2	3	4	5
1	L	4	5	6	8	10	4	4	5	7	9
2	L	3	4	6	6	9	3	4	6	8	10
3	L	6	7	9	10	12	3	4	6	7	9
4	L	5	7	8	10	12	3	4	6	8	10
5	L	5	6	7	8	10	5	5	6	7	10
6	H	4	6	9	9	11	4	4	5	7	10
7	H	3	5	7	10	12	5	5	6	8	11
8	H	6	8	11	10	14	4	4	6	7	10
9	H	5	7	9	10	12	3	4	5	7	9
10	H	5	8	9	11	11	5	5	7	9	11

[a]Number of leaves per plant over five weeks at low (L) and high Nutrient (H) levels for Years 1 and 2. Plants add leaves for a 5- to 6-week period each spring, with an initial flush of leaves within first week of emergence. Leaves die back during the winter.

In fact, it is the three factor interaction (Nutrient × Year × Time) that one should examine first. It should be emphasized that designs with greater than one within-subject factor can be analyzed both by repeated-measures ANOVA (Winer et al., 1991) and profile analysis (MANOVA) (O'Brien and Kaiser, 1985).

Although the analysis at first may seem complicated, it really is not if one keeps track of between- and within-subject factors (Table 6.9). In the MODEL statement in the SAS program each of the Weeks is treated as a different dependent variable for a total of 10 when data for both years are combined (Appendix 6.1B). Also, the number of levels of each of the within-subject factors is listed in the REPEATED command. I have included only the results of the profile analysis (MANOVA) in the results of the analysis (Table 6.9) and have condensed the SAS output. Under the within-subject effect, the three-way interaction (Nutrient × Week × Year) was statistically significant. The interpretation of that significance is that the high Nutrient plants produced more leaves than the low Nutrient plants during Year 1, but not during Year 2, so the treatment produced different effects in different years. As in the analysis of the first year's data, the Nutrient main effect was not statistically significant. The analysis by ANOVA or orthogonal polynomials analyses would result in the same conclusions. See O'Brien and Kaiser (1985) for a detailed example with two between-subjects factors and two within-subject factors.

6.5 Other Ecological Examples of Repeated Measures

I have concentrated on repeated-measures analyses in which Time is the repeated (within-subject) factor. Other repeated-measures situations arise in ecological studies in which individual organisms or experimental units are exposed to several

Table 6.9. Results of Profile Analysis (MANOVA) of Data in Table 6.8[a]

	Between-Subjects		
Source	MS	df	F	P>F
Nutrient (N)	13.69	1	3.83	0.0861
Error	3.58	8		
	Within-Subject		
Source	F	Num df	Den df	P>F
Year (Y)	14.47	1	8	0.0052
Week (W)	135.14	4	5	0.0001
Year × Week	14.34	4	5	0.0060
Nutrient × Year	0.97	1	8	0.3530
Nutrient × Week	0.52	4	5	0.7301
N × Y × W	5.66	4	5	0.0425

[a]The F statistics for all four MANOVA test criteria were identical for all treatment combinations. Num df and Den df are numerator and denominator degrees of freedom respectively.

different conditions and some response is measured. In the example used by Potvin et al. (1990b), the photosynthetic rate of the same plants exposed to different concentrations of CO_2 was compared (within-subject factor). There were two between-subjects factors: plants were from two different populations and were either chilled or kept at the ambient temperature before measurement of photosynthetic rates.

Investigators of plant-herbivorous insect interactions often are interested in the fitness of different populations of an insect species on different plant species or on different plant populations. A typical protocol is to raise sibs from the same egg masses (subjects) from different populations (between-subjects factor) on leaves of different plant species or plant populations (within-subject factor). Horton et al. (1991) give a detailed analysis of the increase in power when such data are analyzed as a repeated-measures design. The egg mass is considered the block or subject, and there are replicate egg masses from each insect population.

In some frog species sexes are different colors, and many also change color on different colored backgrounds. King and King (1991) examined differences in color between male and female wood frogs (*Rana sylvatica*) (between-subjects) in which the same individuals were exposed to different colored backgrounds (within-subject). Because color was assessed in terms of three response variables, this was a doubly multivariate design.

6.6 Alternatives to Repeated-Measures ANOVA and Profile Analysis

The power of MANOVA decreases as the number of dependent variables increases and sample size decreases. One reason for the low power of MANOVA is the lack of restrictions on the structure of the variance-covariance matrix (Crowder and Hand, 1990). Repeated-measures ANOVA may have more power in some circumstances, but also has much more restrictive conditions on the variance–covariance matrix. An alternative has been to fit curves using the general linear model and then to compare parameters of these curves by multivariate procedures, what statisticians regard as the "analysis of growth curves" (Morrison, 1990). An obvious advantage of fitting data to a curve is that it reduces drastically the number of parameters to be compared for within-subject effects. Also, neither repeated-measures ANOVA nor profile analysis can handle missing data for subjects, whereas curve-fitting can. Since the observations usually are highly correlated, this has lead to the development of methods for taking the correlation structure into account in the analysis. The development of analyses based on particular forms of the variance-covariance matrices has been an active area of statistical research on repeated measures (Crowder and Hand, 1990, Chapter 5). These matrices usually fall somewhere between the restrictive "compound symmetry" and unrestricted "unstructured" variance-covariance matrices.

Mixed Linear Models is a rather new (to ecologists), and apparently promising methodology, that is statistically sophisticated (Crowder and Hand, 1990, Chapter 6; McLean et al., 1991; SAS Institute Inc., 1992, procedure MIXED, Release 6.07). These models can handle missing data and many levels of a repeated factor. Just as the General Linear Model is a generalization of the basic fixed-effect ANOVA model, the Mixed Linear Model is a generalization of the mixed-model ANOVA, which is directly applicable to repeated-measures ANOVA since subjects usually are considered a random effect and the within-subject effects a fixed factor (Winer et al., 1991). The analyses can be done assuming different forms for the variance-covariance matrix. It would be wise to consult a statistician when using this approach.

Potvin et al. (1990b) also used nonlinear curve fitting for repeated-measures analysis, and discussed the pros and cons of the procedure. A problem with curve fitting is how sensitive the fitted parameters are to significant changes in the data. See Certato (1990) for an analysis of parametric tests for comparing parameters in the von Bertalanffy growth equation. Potvin et al. also used a nonparametric approach suggested by Koch et al. (1980), which avoids assumptions of normality, but generally has less power. In a comparison of several different methods of analyzing growth curves, Kokoska and Johnson (1987) use a randomization test proposed by Zerbe (1979) for comparing growth curves over a particular interval of time for completely randomized designs.

Muir (1986) presents a univariate test for the linear increase in a response variable with increasing levels of the within-subject factor (e.g., selection over time). The attractiveness of this test is that it increases in power as the number of levels of the within-subject factor increases, and does not have to meet the assumption of circularity because it is testing a specific contrast [see Section 6.2.1 and O'Brien and Kaiser (1985)]. Because the analysis is not straightforward using SAS, a statistician should be consulted when using this technique.

Crossover experiments are considered by some to be related to repeated-measures experiments (Crowder and Hand, 1990, Chapter 7; Mead, 1988). As a very simple example, a group of subjects is first exposed to condition A and then to condition B. A second group is treated in just the opposite sequence. Feinsinger and Tiebout (1991) used such a design in a bird pollination study. Categorical data also can be analyzed as repeated measures (Agresti, 1990, Chapter 11). Finally, time series analysis is a special kind of repeated-measures analysis in which there are many repeated measurements and may be appropriate in some cases (Chapter 7).

6.7 Statistical Packages

Repeated-measures analyses can be done with the major mainframe computer statistical packages: SAS, SPSS, and BMDP. Crowder and Hand (1990) summarize the methods and capabilities of the routines. As mentioned in section 6.4,

SAS recently has added procedure MIXED in release 6.07 for using Mixed Linear Models. It uses maximum likelihood and restricted maximum likelihood methods to handle a variety of different variance-covariance matrices, as do BMDP 3V and 5V. All three packages automatically generate the appropriate contrasts needed for multiple within-subject factor designs (O'Brien and Kaiser, 1985).

Personal computer versions are available for all these programs except procedure MIXED. The amount of RAM available may limit the ability to run more complex analyses. Statistica, SYSTAT, SuperANOVA, and JMP are other statistical packages that run on personal computers and do both univariate and multivariate repeated-measures analyses. The SPSS manual (Norusis, 1990) has a particularly clear introduction and discussion of repeated-measures analyses (ANOVA and profile analysis).

6.8 Conclusions

6.8.1 Texts

Many of the texts and much of the statistical research on repeated-measures analysis over the past several decades have been associated with the psychological literature. The current texts span the spectrum in terms of their coverage of the topic: Winer et al. (1991) present primarily the univariate approach and Maxwell and Delaney (1990) give an extensive discussion of both the univariate and the multivariate approaches. Harris (1985) and Stevens (1992) concentrate on the multivariate approach, but give summaries of the univariate. Hand and Taylor (1987) introduce the multivariate approach with an extensive discussion of contrasts and detailed analyses of real examples. Crowder and Hand (1990) present a statistician's review of the topic at a more technical level than the others and cover techniques not covered in the other texts. The introductions to the topic in the papers by O'Brien and Kaiser (1985), Gurevitch and Chester (1986), and Potvin et al. (1990b) have been described above. See also Koch et al. (1980). With the exception of Winer et al. (1991), the trend in the psychologically oriented texts is toward greater reliance on the multivariate approach. Crowder and Hand (1990), however, do not show this same preference.

6.8.2 Recommendations—Repeated-Measures ANOVA vs Profile Analysis

There is not unanimity among statisticians in recommending which approach to take in repeated-measures analyses. I summarize the alternative views below.

1. The general opinion seems to be not to use Mauchly's sphericity test as a basis of decision for using univariate repeated measures (O'Brien and Kaiser, 1985; Winer et al., 1991; Stevens, 1992).

2. If one agrees with Mead's (1988) criticism of using Time as the within-subject factor (6.2), the multivariate approach seems preferable (profile analysis) [O'Brien and Kaiser (1985), Maxwell and Delaney (1990)]:

   a. Maxwell and Delaney (1990) recommend using MANOVA only when $(N - M > k + 9)$ to have sufficient power. These conditions may be hard to meet in some ecological systems.

   b. If one cannot meet these criteria, and cannot increase replication, or reduce the number of levels of the within-subject factor, one can follow the recommendation of Maxwell and Delaney (1990) and set up individual multivariate contrasts.

   c. Alternatively, the multivariate analysis can be done with low replication, but α set at 0.1 or 0.2 to compensate for the low power, especially if it is an exploratory study (Stevens, 1992).

   d. If neither of these is feasible, then a nonparametric approach (Potvin et al., 1990b) can be used.

   e. As alternatives to repeated-measures ANOVA when time is the within-subject factor, Mead (1988) suggests individual contrasts on the different time intervals, MANOVA, or curve fitting.

3. If one does not accept Mead's criticism and there is sufficient replication to use the multivariate approach, then there are three different opinions on what to do. The respective recommendations depend primarily on the authors' views of the respective overall power of the univariate and multivariate approaches for different experimental designs and tests:

   a. O'Brien and Kaiser (1985) prefer, and Maxwell and Delaney (1990) give a "slight edge" to, the multivariate approach over the adjusted univariate approach, including doing individual multivariate contrasts if there is trouble meeting the criteria for MANOVA.

   b. Stevens (1992) recommends doing both the multivariate and the adjusted univariate tests ($\alpha = 0.025$ for each test), especially if it is an exploratory study.

   c. Finally, Crowder and Hand (1990) say to use either, if it is possible to do MANOVA.
   Given the variation in opinion above, Stevens' (1992) recommendation seems a safe option.

4. There also are several options when one does the univariate repeated-measures analysis:

   a. The conservative approach is to use the Greenhouse-Geisser corrected probability. Maxwell and Delaney (1990) recommend this because the Huynh-Feldt adjustment occasionally can fail to properly control type I error (overestimates ϵ).

   b. The more liberal approach is to use the Huynh-Feldt adjustment.

   c. Stevens (1992) suggests averaging the ϵs of the two corrections.

5. Because there are limitations of both the standard univariate and multivariate techniques one should consider fitting response curves and nonparametric analysis. Also, ecologists should keep an eye on new developments in "growth models" and mixed linear models (Section 6.6), especially if one has a study in which there are many levels of the within-subject factor(s).

Acknowledgments

I thank Jessica Gurevitch and Sam Scheiner for their indulgence and especially helpful editorial comments, Ann Hedrick for first introducing me to O'Brien and Kaiser (1985), an anonymous reviewer for highlighting inconsistencies and providing the Muir (1986) reference, and Stephen F. O'Brien and Donald R. Brown for answering my naive statistical questions. Any mistakes remain my own.

Appendix 6.1

SAS program code for ANOVA and MANOVA (profile analysis) repeated-measures analyses.

A. Analyses of data in Table 6.1

Output in Tables 6.2, 6.3, 6.5, 6.6, and 6.7.

```
PROC ANOVA;                      /use procedure ANOVA for balanced designs/
  CLASS NUTRIENT;
  MODEL WK1 WK2 WK3 WK4 WK5                      /NOUNI suppresses ANOVAs
    = NUTRIENT/NOUNI;                              of dependent variables/
  REPEATED TIME 5 (1 2 3 4 5)    /REPEATED = repeated-measures factor label
    PROFILE/SUMMARY;             TIME. 5 indicates levels of TIME. (1 2 3 4 5) are
                                 intervals of TIME, which can be unequally spaced,
                                 default is equally spaced intervals. PROFILE = profile
                                 transformation. SUMMARY prints the ANOVA
                                                            for each contrast/
  REPEATED TIME 5 (1 2 3 4 5)    /POLYNOMIAL = polynomial transformation/
    POLYNOMIAL/SUMMARY;
```

B. Repeated-Measures Analysis Data in Table 6.8 (Two Within-Subject Factors)

Output in Table 6.9.

```
PROC ANOVA;
  CLASS NUTRIENT;
  MODEL Y1W1 Y1W2 Y1W3           /Y1W1 ... Y1W5, Y2W1 ... Y2W5 identifies the
    Y1W4 Y1W5 Y2W1 Y2W2             10 dependent variables for the five weeks
    Y2W3 Y2W4 Y2W5 =                             in the two years/
    NUTRIENT/NOUNI;

  REPEATED YEAR 2, WEEK 5        /YEAR = first within-subject factor with 2 levels,
    POLYNOMIAL/SUMMARY;         WEEK = second within-subject factor with 5 levels.
                                  See SAS Institute Inc. (1989b, p. 926) for an
                              explanation of how to define dependent variables
                                 when there are two within-subject factors./
```

7

Time-Series Intervention Analysis: Unreplicated Large-Scale Experiments

Paul W. Rasmussen, Dennis M. Heisey,
Erik V. Nordheim, and Thomas M. Frost

7.1 Introduction

Some important ecological questions, especially those operating on large or unique scales, defy replication and randomization (Schindler, 1987; Frost et al., 1988; Carpenter, 1990). For example, suppose one is interested in the consequences of lake acidification on rotifer populations. Experimental acidification of a single small lake is a major undertaking, so it may be possible to manipulate only one lake. Even with baseline data from before the application of the acid, such a study would have no replication, and thus could not be analyzed with classical statistical methods such as analysis of variance (Chapter 3).

An alternative approach might be to use some biological/physical model of the system of interest that allows for replication. For example, one could construct small replicated enclosures in lakes, and acidify some of them. While this would (with proper execution) permit valid statistical analysis of differences among units the size of the enclosures, it is questionable whether this model allows valid ecological generalization to the lake ecosystem. For large-scale phenomena, experiments on small-scale models may not be a trustworthy ecological substitute for large-scale studies (although they may provide valuable supplementary information).

In this chapter, we examine how certain types of unreplicated studies can be analyzed with techniques developed for time series data. Time series are repeated measurements, or subsamples, taken on the same experimental unit through time. Time series analysis techniques include methods for determining whether nonrandom changes in the mean level of a series have occurred at prespecified times. The results of such analyses can help determine whether other changes or manipulations occurring at those prespecified times may have caused changes in the observed series.

The methods we describe here take advantage of the long term data available for many large-scale studies (see also Jassby and Powell, 1990). Previous authors have proposed a variety of techniques for assessing such studies, ranging from graphic approaches (Bormann and Likens, 1979) to more sophisticated statistical analyses (Matson and Carpenter, 1990). These techniques share an emphasis on time series data and usually involve comparing a series of pre- and posttreatment measurements on a treatment and a reference system (e.g., Stewart-Oaten et al., 1986; Carpenter et al., 1989). We will confine our discussion of time series methods to the use of ARIMA modeling techniques, although many other time series methods are available.

We first consider how time series data affect the use and interpretation of classical statistical analysis methods (Section 7.2). Section 7.3 presents some examples of unreplicated time series designs. Section 7.4 introduces some of the key ideas of time series analysis. Section 7.5 describes intervention analysis, an extension of time series analysis useful for examining the impact of a treatment or natural perturbation. Section 7.6 illustrates the application of intervention analysis to data from a whole-lake acidification experiment. (If you are not an aquatic ecologist, mentally replace "rotifer" and "lake" with your favorite organism and ecosystem, respectively, in our examples. The principles and potential applications are general.) We discuss some general issues in Section 7.7.

7.2 Replication and Experimental Error

We first return briefly to the lake acidification example. The prescription in classical experimental design for testing the effect of lake acidification on rotifer densities would be similar to the following:

1. Identify all lakes to which we want our inferences to apply.
2. Select, say, four lakes at random from this population.
3. Randomly select two of the four lakes to be acidified.
4. Use the other two lakes as reference, or control, lakes.

Suppose estimates of pre- and posttreatment rotifer densities are available for all four lakes; for each lake one can compute the change (difference) between the pre- and posttreatment periods for further analysis. Classical analysis then proceeds to examine the question of whether the changes in the acidified lakes are consistently different from the changes in the reference lakes, judged in light of the typical lake variability. This typical lake variability, or experimental error, can be estimated only by having replicate lakes within each group. [Two lakes per group is the minimum that permits estimation of experimental error, although a design with only two lakes per group would normally return a rather poor estimate of experimental error (Carpenter, 1989).] An important feature of the classical approach is that one can specify the probability of mistakenly declaring an acidification effect when none exists.

While it may be difficult to acidify more than one lake, getting a long time series of pre- and posttreatment data from a single acid-treated lake may be feasible. With such a series, one can examine the related question of whether there was an unusual change in the series at the time of the treatment, judged in light of normal variation in the series through time. The hypothesis tested here is weaker than the classical hypothesis. A test of this hypothesis can answer the question only of whether a change occurred at the time of the treatment; it cannot resolve the issue of whether the change was due to the treatment rather than some other coincidental event (Frost et al., 1988). Making the case for the change being due to the treatment in the absence of replication is as much an ecological issue as a statistical one; such arguments can often be supported by corroborating data, such as from enclosure experiments. In the next section we discuss time series designs that to some extent guard against the detection of spurious changes.

Determining the "normal variation through time" of a time series is a more difficult issue than determining experimental error in a replicated experiment. In a replicated, randomized experiment, the experimental units are assumed to act independently of one another, which greatly simplifies the statistical models. On the other hand, measurements in time series are usually serially dependent or autocorrelated; the future can (at least in part) be predicted from the past. The variation of a system through time depends on its autocorrelation structure, and much of time series analysis focuses on constructing and evaluating models for such structure.

We caution the reader to guard against treating subsampling through time as genuine replication and then analyzing the unreplicated experiment as if it were replicated. In some cases, it may turn out that a classical test, such as a *t*-test, can be applied to a time series. Such a test could be used if analysis indicated that the measurements were not autocorrelated (Stewart-Oaten et al., 1986). Even then it must be remembered that the hypothesis tested has to do with a change in level at the time of the treatment application, and that such a change may not be due to the treatment. These issues are addressed by Hurlburt (1984) and Stewart-Oaten et al. (1992), and a basic understanding of them is necessary for conducting effective ecological experimentation.

We do not advocate unreplicated designs as a matter of course but they are sometimes necessary. Even with the detailed machinery of time series, some caution will be required in the interpretation of the results. Without replication, one cannot be sure that one has an adequate understanding of the underlying error against which treatment effects should be judged. Increasing confidence in making such an interpretation can be developed in several ways, however. These include (1) considering information on the natural variability that is characteristic of the types of systems being manipulated, (2) integrating smaller-scale experiments within the large-scale manipulations, and (3) developing realistic, mechanistic models of the process being evaluated (Frost et al., 1988).

7.3 Unreplicated Time Series Designs

Suppose an ecologist monitors rotifer densities in a lake biweekly for 5 years, then acidifies it, and continues to monitor it for an additional 5 years. In this example, there is a sequence of observations over time (a time series) on a single experimental unit (the lake). We refer to such a single series with an intervention as a before-after design. (Acidification is the intervention.)

Sometimes nature, perhaps unexpectedly, applies a perturbation to a system. This can be called a "natural experiment." In such a case there may be no choice but to use a simple before-after design. Such designs are most prone to having coincidental "jumps" or trends spuriously detected as intervention effects. When the treatment is under the investigator's control, improvements in the before-after design are possible. One simple modification is to have multiple interventions, or to switch back and forth between the treatment and control conditions. If each intervention is followed by a consistent response, the likelihood that the observed responses are merely coincidental is reduced.

A design with paired units is also useful. One unit receives a treatment (intervention) during the course of the study, and the other serves as a reference, or baseline, and is never treated. Both units are sampled repeatedly and at the same times before and after the treatment is applied. This design has been discussed by Stewart-Oaten et al. (1986), and earlier by Eberhardt (1976), although it had been in use among field biologists previous to either of these discussions (see Hunt, 1976). We will refer to this design as the before–after–control–impact (BACI) design, following Stewart-Oaten et al. (1986). In a BACI design, the reference unit provides a standard against which to compare the treatment unit. This helps to determine whether changes seen in the treatment unit are due to the treatment itself, long-term environmental trends, natural variation in time, or some other factor.

Two important decisions involved in implementing a BACI design are the selection of experimental units, and the timing and number of samples per unit. The treatment and reference units should be representative of the systems to which an observed treatment effect is to be generalized, and they should be similar to one another in their physical and biological characteristics. Typically samples are evenly spaced with the time between samples depending on the rate of change in the population or phenomenon studied (Frost et al., 1988). For instance, rapidly changing populations of insects or zooplankton are often sampled many times a year, while slowly changing vertebrate or plant populations are usually sampled only once a year. The sampling frequency will depend on both the system studied and the specific questions asked. The duration of each evaluation period (before or after treatment) is determined by the periodicity of natural cycles, the magnitude of random temporal fluctuations, and the life span of the organisms studied. Longer sampling periods are needed when random fluctuations are greater, cycles have longer periods, or study organisms are longer lived.

We will use an example of a BACI design with multiple interventions to demonstrate in some detail the analysis and interpretation of a "real" unreplicated design. This concerns the effect of acidification on Little Rock Lake, a small, oligotrophic seepage lake in northern Wisconsin (Brezonik et al., 1986). Since 1983, it has been the site of an ecosystem-level experiment to investigate the responses of a seepage lake to gradual acidification (Watras and Frost, 1989). Little Rock Lake was divided into a treatment and a reference basin with an impermeable vinyl curtain. Following a baseline period (from August 1983 through April 1985), the treatment basin was acidified with sulfuric acid in a stepwise fashion to three target pH levels, 5.6, 5.1, 4.7, each of which was maintained for 2 years. The reference basin had an average pH of 6.1 throughout the experiment. Details on the lake's limnological features are provided in Brezonik et al. (1986). Our analysis here will focus on populations of the rotifer *Keratella taurocephala*, which exhibited marked shifts with acidification (Gonzalez et al., 1990).

Creating good unreplicated designs requires some creativity. So does the analysis of the resulting data. The analytical methods for most studies with genuine replication (e.g., analysis of variance) are quite straightforward. However, much of the analysis of unreplicated experiments is less clearcut and tends to be more subjective. With time series data of sufficient length (at least 50 observations), it may be possible to use developed statistical techniques that have a rigorous theoretical underfooting. We discuss some of these techniques in Sections 7.4 and 7.5. The general goal of such statistical modeling is to find a parsimonious model, a simple model that fits the data well. By simple, we mean a biologically and physically reasonable model with a small number of parameters to be estimated. We will use analytical techniques appropriate to time series with an intervention. As will be seen in the next two sections, proper use of these techniques requires mastery of certain fundamentals and considerable care. Unless you are quite skilled in the area, we recommend consultation with a statistician when conducting the analysis. (Such consultation at the design phase is also strongly recommended.)

7.4 Time Series

Time series analysis encompasses a large body of statistical techniques for analyzing time-ordered sequences of observations with serial dependence among the observations. We will focus on a subset of time series techniques developed for autoregressive integrated moving average (ARIMA) models. The classic reference for ARIMA modeling is Box and Jenkins (1976). Other references include McCleary and Hay (1980), Cryer (1986), Wei (1990), and Diggle (1990).

ARIMA models are valuable because they can describe a wide range of processes (a process is a sequence of events generated by a random model) using only a few parameters. In addition, methods for identifying, estimating, and checking the fit of ARIMA models have been well studied. These models are appropriate for modeling time series where observations are available at discrete, (usually) evenly spaced, intervals.

ARIMA models include two basic classes of models: autoregressive (AR) and moving average (MA) models. AR models are similar to regression models in which the observation at time t is regressed on observations at earlier times. First, we need some notation. Let y_t be the original time series, say rotifer density at time t. It is convenient to work with the series centered at its mean, say $z_t = y_t - \mu$. z_t is the deviation of the density at time t from the long term mean μ. An AR model of order $p = 2$ [denoted AR(2)] for the centered rotifer density is of the following form

$$z_t = \phi_1 z_{t-1} + \phi_2 z_{t-2} + \varepsilon_t \tag{7.1}$$

where the ϕs are coefficients (like regression coefficients) and ε_t is a random error at time t (usually assumed to be uncorrelated with mean 0 and variance σ^2). This model states the present rotifer density is a linear function of the densities at the previous two sampling times plus random error.

An MA model relates an observation at time t to current and past values of the random error ε_t. For example, an MA model of order $q = 2$ [denoted MA(2)] has the form

$$z_t = \varepsilon_t + \theta_1 \varepsilon_{t-1} + \theta_2 \varepsilon_{t-2} \tag{7.2}$$

where the θs are coefficients.

More general models can have both AR and MA terms, so-called ARMA models (note that ARIMA models include ARMA models, as described below). The goal of fitting an ARMA model is to describe all of the serial dependence with autoregressive and moving average terms so that the residuals, or estimated error terms, look like uncorrelated random errors, or "white noise."

ARMA processes have the same mean level, the same variance, and the same autocorrelation patterns, over whatever time interval they are observed. This is an intuitive description of a fundamental property known as stationarity. (More rigorous definitions can be found in the references.) Stationarity allows valid statistical inference and estimation despite a lack of genuine replication. A stationary process exhibits the same kind of behavior over different time intervals, as if replication were built into the process. Note that because the autocorrelation between observations depends on the number of time steps between them, the observations usually must be equally spaced. For some biological processes, however, it is not clear that this is necessary. For instance, the autocorrelation among zooplankton abundance estimates 2 weeks apart may be larger in winter when the population is changing slowly than in the summer when it changes more quickly.

Because many processes observed in nature are clearly not stationary, it is desirable to find a way to modify an observed time series so that the modified series is stationary; then ARMA models with few parameters can be fit to the modified data. There are two primary ways to do this. One is to incorporate deterministic functions into the model, such as a linear trend over time, a step increase at a known time, or periodic functions such as sines and cosines to represent seasonal behavior. The other way is to difference the series, that is, to compute new observations such as $z_t - z_{t-1}$ (first difference) or $z_t - z_{t-12}$ (seasonal difference with period 12). Differencing is a more general approach since it can represent both deterministic and random trends. On the other hand, when deterministic trends are of special interest, it may make more sense to fit them than simply to remove them by differencing. Because of its greater generality, we will focus on achieving stationarity by differencing.

Describing some processes may require both AR and MA parameters as well as differencing. This leads to ARIMA models with p AR parameters, q MA parameters, and d differences [denoted ARIMA(p,d,q)]. For instance, an ARIMA(1,1,1) model is of the form

$$x_t = \phi_1 x_{t-1} + \varepsilon_t + \theta_1 \varepsilon_{t-1} \qquad (7.3)$$

where $x_t = z_t - z_{t-1}$ and ε_t is uncorrelated error as before.

Box and Jenkins (1976) developed methods for identifying appropriate forms for ARIMA models from data. Their methods require computing sample autocorrelations and related functions from the data, and using the properties of these functions to identify possible ARIMA models. Each sample autocorrelation, r_k, at lag k (i.e., the correlation among observations k time steps apart) is computed for lags 1,2, . . . to obtain the sample autocorrelation function (ACF). A related function, the sample partial autocorrelation function (PACF), represents the sample autocorrelation among observations k time steps apart, when adjusted for autocorrelations at intermediate lags.

The model identification process involves identifying from the ACF, the PACF, and plots of the original series, a subset of ARIMA models for fitting. The original series may show a long-term trend or seasonal behavior and thus suggest nonstationarity. The ACF for nonstationary series also declines very slowly to zero. If the series appears nonstationary, it should be differenced until it is stationary before examining the ACF and PACF further.

The theoretical ACF and PACF give distinctive signatures for any particular ARIMA model. In ARIMA modeling, the sample ACF and PACF are estimated from the (perhaps differenced) data with the hope of recognizing such a signature, and thereby identifying an appropriate ARMA parametrization. The theoretical ACF for pure MA models has large values only at low lags, while the PACF declines more slowly. The number of lags with autocorrelations different from zero suggests q, the order of the MA model. The opposite is true for pure AR models, where the PACF has large values at low lags only, and the ACF declines more slowly. In this case, the order, p, of the AR model is suggested by the number of partial autocorrelations different from zero. Often AR or MA models of low order (1 or 2) provide good fits to observed series, but models with both AR and MA terms may be required. The adequacy of fit of the model can be checked by examining the residuals. The sample ACF and PACF of the residuals should not have large values at any lags, and should not show strong patterns of any sort. Box and Pierce (1970) suggested an overall test for lack of fit, and that test has since been refined (Ljung and Box, 1978).

7.5 Intervention Models

Intervention analysis extends ARIMA modeling methods to investigate the effects of known events, or interventions, on a time series. The response to a treatment, or intervention, can have a number of forms. The two simplest forms are a permanent jump to a new level after the intervention, or a temporary pulse, or spike, at the intervention. For example, a step change in an ARIMA model can be represented by a model of the form

$$z_t = \omega S_t + N_t \tag{7.4}$$

where $S_t = 0$ before intervention and $S_t = 1$ at and after intervention, ω is a coefficient, and N_t is an ARIMA model. In this representation, z_t is the suitably differenced series if the original series was not stationary. To model a spike or pulse response, $S_t = 1$ at the intervention, and 0 otherwise. Box and Tiao (1975) discuss more complicated models such as linear or nonlinear increases to new levels. Examples are illustrated in McCleary and Hay (1980). The lag between the intervention and the response can also be examined by fitting models with different lags.

There is usually no direct way to identify the ARIMA form of the intervention model from the observed time series itself. Typically a plausible ARIMA model is developed, perhaps for the series as a whole, or separately for the series before and after the intervention. The form of the response to intervention may be suggested by examination of residuals from the ARIMA model without the intervention, or from theoretical knowledge about the expected response. Usually a number of plausible models must be examined and the best chosen from among them. If the intervention form is unknown, McCleary and Hay (1980) suggest first trying the spike model, then the gradual increase model, and finally the step model. Box and Jenkins (1976) have emphasized the iterative approach to model building, where the goal is to obtain a model that is simple but fits the data adequately, and that has a reasonable scientific interpretation. Note that in following this iterative model building philosophy a series of decisions are made, and analyses of different data sets will proceed in different ways. While we can give a general outline of the steps involved in intervention analysis, we cannot exactly prescribe a method that will work for all data sets.

Although each analysis is unique, the following sequence of steps is useful in carrying out intervention analysis:

1. Plot the time series.
2. Choose a transformation of the original series, if necessary.
3. Determine whether the series must be differenced.
4. Examine the sample ACF and PACF of the (possibly) transformed and differenced series to identify plausible models.
5. Iterate between fitting models and assessing their goodness of fit until a good model is obtained.

We emphasize that the way an analysis proceeds depends on the data set. The following example should not be viewed as a prescription for intervention analysis, but as an example of a general process. Other examples of the use of intervention analysis in ecology and environmental monitoring have been discussed by Pallesen et al. (1985), van Latesteijn and Lambeck (1986), Madenjian et al. (1986), Noakes (1986), and Bautista et al. (1992).

7.6 Example

7.6.1 Data

We will analyze abundance data (animals/liter) for the rotifer *Keratella tauroceph-ala* from Little Rock Lake, Wisconsin. Sampling methods are described in Gonzalez et al. (1990). Data were collected every 2 weeks when the lake was ice free, and every 5 weeks when the lake was ice covered. There were slight variations from this schedule during the first years of monitoring; we have dropped extra observations or used an observation for two consecutive sampling dates when necessary to produce the same schedule in all years (we dropped a total of eight observations, and used six observations twice). We will discuss the effects of the spacing of observations, and ways to obtain equal spacing, in Section 7.6.5. The example data set includes 19 observations per year (106 total), with observations on approximately the same dates in all years. We will examine the effect of two interventions: the pH drop from 6.1 to 5.6 on 29 April 1985 and from 5.6 to 5.1 on 27 April 1987. We have not used data collected after the third intervention on 9 May 1989.

One of the early steps in any data analysis should be careful examination of plots of the data. The plot of the abundance series for both basins (Fig. 7.1A) suggests that there is little change in *K. taurocephala* abundance in the reference basin over time, but an increase in both the level and variability of abundance in the treatment basin after the interventions.

7.6.2 Derived Series

The first decision to make is whether to analyze the series for the two basins separately, or to analyze a single series derived from the two original series. There are advantages to both approaches. The single derived series may be simpler to analyze: not only is there only one series, but the derived series may have less serial autocorrelation, and less pronounced seasonal behavior than the original series (Stewart-Oaten et al., 1986). In addition, the analysis of the derived series may lead to a single direct test of the intervention effect, while separate analyses of the treatment and reference series would have to be considered together in evaluating the effect of the intervention. On the other hand, the derived series is more remote from the observed data, and it may be of interest to model the serial dependence, seasonal behavior, and effects of interventions in the original series themselves. If temporal patterns in the two units are quite different, it may be essential to analyze the two series separately. In many cases, both approaches may be appropriate. We will discuss both here.

Two possible forms of derived series are the series of differences between the treatment and reference observations at each sampling date, and the series of ratios of the treatment to reference observations at each date. Most discussions of this topic in the ecological literature have argued that the log transformed ratio series (equivalent to the difference of the two log series) best meets the assumptions of statistical tests (Stewart-Oaten et al., 1986, Carpenter et al., 1989, Eberhardt and Thomas, 1991). The issue in part revolves around the assessment of whether factors leading to change in the response are additive or multiplicative. Because factors affecting populations often have multiplicative effects, the series of ratios may work best for abundance data. It is usually best to examine both derived series. Both derived series showed similar patterns for the *K. taurocephala* data, with the mean level and variability of the series increasing after each intervention. The increase in variability was much greater for the series of differences; we chose to work with the ratio series because we could find a transformation that more nearly made the variance of that series constant. Note that we added one to each value to avoid dividing by zero.

7.6.3 Transformations

A fundamental assumption of ARIMA models, as well as of many standard statistical models, is that the variance of the observations is constant. The most common violation of this assumption involves a relationship between the variance and the mean of the observations. We will use a simple method for determining a variance stabilizing transformation when the variance in the original series is proportional to some power of the mean (Box et al., 1978). Poole (1978) and Jenkins (1979) described the use of this procedure in time series analysis. Jenkins (1979) recommends dividing the series into subsets of size 4 to 12, related to the length of the seasonal period (we used 6-month intervals). Then the mean and variance are calculated for each subset and the log standard deviation is regressed on the log mean. The value of the slope (b) from this regression suggests the appropriate transformation (see SAS commands in Appendix 7.1). The assumption that the variance is proportional to some power of the mean determines the form of the transformation:

$$z = y^{1-b}, (1-b) \neq 0 \qquad (7.5)$$
$$= \log y, \qquad b = 1$$

For the ratio series the slope from the regression of log standard deviation on log mean was 1.33. The estimate 1.33 has some associated error; one traditionally uses a value near the estimate that corresponds to a more interpretable transformation rather than the exact estimate. Thus, we might consider either the reciprocal square-root (the reciprocal of the square-root of the original value; appropriate for a slope of 1.5) or the log transformation (for a slope of 1.0). We carried out analyses on both scales and found that conclusions were similar although the reciprocal square-root scale did a slightly better job of stabilizing the variance. We will report analyses of the log ratio series (we used logs to the base 10), since that transformation is more familiar and somewhat easier to interpret.

7.6.4 ARIMA Models

The plot of the log abundance series for both basins (Fig. 7.1B) presents a very different appearance than that of the abundance series (Fig. 7.1A). The large increase in both level and variability of the treatment basin series on the original scale is much reduced on the log scale. The reference basin series may show a slight increasing trend on the log scale. The log ratio series (Fig. 7.1C) appears to have increased in level after the first intervention, and perhaps in variability after the second intervention. These subjective impressions should be kept in mind while carrying out the statistical analyses.

Methods for identifying ARIMA models require that the series be stationary. If an intervention changes the level of a series, that series is no longer stationary,

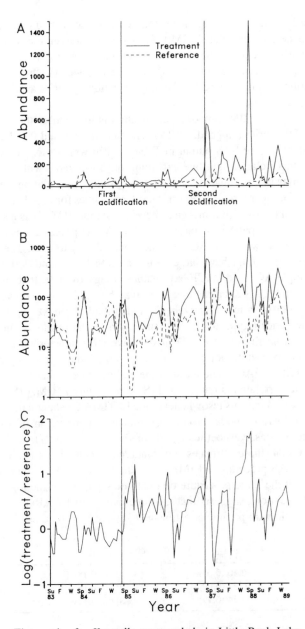

Figure 7.1. Time series for *Keratella taurocephala* in Little Rock Lake, Wisconsin, during acidification treatments. (A) The abundance series for both the treatment and reference basins shown on the original scale, animals/liter. (B) The same series shown on a $\log_{10}$ scale. (C) The series derived as the $\log_{10}$ of the ratio of the treatment series over the reference series.

and standard model identification procedures may be difficult to apply. One approach to determining the ARIMA form for an intervention series is to fit models to the segments of the series between interventions. Before discussing the results of this process for the log ratio series, we will first demonstrate model identification procedures on the longest stationary series available, the log reference series.

The first step in ARIMA model identification is to examine the sample ACF and PACF of the series (Appendix 7.2A). The sample ACF and PACF are usually presented as plots (see SAS output in Table 7.1) in which each autocorrelation or partial autocorrelation is a line extending out from zero, with two standard error (SE) limits (approximate 95% confidence limits) indicated on the plot. The sample ACF for the log reference series has large values for lags 1, 2, and 3, and values less than the two SE limits elsewhere; the sample PACF has a single large value at lag 1. This pattern is suggestive of an AR(1) process, since the PACF "cuts off" at lag 1, while the ACF declines more slowly to zero. The PACF for an AR(2) process would have large values at both lags 1 and 2, for instance. Both the sample ACF and PACF have values at lags 18 or 19 that almost reach the two SE limits; this suggests that there may be a weak annual cycle.

Once a model form has tentatively been identified, the next step is to fit that model to the series (Appendix 7.2B), and examine the residuals. In this case, the residuals from fitting an AR(1) model have the characteristics of random, uncorrelated error (white noise): there are no large values or strong patterns in the sample ACF or PACF, and a test of the null hypothesis that the residuals are random and uncorrelated is not rejected [SAS procedure ARIMA (SAS Institute Inc., 1988) carries out this test whenever the ESTIMATE statement is used to fit a model]. The AR(1) model produced estimates for the mean of 1.44 (SE = 0.08), an autoregressive coefficient ($\hat{\phi}$) of 0.65 (SE = 0.07), and a residual variance of 0.084 (these estimates and standard errors are included in the output produced by SAS procedure ARIMA when the ESTIMATE statement is used).

Carrying out the same procedure on the three segments of the log ratio series between interventions suggests that all three can be fit by AR(1) models, and leads to the following parameter estimates (SEs are in parentheses):

Segment	$\hat{\mu}$	$\hat{\phi}$	Variance
1: pH 6.1	−0.10 (0.05)	0.28 (0.18)	0.04
2: pH 5.6	0.37 (0.09)	0.47 (0.15)	0.09
3: pH 5.1	0.54 (0.19)	0.62 (0.13)	0.21

Although all three segments have the same AR(1) form, there may be differences in parameter estimates and variances. These estimates may be poor because the series are short (between 31 and 38 observations), but the differences suggest some caution when interpreting the intervention model. Differences between the

Table 7.1. Plots of the Autocorrelation Function (ACF) and Partial Autocorrelation Function (PACF) from the Program Listed in Appendix 7.2A.

ARIMA Procedure

Name of variable = LOGREF.

Mean of working series = 1.443589
Standard deviation   = 0.379975
Number of observations =   106

Autocorrelations

Lag	Covariance	Correlation	−1 9 8 7 6 5 4 3 2 1 0 1 2 3 4 5 6 7 8 9 1	Std
0	0.144381	1.00000	|********************	0
1	0.093730	0.64918	a |*************	0.097129
2	0.049058	0.33978	. |*******	0.131855
3	0.028400	0.19670	. |**** .	0.139871
4	0.010254	0.07102	. |* .	0.142457
5	0.0034776	0.02409	. | .	0.142791
6	0.0010632	0.00736	. | .	0.142829
7	−0.0067425	−0.04670	. *| .	0.142833
8	−0.0084151	−0.05828	. *| .	0.142977
9	−0.0005594	−0.00387	. | .	0.143201
10	−0.0014308	−0.00991	. | .	0.143202
11	−0.0089326	−0.06187	. *| .	0.143208
12	−0.0090990	−0.06302	. *| .	0.143460
13	−0.0028932	−0.02004	. | .	0.143721
14	0.00078262	0.00542	. | .	0.143747
15	0.0053034	0.03673	. |* .	0.143749
16	−0.0010200	−0.00706	. | .	0.143838
17	0.00089464	0.00620	. | .	0.143841
18	0.022174	0.15358	. |*** .	0.143844
19	0.026635	0.18448	. |**** .	0.145382
20	0.010681	0.07398	. |* .	0.147574
21	−0.0084435	−0.05848	. *| .	0.147924
22	−0.026879	−0.18617	. ****| .	0.148142
23	−0.026095	−0.18074	. ****| .	0.150333
24	−0.023388	−0.16199	. ***| .	0.152369

Partial Autocorrelations

Lag	Correlation	−1 9 8 7 6 5 4 3 2 1 0 1 2 3 4 5 6 7 8 9 1
1	0.65918	. |*************
2	−0.14114	***| .
3	0.06457	. |* .
4	−0.09598	. **| .
5	0.04833	. |* .
6	−0.01962	. | .
7	−0.07093	. *| .
8	0.01676	. | .
9	0.07100	. |* .
10	−0.06325	. *| .
11	−0.07167	. *| .
12	0.02016	. | .
13	0.05950	. |* .
14	−0.00623	. | .
15	0.02763	. |* .
16	−0.09466	. **| .
17	0.10644	. |** .
18	0.20260	. |**** .
19	−0.05757	. *| .
20	13497	. ***| .
21	−0.12215	. **| .
22	−0.11148	. **| .
23	0.08049	. |** .
24	−0.09544	. **| .

a Marks two standard errors.

mean levels for the three segments provide estimates for the effects of the interventions (see further discussion below). Notice that if the reference and treatment series had the same mean level initially, the estimated mean for the log ratio series would be zero at pH 6.1.

The preceding analyses indicate that we can begin fitting intervention models for all series by assuming an AR(1) form. We also need to specify the form for the interventions. The acid additions to the treatment basin were step changes and so may be represented by variables that take the value of zero before, and one after, the intervention. These variables are just like dummy variables used in regression (Draper and Smith, 1981). We have constructed two dummy variables, one for each intervention. The first has the value of one between 29 April 1985 and 27 April 1987, and zero elsewhere; the second has the value of one after 27 April 1987 and zero elsewhere. In this parametrization, the parameter for the mean estimates the initial level of the series, and the intervention parameters estimate deviations from that initial level to the levels after the interventions. Alternative dummy variables can be used, and will lead to the same predicted values, although the interpretation of the intervention parameters will differ. The dummy variables are treated as input variables when using the ESTIMATE statement in SAS procedure ARIMA to fit the model (Appendix 7.3).

We fit intervention models to the log ratio series (Appendix 7.3), and also to the log treatment basin and log reference basin series separately. In all cases the simple AR(1) model fit well, although for the log treatment series there was also some evidence for an annual cycle. We also fit models for more complicated responses to intervention, but the additional parameters were not significantly different from zero. Results for the three series are presented below (SEs are in parentheses after each estimate; $\hat{\omega}_1$ and $\hat{\omega}_2$ are coefficients for the first and second intervention parameters):

Series	$\hat{\mu}$	$\hat{\phi}$	$\hat{\omega}_1$	$\hat{\omega}_2$	Variance
Ratio	−0.06 (0.13)	0.57 (0.08)	0.40 (0.17)	0.60 (0.18)	0.12
Treatment	1.40 (0.12)	0.56 (0.08)	0.23 (0.15)	0.69 (0.16)	0.09
Reference	1.46 (0.12)	0.60 (0.08)	−0.16 (0.16)	0.09 (0.16)	0.08

We can construct statistics similar to t-statistics for the parameter estimates by dividing the estimates by their SE, just as in linear regression. Approximate tests can be based on the standard normal distribution; P-values can be determined from the Z-score using a table or software package.

Models for all three series have similar autoregressive coefficients, and the value of approximately 0.6 indicates that there is strong serial correlation in all series. The intervention coefficients provide estimates of the shift in level from the initial baseline period. Thus, for the log ratio series, 0.4 is the shift from the initial level of -0.06 to the level after the first intervention, and 0.6 is the shift from the initial level to the level after the second intervention. Both of these coefficients are significantly different from zero ($Z = 2.35$, $P = 0.02$; and $Z = 3.33$, $P = 0.001$). Shifts at the time of the second intervention can be computed as the difference, $\hat{\omega}_2 - \hat{\omega}_1$. The largest shift in level for the log ratio series was at the first intervention, while for the log treatment series it was at the second intervention.

Notice that the difference between intervention parameters for the treatment and reference series approximately equals the parameters for the log ratio series [i.e., $0.23 - (-0.16) = .39$, and $0.69 - 0.09 = 0.60$]. The larger shift at the time of the first intervention for the log ratio series is due to a shift up for the log treatment series and a shift down for the log reference series, while the smaller shift in the log ratio series at the time of the second intervention is due to shifts up in both the log reference and treatment series. [Such additivity of the individual series and the derived, transformed series should not be expected in general. It occurs with the log ratio because $\log(x/y) = \log(x) - \log(y)$.]

The intervention parameters are more easily interpreted when converted back to the original abundance scale. For the log ratio series, the initial predicted level is (t is the treatment basin value, r is the reference basin value): $\log_{10}(t/r) = -0.06$, which implies $t = 0.87r$. Thus, the treatment abundance is 0.87 as large as the reference abundance prior to the first intervention. After the first intervention, the level of $-0.06 + 0.40 = 0.34$ implies that $t = 2.19r$ and after the second intervention, the level of $-0.06 + 0.60 = 0.54$ implies that $t = 3.47r$. The shift in level of 0.23 at the time of the first intervention in the log treatment series corresponds to a multiplication of $10^{0.23} = 1.70$ in the abundance of the rotifer. The shift at the second intervention represents a multiplication of $10^{0.69} = 4.9$ times the initial abundance.

7.6.5 Spacing of Observations

We analyzed our series as if the observations were equally spaced when in fact the spacing differed between summer and winter. To investigate how the spacing of observations may have affected our analyses, we constructed two series with equally spaced observations from the log ratio series and fit intervention models to them. Some approaches to constructing series with equally spaced observations include

1. interpolate to fill in missing observations,
2. aggregate data to a longer sampling interval (Wei, 1990), and
3. use a statistical model to fill in missing observations (Jones, 1980).

We used method (1) to construct a series with 24 observations per year by interpolating between winter observations taken at 5-week intervals. We also used method (2) to construct a series with 13 observations per year by averaging all observations within each 4-week time period.

Both of these constructed series were adequately fit with AR(1) models, and estimates of intervention effects were very similar to those from the model for the original series. The variance was smaller for the two constructed series because the process of constructing them involved smoothing. The size of the autoregressive parameter increased as the frequency of observation increased. The important point here is that the intervention parameter estimates were not much affected by the spacing of observations. Thus, we can feel comfortable that the unequal spacing has not distorted our estimation of the intervention parameters. We chose to describe analyses of the data set that corresponded most closely to the actual observations.

7.6.6 Comparison with Other Analytical Procedures

If we ignore the serial dependence among observations over time, we can compare the postintervention data with the preintervention data by t-tests. Recall that a fundamental assumption of the t-test and similar classical statistical procedures is that the observations are independent. The $K.\ taurocephala$ series we analyzed had autocorrelations of about 0.57 for observations one time unit apart, with the autocorrelations decreasing exponentially as the number of time units between observations increased. The t-statistic is positively inflated when observations are positively correlated, and the degree of inflation increases as the autocorrelation increases (Box et al., 1978, Stewart-Oaten et al., 1986). For the log ratio series, the Z-statistics calculated for the intervention coefficients were one-third to one-half the size of the t-statistics calculated by standard t-tests comparing the preintervention period to either of the post-intervention periods. In this case, all tests are significant, but in situations where the intervention has a smaller effect, the t-test could lead to misleading conclusions.

Randomized intervention analysis (RIA) has been proposed as an alternative approach for analyzing data from experimental designs such as we have discussed here (Carpenter et al., 1989). Randomization tests involve comparing an observed statistic with a distribution derived by calculating that statistic for each of a large number of random permutations of the original observations. Although randomization tests do not require the assumption of normality, they may still be adversely affected by lack of independence among observations (Chapter 13). Empirical evaluations suggest that RIA works well even when observations are moderately autocorrelated (Carpenter et al., 1989), although Stewart-Oaten et al. (1992) question the generality of these evaluations. Conclusions drawn from RIA for the Little Rock Lake $K.\ taurocephala$ data parallel those we found (Gonzalez et al., 1990).

Although his approach is not possible when there is only one reference unit, Schindler et al. (1985) used conventional statistical methods to compare a single manipulated system with many reference systems. This may be useful when standard monitoring can provide data from a number of reference systems.

7.6.7 Ecological Conclusions

Based on our intervention analysis, we would conclude that there have been nonrandom shifts in the population of *K. taurocephala* in Little Rock's treatment basin with acidification. The question remains, however, as to whether these shifts can be attributed to the acidification or if they reflect natural population shifts coincident with the pH manipulations.

Information on the ecology of *K. taurocephala* provides strong support for the conclusion that the increases with acidification in Little Rock Lake reflect a response to the pH manipulation. In surveys across a broad sample of lakes, *K. taurocephala* showed a strong affinity with low pH conditions (MacIssac et al., 1987). Such previous information on the distribution of *K. taurocephala* led to the prediction, prior to any acidification, that its population would increase at lower pH levels in Little Rock Lake (Brezonik et al., 1986). Detailed mechanistic analyses of the response of *K. taurocephala* to conditions in Little Rock Lake's treatment basin revealed that its increased population levels were linked with reductions in the populations of its predators under more acid conditions (Gonzalez, 1992).

7.7 Discussion

Intervention analysis provides three major advantages over other analytical methods that have been used to analyze similar time series data sets. First, it provides good estimates of intervention effects and standard errors even in the presence of autocorrelation among observations. Second, it requires that the autocorrelation structure of the data is modeled explicitly, and this may provide additional information about underlying processes. Third, the iterative model building approach that is inherent in intervention analysis is itself a way of learning about characteristics of the data that may otherwise be difficult to discover. It is certainly a very different approach from that of simply carrying out a standard test to compute a P-value. The disadvantages of intervention analysis are that it requires a long series of observations (usually 50 or more) to determine the pattern of serial dependence, and it involves a large investment of time on the part of the data analyst, both in becoming familiar with the methodology and in carrying out each analysis.

There are many additional complexities of intervention analysis that we have not covered. We analyzed time series with weak seasonal behavior, where nonseasonal models were adequate. In many data sets, seasonal behavior may be pronounced, and will have to be incorporated into ARIMA intervention models. We have used methods appropriate for univariate time series. Our series was initially bivariate with one series for the treatment, another for the reference. We transformed the problem to a univariate one by taking the ratio of the two series at each sampling period. In general, such combination of series requires that the series be synchronous; that is, one series does not lag or lead the other. Such combination also requires that the two processes are operating on the same scale; Stewart-Oaten et al. (1986) discuss this in some detail within the context of achieving additivity. There are bivariate approaches that allow these assumptions to be relaxed to some extent, but at the expense of ease of model identification and interpretation. One such approach is to use a general transfer function model (Box and Jenkins, 1976), of which intervention models are a special type. The methods of identifying and estimating such models can be quite involved and are well beyond the scope of this chapter, but such models are worth considering if the assumptions required for the univariate approach are suspect.

Despite our emphasis on unreplicated studies, replication should always be employed when possible. Replication is the only way of reducing the probability of detecting a spurious treatment effect in a manner that is readily quantifiable. A detailed treatment of replication is beyond the scope of this chapter, but we will mention some basics. A design with replicated time series is referred to as a repeated-measures design; repeated, or sequential, measurements are taken over time within each experimental unit (in an unreplicated time series, there is just one experimental unit). The usual approach to analyzing repeated-measures designs relies on applying a standard multivariate analysis of variance (Chapter 6). Such an approach requires estimating a moderate number of parameters, which might be a problem if the amount of replication is not fairly substantial. This problem can be reduced to some extent by using repeated-measures models with structured covariance matrices; such models are "hybrids" of ARIMA models and usual multivariate analysis of variance models (Laird and Ware, 1982; Jennrich and Schluchter, 1986). The advantages of such repeated-measures models over standard multivariate ANOVA models are (1) they allow testing hypotheses about the nature of the serial dependencies within units just as in ARIMA modeling, and (2) they require the estimation of fewer parameters, and hence demand fewer data for estimation. Such procedures are relatively new; software implementations are the BMDP procedure 5V (1990 release) and SAS procedure MIXED (version 6.07, SAS Institute Inc., 1992).

Time series methods have broad potential applicability in ecology; intervention analysis is a relatively small facet. Poole (1978) gives an interesting review and summary of time series techniques in ecology, including transfer function models. Because of their underlying statistical similarities, a knowledge of time series models is also useful in understanding spatial statistical techniques. Time series techniques typically receive little or no attention in the basic statistics courses that most graduate ecology programs require. Many universities offer applied time series courses that are accessible to students with an introductory statistics background. Considering the frequency with which ecologists deal with time series data, graduate ecology programs should encourage such formal training.

We also encourage ecologists who are designing and/or analyzing time series studies to seek the collaboration of their statistical colleagues.

Acknowledgments

We thank Pamela Montz for compiling the rotifer data evaluated here. Comments by Steve Carpenter on an earlier version of this manuscript substantially improved it. This work was supported in part by funds from the U.S. Environmental Protection Agency and the National Science Foundation, Long-term Ecological Research Program.

Appendix 7.1.

SAS program code for data input and determination of value for transformation.

```
DATA ABUN;                                  /input data/
   INPUT MONTH                              /month sampled/
      DAY                                   /day sampled/
      YEAR                                  /year sampled/
      PERIOD          /six-month period dating from the start of monitoring/
      S1                      /dummy variable - has value of 1 while treatment
                                    basin pH is 5.6, value of 0 otherwise/
      S2                      /dummy variable - has value of 1 while treatment
                                    basin is pH 5.1, value of 0 otherwise/
      REF                     /abundance of K. taurocephala in reference basin/
      TRT;                    /abundance of K. taurocephala in treatment basin/
   RATIO = (TRT+1)/(REF+1);   /compute ratio of treatment to reference abundance/
   LOGRATIO = LOG10(RATIO);         /compute log transformation of ratio/
   LOGREF = LOG10(REF);          /compute log transformation of original data/

PROC SUMMARY NWAY;              /calculate mean and SD of ratios for each year/
   CLASS YEAR; VAR RATIO;
   OUTPUT OUT=SUMRY                         /output SDs/
      MEAN=MRATIO STD=SD;
DATA SUMRY; SET SUMRY;                   /calculate log of mean and SD/
   LOGRATIO = LOG10(MRATIO);
   LOGSD = LOG10(SD);
PROC PLOT;                               /plot log SD vs log mean/
   PLOT LOGSD*LOGRATIO;

PROC REG;                                /regress log SD on log mean/
   MODEL LOGSD = LOGRATIO;
```

Appendix 7.2

SAS program code for fitting the ARIMA model to the log reference series. Note that the data must be sorted by sample date.

A. Initial Analysis for Examining the ACF and PACF Ratios.

```
PROC ARIMA;                              /compute the ACF and PACF/
   IDENTIFY VAR = LOGREF;        /ACF and PACF plots produced automatically/
```

B. Based on the Initial Analysis, A Model of Order p = 1 *Is Fit to the Data.*

```
PROC ARIMA;                                        /fit an AR(1) model/
   IDENTIFY VAR = LOGREF;
   ESTIMATE P = 1                                  /P = order of model/
      MAXIT = 30                      /maximum number of iterations = 30/
      METHOD = ML                            /use maximum likelihood/
      PLOT;               /plot ACF and PACF residuals to test for model fit/
```

*** Similar analyses of reference and treatment basins not shown for brevity ***

Appendix 7.3

SAS program code for fitting an intervention model to the log ratio series.

```
PROC ARIMA;                                    /fit intervention model/
   IDENTIFY VAR = LOGRATIO
      CROSSCOR = (S1 S2);              /indicate intervention variables/
   ESTIMATE P = 1 INPUT = (S1 S2)              /list input variables/
      MAXIT = 30 METHOD = ML
      PLOT;
```

8

Nonlinear Curve Fitting: Predation and Functional Response Curves

Steven A. Juliano

8.1 Ecological Issues

8.1.1 Predation and Prey Density

The number of prey that an individual predator kills (or a the number of hosts a parasitoid attacks) is a function of prey density, and is known as the functional response (Holling, 1966). In general, the number of prey killed in a fixed time approaches an asymptote as prey density increases (Holling, 1966). There are at least three types of curves that can be used to model the functional response (Holling, 1966; Taylor, 1984; Trexler et al., 1988) that represent differences in the proportion of available prey killed in a fixed time. In type I functional responses, number of prey killed increases linearly up to a maximum, then remains constant as prey density increases (Fig. 8.1A). This corresponds to a constant proportion of available prey killed up to the maximum (density independence), followed by a declining proportion of prey killed (Fig. 8.1B). In type II functional responses, the number of prey killed approaches the asymptote hyperbolically as prey density increases (Fig. 8.1C). This corresponds to an asymptotically declining proportion of prey killed (inverse density dependence) (Fig. 8.1D). In type III functional responses, the number of prey killed approaches the asymptote as a sigmoid function (Fig. 8.1E). This corresponds to an increase in the proportion of prey killed (density dependence) up to the inflection point of the sigmoid curve, followed by a decrease in the proportion of prey killed (Fig. 8.1F).

Most ecological interest in functional responses has involved types II and III. At a purely descriptive level it is often desirable to be able to predict the number of prey killed under a given set of circumstances. Type II functional responses especially have figured prominently in behavioral ecology, serving as the basis for foraging theory (Stephens and Krebs, 1986; Abrams, 1990). Mechanistic

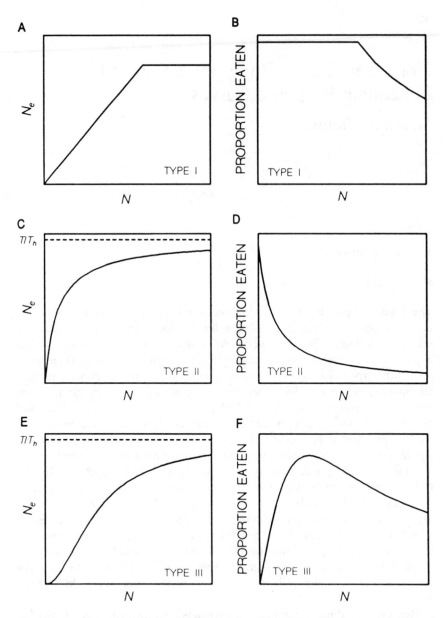

Figure 8.1. Three types of functional responses. The relationships between number of prey eaten (N_e) and number of prey present (N) are depicted in **A**, **C**, and **E**. The corresponding relationships between proportion eaten (N_e/N) and number of prey present (N) are depicted in **B**, **D**, and **F**.

models of population dynamics of resource limited consumers (e.g., Williams, 1980) and predation limited prey (e.g., Hassell, 1978) have also made extensive use of type II functional responses. Biological investigations of functional responses may address different questions:

1. What is the shape (type) of the functional response? This question is often of interest in attempting to determine whether density dependent predation is a stabilizing factor in predator–prey population dynamics, and should also be answered prior to fitting any specific mathematical model to the functional response (van Lenteren and Bakker, 1976; Hassell et al., 1977; Trexler et al., 1988).

2. What are the best estimates of parameters of a mechanistic model of the functional response? This question must be answered by investigators who wish to use functional responses in mechanistic models of resource competition or predation (e.g., Hassell, 1978; Williams, 1980).

3. Are parameters of models describing 2 (or more) functional responses significantly different? Such a question may arise when different predator species or age classes are being compared, perhaps in an attempt to determine which is most effective at killing a particular prey (Thompson, 1975; Russo, 1986), or when different species or populations of prey are being compared, perhaps in an attempt to find evidence of differential predator–prey coevolution (Livdahl, 1979; Houck and Strauss, 1985; Juliano and Williams, 1985).

These three questions require different methods of statistical analysis, and it is these analytical methods that are the subject of this chapter. The general approach described in this chapter involves two steps: model selection and hypothesis testing. Model selection involves using a logistic regression of proportion of prey killed vs number of prey to determine the general shape of the functional response (Trexler et al., 1988; Trexler and Travis, 1993). Hypothesis testing involves using nonlinear least-squares regression of number of prey eaten vs number offered to estimate parameters of the functional response and to compare parameters of different functional responses (Juliano and Williams, 1987).

Both of these statistical methods are related to more typical regression models and I will assume some familiarity with regression analyses and assumptions (see Neter and Wasserman, 1974; Glantz and Slinker, 1990). Although the methods described in this chapter may have other applications in ecology (Section 8.4), the descriptions of procedures given in this chapter are fairly specific to analyses of functional responses. General descriptions of methods for logistic regressions are given by Neter and Wasserman (1974) and Trexler and Travis (1993), and descriptions of methods for nonlinear regressions are given by Glantz and Slinker (1990) and Bard (1974).

8.1.2 Mathematical Models of Functional Responses

Because most interest has centered on types II and III functional responses, the remainder of this chapter will concentrate on these types. Numerous mechanistic and phenomenological models have been used to describe functional responses. Holling's (1966) original model assumed that asymptotes were determined by time limitation with predators catching and eating prey as rapidly as they can. Holling also assumed that encounter rates with prey were linearly related to prey density, that while predators were handling prey they could not make additional captures, and that prey density was constant. He modeled type II functional responses using the "Disc Equation":

$$N_e = \frac{aNT}{1 + aNT_h} \tag{8.1}$$

where N_e = number eaten, a = the attack constant (or instantaneous search rate), which relates encounter rate with prey ($= aN$) to prey density, N = prey density, T = total time available, and T_h = handling time per prey, which includes all time spent occupied with the prey and unable to attack other prey. This model is the most widely used representation of the type II functional response. Although the ecological literature contains numerous citations of this model with alternative symbols, Holling's symbols are used throughout this chapter.

Type III functional responses can be modeled using Eq. 8.1 if the attack constant a is made a function of prey density N (Hassell, 1978). In the most general useful form, a is a hyperbolic function of N:

$$a = \frac{d + bN}{1 + cN} \tag{8.2}$$

where b, c, and d are constants. This is a more general form of the hyperbolic relationship postulated by Hassell (1978), which was equivalent to Eq. 8.2 with $d = 0$. Substituting Eq. 8.2 into Eq. 8.1 and rearranging yields

$$N_e = \frac{dNT + bN^2T}{1 + cN + dNT_h + bN^2T_h} \tag{8.3}$$

In general, type III functional responses can arise whenever a is an increasing function of N. Numerous mathematical forms could describe the relationship between a and N, and it is difficult to know beforehand which form will be best for a given data set. The form given in Eq. 8.2 has the advantage that if certain parameters are equal to 0, the expression reduces to produce other likely relationships of a and N (Table 8.1). Note, however, that if both b and d are 0, then a is 0 and there is no functional response.

These models describe predation occurring at a constant prey density N. For experimental studies of predation, this assumption is often not realistic. In typical functional response experiments, individual predators in a standardized state of hunger are offered a number of prey (initial number $= N_0$) under standardized conditions of temperature, light, container size and shape, etc. Usually prey are not replaced as they are eaten, or they are only replenished at some interval. Although it would be desirable for experimenters to replace prey as they are eaten (Houck and Strauss, 1985), this would obviously become prohibitive because it would require that someone observe each replicate continuously. After a set time (T) the number of prey eaten (N_e) is determined. This process is repeated across a range of numbers of prey, usually with replication at each number of prey. The resulting data set contains paired quantitative values, usually (N_0, N_e). Investigators then attempt to answer one or more of the three questions outlined in the first section by statistical analysis of these data. Thus in typical functional response experiments, prey density declines as the experiment proceeds. Under such circumstances, Eqs. 8.1 and 8.3 do not accurately describe the functional response, and parameters estimated using these models will be subject to errors dependent upon the degree to which prey are depleted. When prey depletion occurs, the appropriate model describing the functional response is the integral of Eq. 8.1 or 8.3 over time to account for changing prey density (Rogers, 1972). For the type II functional response (Eq. 8.3), integrating results in the "random predator equation":

$$N_e = N_0 \{1 - \exp[a(T_h N_e - T)]\} \tag{8.4}$$

where $N_0 =$ the initial density of prey.

For type III functional responses, the precise form of the model incorporating

Table 8.1.   *Various Forms of the Relationships between* a *and* N *produced by Eq. 8.2*[a]

Parameters equal to 0	Resulting equation for a	Relationship of a to N
None	$\dfrac{d + bN}{1 + cN}$	Increasing to asymptote; intercept $\neq 0$
c	$d + bN$	Linear increase; intercept $\neq 0$
d	$\dfrac{bN}{1 + cN}$	Increasing to asymptote; intercept $= 0$
d and c	bN	Linear increase; intercept $= 0$

[a]It is assumed that the functional response is known to be type III, hence that at least $b > 0$.

depletion depends on whether the attack constant a is a function of initial density (N_0) or current prey density (N) (Hassell et al., 1977; Hassell, 1978). The simplest form arises when a is a function of initial density, as in Eq. 8.2:

$$N_e = N_0 \{1 - \exp[(d + bN_0)(T_h N_e - T)/(1 + cN_0)]\} \qquad (8.5)$$

This will be the form used throughout this chapter.

Other mathematical forms have been used to model functional responses. Ivlev (1961) used an exponential model based on the assumption of hunger limitation of the asymptotic maximum number eaten to generate type II functional responses. Several authors (e.g., Taylor, 1984; Williams and Juliano, 1985) have used phenomenological models derived from Michaelis-Menten enzyme kinetics to model type II functional responses (note that Eq. 8.2 is such a model). Trexler et al. (1988) showed that several phenomenological models could generate type II and III functional responses. Although different models may be employed, the general issues involved in statistical analyses of functional responses are similar. In this chapter I will concentrate on models derived from Holling's handling time limited case.

8.2 Statistical Issues

Many investigators have analyzed experiments done without replacement of prey, but employed the models appropriate for predation with constant prey density (Eqs. 8.1 and 8.3) (e.g., Livdahl, 1979; Juliano and Williams, 1985; Russo, 1986). Such analyses are of questionable value, as parameter estimates and comparisons are sure to be biased and the shape of the functional response may be incorrectly determined (Rogers, 1972; Cock, 1977). Fitting models such as Eqs. 8.2 and 8.3 to data is a simpler task than is fitting models such as Eqs. 8.4 and 8.5 (see Williams and Juliano, 1985; Houck and Strauss, 1985; Glantz and Slinker, 1990). Such models involving no prey depletion have the desirable property of giving the correct expectation for stochastic predation (Houck and Strauss, 1985; Sjoberg, 1980). However, if experiments are conducted with prey depletion, this prey depletion must be incorporated into the statistical analysis by using Eqs. 8.4 and 8.5 in order to reach valid conclusions.

The most common method used to analyze functional responses is some form of least squares regression involving N_0 and N_e (e.g., Livdahl, 1979; Livdahl and Stiven, 1983; Juliano and Williams, 1985, 1987; Williams and Juliano, 1985; Trexler et al., 1988). Some authors have used linear regressions, especially on linearized versions of Eqs. 8.1 and 8.4 (Rogers, 1972; Livdahl and Stiven, 1983). However, these linearizations produce biased parameter estimates and comparisons of parameters may not be reliable (Cock, 1977; Williams and Juliano, 1985; Juliano and Williams, 1987). Linearized expressions are also unsuitable for distinguishing type II from type III functional responses.

Nonlinear least squares has proved more effective for parameter estimation and comparison (Cock, 1977; Juliano and Williams, 1987), but seems to be a less desirable choice for distinguishing types II and III functional responses (Trexler et al., 1988), and one reason for this conclusion is apparent when Fig. 8.1C and E are examined. Distinguishing the curves in Fig. 8.1C and E using nonlinear least squares could be done by testing for significant lack of fit for models using Eqs. 8.4 and 8.5 (Trexler et al., 1988). Significant lack of fit for eq. (8.4) but not for eq. (8.5) would indicate a type III functional response. These two types could also be distinguished by fitting a nonlinear model using Eq. 8.5, and testing H_0: $b = 0$ and $c = 0$. Rejecting the null hypothesis that $b \leq 0$ is sufficient to conclude that the functional response is of a type III form. However, comparison of Fig. 8.1C and E illustrates why nonlinear regressions involving N_0 and N_e are unlikely to distinguish the two types. In many type III functional responses, the region of increasing slope is quite short, hence the difference between the sigmoid curve of Fig. 8.1E and the hyperbolic curve of Fig. 8.1C is relatively small and will be difficult to detect in variable experimental data (Porter et al., 1982; Trexler et al., 1988; Peters, 1991).

Another difficulty in using Eq. 8.5 to distinguish type II and III functional responses is that the parameter estimates for b, c, and d may all be nonsignificant when incorporated into a model together despite the fact that a model involving only one or two of these parameters may yield estimates significantly different from 0. This may occur because of reduced degrees of freedom for error in more complex models and inevitable correlations among the parameter estimates.

8.3 Statistical Solution

8.3.1 Experimental Analysis

Analyzing functional responses and answering questions 1, 2, and 3 posed above requires two distinct steps. The methods best suited for answering question 1 are distinct from those best suited for answering questions 2 and 3. Further, question 1 must be answered before questions 2 or 3 can be answered.

Model selection: determining the shape of the functional response. Trexler et al. (1988) demonstrated that the most effective way to distinguish type II and III functional responses involves logistic regression (Neter and Wasserman, 1974; Glantz and Slinker, 1990) of proportion of prey eaten vs number of prey present. This focuses the comparison on the curves in Fig. 8.1D and F, which are clearly much more distinct than their counterparts in Fig. 8.1C and E. Logistic regression involves fitting a phenomenological function predicting proportion of individuals in a group responding (in this case, being eaten) from one or more continuous variables (Neter and Wasserman, 1974; SAS Institute Inc., 1989b; Glantz and

Slinker, 1990). The dependent variable Y is dichotomous, representing individual prey alive ($Y = 0$) or dead ($Y = 1$) at the end of the trial. In this case the continuous variable is N_0. The general form of such a function is:

$$\frac{N_e}{N_0} = \text{Prob}\{Y = 1\} = \frac{\exp(P_0 + P_1 N_0 + P_2 N_0^2 + P_3 N_0^3 + \ldots + P_z N_0^z)}{1 + \exp(P_0 + P_1 N_0 + P_2 N_0^2 + P_3 N_0^3 + \ldots + P_z N_0^z)} \quad (8.6)$$

where P_0, P_1, P_2, P_3, $\ldots$ P_z are parameters to be estimated. In most statistical packages, the parameters are estimated using the method of maximum likelihood, rather than least squares (e.g., SAS Institute Inc., 1989a, procedure CATMOD; see also Chapter 14). In this method, rather than minimizing sums of squared deviations of observed from expected, the probability of the observed values arising from the parameter estimates is maximized (Glantz and Slinker, 1990). This involves iterative search for the parameter values that maximize the likelihood function L:

$$L = \prod_k \frac{[\exp(P_0 + P_1 N_{0k} + P_2 N_{0k}^2 + P_3 N_{0k}^3 + \ldots + P_z N_{0k}^z)]^{Y_k}}{1 + \exp(P_0 + P_1 N_{0k} + P_2 N_{0k}^2 + P_3 N_{0k}^3 + \ldots + P_z N_0^z)} \quad (8.7)$$

where k is the subscript designating individual observations in the data set, and Y_k is the value of the dichotomous variable (1 for prey eaten, 0 for prey surviving). For further details consult Glantz and Slinker (1990, pp. 520–522).

The strategy is to find the polynomial function of N_0 that describes the relationship of N_e/N_0 vs N_0. The curves depicted in Fig. 8.1D and F may both be fit by quadratic (or higher order) expressions. However, in Fig. 8.1D the linear term would be negative (initially decreasing) whereas in Fig. 8.1F the linear term would be positive (initially increasing). Thus, one criterion for separating type II and III functional responses by analyzing proportion of prey eaten is to test for significant positive or negative linear coefficients in the expression fit by the method of maximum likelihood to data on proportion eaten vs N_0.

One potential problem with this approach is that a simple quadratic expression may not adequately fit the data (see below). In many cases, a cubic expression will provide a better fit to a type III functional response (Trexler et al., 1988), and cubic expressions provide a good starting point for fitting a logistic regression (Trexler and Travis, 1993). Still higher order expressions will, of course, give even better fits, but this improved fit is usually the result of better fit to points at higher values of N_0. It is likely that the initial slope of the curve will have the same sign. Whatever the order of the expression that is fit, plotting observed and predicted values vs N_0 and observing the slope near the origin is desirable for distinguishing type II and type III functional responses (see Chapter 2). At the model selection stage, significance testing is not as important as is obtaining a good description of the relationship of proportion eaten vs N_0. It is particularly important to determine whether the slope near N_0 is positive or negative (Fig. 8.1).

It is apparent that distinguishing between type II and III functional responses requires information about the trend in proportion of prey eaten near $N_0 = 0$. In designing experiments, this is an important consideration in the choice of values for N_0. If a set of N_0 values that is too large is chosen, any region of density-dependent predation may be undetected. Because the dependent variable is a proportion, relative variability of observations will necessarily increase at low N_0. Replication of observations at very low N_0 should therefore be greater than replication of observations at higher N_0.

One alternative method for determining the shape of the functional response curve is LOWESS (LOcally WEighted regrESSion, Chapter 2). This (and other) computer intensive smoothing technique has the advantage of less restrictive assumptions than regression techniques.

Hypothesis testing: estimating and comparing functional response parameters. Although logistic regression is the most useful technique for distinguishing the types of functional responses, and so should be the first step in analyzing a functional response experiment, many investigators want to fit the mechanistic models given in Eqs. 8.1, 8.3, 8.4, or 8.5 to data to obtain estimates of the parameters. These estimates may be useful in population models or may be used for comparing effects of different experimental conditions on predation. Nonlinear least squares is the preferred method for obtaining such estimates (Cock, 1977; Williams and Juliano, 1985; Juliano and Williams, 1987). Because most experiments are done without prey replacement, Eqs. 8.4 and 8.5 are appropriate. Both of these equations give only an implicit function relating N_e to N_0 (with N_e on both sides of the expressions). Thus, iterative solutions are necessary in order to find the predicted values of $\hat{N}_e$ for any set of parameters. This can be done using Newton's method for finding the root of an implicit equation (Turner, 1989). Consider Eq. 8.4 above. This can be rewritten as

$$0 = N_0 - N_0 \exp[a(T_h N_e - T)] - N_e = f(N_e) \tag{8.8}$$

The problem is to find the value of $\hat{N}_e$ that satisfies Eq. 8.8. Graphically, this problem is represented in Fig. 8.2, in which the value of $\hat{N}_e$ at which $f(N_e) = 0$ is the value that satisfies Eq. 8.8. Newton's method approaches this value iteratively, by using the first derivative of $f(N_e)$ [$= f'(N_e)$]. This derivative defines the slope of a line that crosses the horizontal axis at some point (Fig. 8.2). The slope of this line can also be estimated from any two points $(i, i+1)$ on the line by the difference in the vertical coordinates over the difference in the horizontal coordinates, or in this case

$$f'(N_{ei}) = \frac{0 - f(N_{ei})}{N_{ei+1} - N_{ei}} \tag{8.9}$$

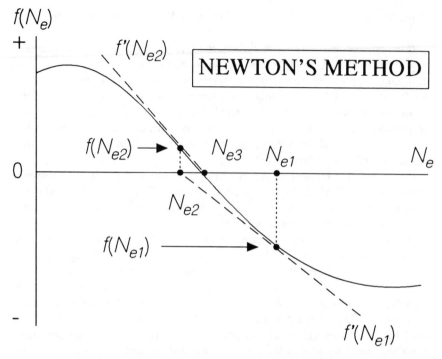

Figure 8.2. Graphic representation of how Newton's method is used to find N_e, the root of the implicit equation describing the functional response with prey depletion (Eqs. 8.4 or 8.5). The point at which the function $f(N_e)$ crosses the horizontal axis is the true solution. Starting from N_{e1}, new approximations of N_e are made using the derivative of $f(N_{e1})$: $N_{ei+1} = N_{ei} - \{f(N_{ei})/f'(N_{ei})\}$. The process is repeated at each new approximate N_e until the approximation is arbitrarily close to the true value [i.e., $f(N_e) = 0$]. See text for details.

where N_{ei+1} is the next value of N_e that serves to start an iteration. Each successive step in the iterative process brings N_{ei+1} closer to the true value that makes Eq. 8.8 true (Fig. 8.2). $\hat{N}_e$ is then the value of N_{ei+1} at which the value of the function $f(N_e)$ is sufficiently close to 0.

Using this procedure to run a nonlinear regression on an implicit function using a computer package such as SAS requires first substituting initial parameter estimates into Eq. 8.8 and obtaining an initial estimate of $\hat{N}_e$. The nonlinear least-squares routine then computes sums of squares, and modifies parameter estimates so that the residual sum of squares is minimized. With each new set of parameter estimates, Newton's method is reemployed in order to determine $\hat{N}_e$.

Using nonlinear least squares assumes normally distributed, homogeneous error variance over the range of N_0 (Bard, 1974; Glantz and Slinker, 1990). Most, if not all, real data fail to meet these assumptions, as variance typically increases with N_0 (Houck and Strauss, 1985; Juliano and Williams, 1987; Trexler et al.,

1988). Violating these assumptions in analyses of functional responses may result in confidence intervals that miss the true parameter values more often than expected (Juliano and Williams, 1987). It appears, however, that even fairly severe departures from these assumptions distort the results only slightly and nonlinear least squares is thus relatively robust (Juliano and Williams, 1987). Residual plots (Neter and Wasserman, 1974) can be used to examine these assumptions graphically. In extreme cases, other remedial measures may be necessary, such as using a gamma distribution of residuals (Pacala and Silander, 1990), nonparametric regressions (Juliano and Williams, 1987), or weighted nonlinear least squares, which may be implemented in the SAS procedure NLIN (SAS Institute Inc., 1989b).

8.3.2 Examples and Interpretation

What follows are two example problems illustrating how to go about answering the three questions described above. Both are based on real functional response data culled from the literature.

Example 1: *Notonecta* preying on *Asellus*. *Notonecta glauca* is a predatory aquatic hemipteran. *Asellus aquaticus* is an aquatic isopod crustacean. This original experiment was described by Hassell et al. (1977). Raw data are given (graphically) by Trexler et al. (1988). Individual *Notonecta* were exposed to a range of densities of *Asellus* from 5 to 100, and allowed to feed for 72 hours, with prey replaced every 24 hours. Substantial prey depletion occurred over the 24-hour periods. At least eight replicates at each density were run. The questions of interest were questions 1 and 2 above: What is the shape of the functional response? and What are the best parameter estimates for the function describing the functional response?

The first step in analyzing the data is to use logistic regression to determine whether the functional response is type II or type III. Data input structure for conducting this analysis using the SAS procedure CATMOD is shown in Appendix 8.1A. The important features to note are: (1) two data lines are entered for each replicate, one for prey eaten (FATE=0) and one for prey alive (FATE=1), (2) the variable NE is equivalent to N_e for observations with FATE=0 (prey eaten), and (3) for a given replicate, NE sums to N0. This data structure is one that allows both analysis of proportions using logistic regression in CATMOD, and with minor modification, analysis of number eaten using nonlinear least squares, along with computation of mean ± SE numbers of prey eaten at each prey density.

The first step in evaluating the shape of the functional response is to fit a polynomial logistic model. Cubic models are likely to be complex enough to describe most experimental data, hence the specific form of the logistic equation is

Table 8.2.   SAS Output for Example 1: Analysis of Notonecta-Asellus *Data[a]*

A. Procedure CATMOD: Maximum Likelihood Analysis of Variance Table

Source	DF	Chi-Square	Prob
INTERCEPT	1	28.68	0.0000
N0	1	6.66	0.0098
N02	1	7.52	0.0061
N03	1	5.67	0.0172
LIKELIHOOD RATIO	85	206.82	0.0000

B. Procedure CATMOD: Analysis of Maximum Likelihood Estimates.

Effect	Parameter	Estimate	Standard Error	Chi-Square	Prob
INTERCEPT	1	−1.3457	0.2513	28.68	0.0000
N0	2	0.0478	0.0185	6.66	0.0098
N02	3	−0.00103	0.000376	7.52	0.0061
N03	4	5.267E−6	2.211E−6	5.67	0.0172

C. Procedure NLIN: Full Model

Non-Linear Least Squares Summary Statistics				Dependent Variable NE
Source	DF	Sum of Squares	Mean Square	
Regression	4	9493.754140	2373.438535	
Residual	85	1766.245860	20.779363	
Uncorrected Total	89	11260.000000		
(Corrected Total)	88	4528.808989		

Parameter	Estimate	Asymptotic Std. Error	Asymptotic 95 % Confidence Interval	
			Lower	Upper
BHAT	0.000608186	0.0246696938	−0.0484419866	0.0496583578
CHAT	0.041999309	2.6609454345	−5.2486959321	5.3326945493
DHAT	0.003199973	0.0591303799	−0.1143675708	0.1207675172
THHAT	3.157757984	1.7560019268	−0.3336593588	6.6491753258

[a]Only the analysis and final parameter estimates are shown; output from the iterative phases of NLIN is omitted for brevity.

$$N_e/N_0 = \frac{\exp(P_0 + P_1 N_0 + P_2 N_0^2 + P_3 N_0^3)}{1 + \exp(P_0 + P_1 N_0 + P_2 N_0^2 + P_3 N_0^3)} \tag{8.10}$$

where P_0, P_1, P_2, and P_3 are parameters to be estimated. Program steps to fit this model to data are given in Appendix 8.1B. Note that the input statements in Appendix 8.1B include estimated parameters from the logistic regression (see section on Generating predicted proportions eaten), which would be available only after a preliminary run of the procedure CATMOD. Output from these procedures appears in Table 8.2.

After information concerning the data set and the iterative steps toward the

Table 8.2.   Continued.

D. Procedure NLIN: Reduced Model 1

Non-Linear Least Squares Summary Statistics			Dependent Variable NE	
Source	DF	Sum of Squares	Mean Square	
Regression	3	9526.797240	3175.599080	
Residual	86	1733.202760	20.153520	
Uncorrected Total	89	11260.000000		
(Corrected Total)	88	4528.808989		

Parameter	Estimate	Asymptotic Std. Error	Asymptotic 95 % Confidence Interval	
			Lower	Upper
BHAT	0.000415285	0.00022766485	−0.0000373001	0.0008678694
DHAT	0.000835213	0.00380836014	−0.0067355902	0.0084060158
THHAT	4.063648373	0.35698844155	3.3539756864	4.7733210598

E. Procedure NLIN: Reduced Model 2

Non-Linear Least Squares Summary Statistics			Dependent Variable NE	
Source	DF	Sum of Squares	Mean Square	
Regression	2	9525.757196	4762.878598	
Residual	87	1734.242804	19.933825	
Uncorrected Total	89	11260.000000		
(Corrected Total)	88	4528.808989		

Parameter	Estimate	Asymptotic Std. Error	Asymptotic 95 % Confidence Interval	
			Lower	Upper
BHAT	0.000460058	0.00010357262	0.0002541949	0.0006659204
THHAT	4.104328418	0.28920091748	3.5295076540	4.6791491819

solution (not shown), CATMOD produces maximum likelihood tests of hypotheses that the parameters are zero, along with a likelihood ratio test, which tests the overall fit of the model (Table 8.2A). A significant test indicates lack of fit. In this example, note that all parameters are significantly different from 0, and that even this cubic model still yields significant lack of fit. Next, the output shows the actual parameter estimates with their standard errors (Table 8.2B). The intercept is not particularly informative. Because the linear parameter (labeled N0 in the output) is positive and the quadratic parameter (labeled N02 in the output) is negative, these results indicate a type III functional response, with proportion eaten initially increasing, then decreasing as N_0 increases, as in Fig. 8.1F. A simple verification of this can be obtained by plotting observed mean proportions eaten along with predicted proportions eaten (Appendix 8.1B). The resulting graph clearly indicates a type III functional response (Fig. 8.3). The scatter of the observed means around the predicted curve is consistent with the significant lack of fit test, and better fit would likely be obtained by fitting quartic, or higher order equations to the data. Fitting a quartic equation (results not shown) to this data set yields a significant fourth-order term, but does not alter the general

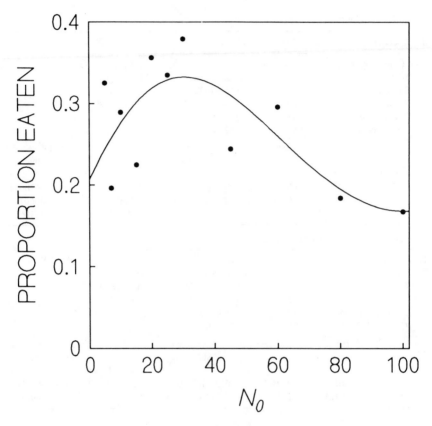

Figure 8.3. Observed mean proportions of prey eaten at each initial prey density in the *Asellus-Notonecta* example, and the fitted relationship produced by logistic regression. Parameter estimates for the logistic regression are given in Table 8.2B.

shape of the curve, and does not alter the conclusion that *Notonecta* shows a type III functional response to *Asellus*. Exploratory analysis using LOWESS also yields the conclusion that *Notonecta* shows a type III functional response (Chapter 2).

If a cubic equation yields a nonsignificant cubic parameter (labeled NO3 in the output), then even if all other parameters are nonsignificant it is desirable to reduce the model by eliminating the cubic term from Eq. 8.10, and to retest the other parameters. Models fit with excess nonsignificant parameters may be misleading.

The utility of determining by logistic regression that a functional response is type II or type III is apparent when attempting to answer the second question posed above, and determining values of the parameters. If logistic regression indicates a type II functional response, a simpler model can be fit. In this case, the more complex type III model is known to be necessary.

To fit a mechanistic model to the data and to estimate parameters, nonlinear least-squares regression is used (SAS procedure NLIN, SAS Institute Inc., 1989b). Several methods are available in this procedure, some of which require the partial derivatives of the nonlinear functions with respect to the parameters of the model. A simpler derivative free procedure (referred to as DUD in SAS) is also available, and because of its simplicity, this procedure will be used. In many cases, different methods within the procedure NLIN will yield similar results, however, investigators may wish to try two or more methods on the same data set to determine if the solution is independent of the method used. NLIN allows programming steps to be incorporated into the procedure, and this makes it possible to use Newton's method to solve the implicit functions given in Eqs. 8.4 or 8.5.

If the experiment had been done with prey replacement (constant N), NLIN could be used to fit one of the explicit nonlinear relationships given in Eqs. 8.1 or 8.3. For this circumstance, the lines from the "Define implicit function" comment through the MODEL statement (Appendix 8.1C, D, E) would be eliminated and replaced with a simple model statement representing Eq. 8.1 or Eq. 8.3.

In this example, because logistic analysis indicates a type III functional response, the general function given in Eq. 8.5 is the starting point. This initial model involves estimation of four parameters. Three parameters (b, c, and d) are defined in Eq. 8.2, and define the relationship between the attack constant, a, and N_0. Because the functional response is type III, at least the parameter b in Eqs. 8.2 and 8.5 must be greater than 0. This will prove important in defining reduced models if the full model is not satisfactory. The fourth parameter is T_h.

The original data set must be modified in order to use nonlinear least squares. One paired observation, with values (N_0, N_e), is required for each replicate. In the data set defined in Appendix 8.1, this corresponds to all observations with FATE = 0, and this subset is selected as the new data set NOTONEC2 (Appendix 8.1B).

NLIN, like other nonlinear procedures, is iterative and requires initial estimates of the parameters. In many cases, the success of nonlinear regression depends upon having reasonably accurate estimates of the parameters. In functional response models, handling time (T_h) can be approximated by T/N_e at the highest N_0. In this model of type III functional responses, initial estimates of the other parameters are less obvious. The parameter d is likely to be small in many cases, hence an initial estimate of 0 usually works. For b and c it is usually best to select several initial estimates varying over as much as two orders of magnitude, and let the procedure find the best initial estimate. This is illustrated in Appendix 8.1C, D, and E for initial values of the parameter b. Multiple initial parameter values may be useful for T_h and d as well, however using multiple initial values for all four parameters may require long computation time. Values of all parameters depend upon the time units chosen.

NLIN allows bounds to be placed on parameters, and this step should always be employed in analysis of functional responses. This will help to avoid nonsensical results, such as negative handling times. In type III functional responses, the parameters T_h and b must be greater than 0, and the parameters c and d must be nonnegative. In Appendix 8.1C, D, and E a BOUNDS statement is used to set these limits.

Use of Newton's method to find $\hat{N}_e$ also requires an initial estimate of $\hat{N}_e$. For simplicity, observed values of N_e may be used, as in Appendix 8.1. Mean N_e at each value of N_0 could also be used, and would likely be closer to the final value of $\hat{N}_e$, hence would reduce the number of iterations necessary to reach a solution.

The strategy for obtaining parameter estimates is to begin with the full model (four parameters) and to eliminate from the model the parameters c and d (one parameter at a time) if these are not significantly different from 0 (Appendix 8.1C–E). For a type III functional response, the minimal model includes T_h and b. In the output from NLIN, significant parameters have asymptotic 95% confidence intervals that do not include 0 (Table 8.2C). For the full model (input in Appendix 8.1C), estimates of b, c, and d were not significantly different from 0 (Table 8.2C). The next step is to eliminate the parameter c, resulting in a linear relationship of a to N_0, and rerun the model (Appendix 8.1D). Again, all parameters except T_h were not significantly different from 0 (Table 8.2D). The final step is to eliminate the parameter d, resulting in the minimal model for a type III functional response, and rerun the model (Appendix 8.1E). In this case the parameters b and T_h were both significant (Table 8.2E), indicating that the relationship of a to N_0 is linear with a slope ($\pm$ SE) of 0.00046 ± 0.00010, and goes through the origin. Handling time is estimated to be 4.104 ± 0.289 h.

Graphic presentation of observed and predicted values (Fig 8.4) indicates a very short region of density dependence and a reasonably good fit of the model to the data. A residual plot (not shown) indicates variance increasing with N_0, as is common in functional response data sets. Weighted least squares may be needed to remedy this problem.

This type III response is described by a two-parameter model. It is interesting to compare this model to the alternative two-parameter model for a type II functional response (constant a). Fitting a type II functional response results in parameter estimates $\pm$ SE of $\hat{a} = 0.0240 \pm 0.0043$ and $\hat{T}_h = 0.908 \pm 0.161$. Both parameters are significantly different from 0. However, residual sum of squares for the type II model is 1813.5, considerably greater than that for the minimal type III model (Table 8.2E). Thus, a type III functional response fits the data better than a type II functional response, even though the number of parameters is the same. This example also illustrates the value of using logistic regression to establish the type of functional response, even if the primary interest is in

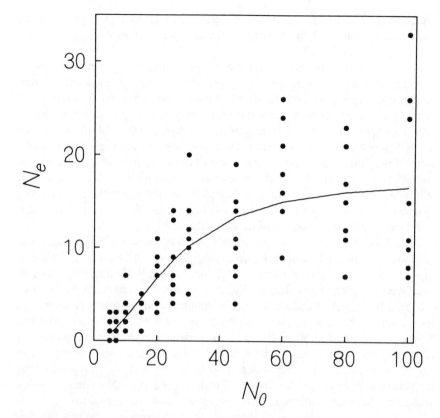

Figure 8.4. Observed numbers of prey eaten in the *Asellus-Notonecta* example, and the functional response curve fit by nonlinear least squares. Parameter estimates for the functional response are given in Table 8.2E.

estimating the parameters. Once it is known that this is a type III functional response, the decision of which two-parameter model to choose is simplified. Finally, this example illustrates how models with excess parameters (the four- and three-parameter models) may give misleading conclusions.

Example 2: *Toxorhynchites* preying on two strains of *Aedes triseriatus*. *Aedes triseriatus* is a tree hole dwelling mosquito from eastern North America. In the southern part of its range, larval *A. triseriatus* are preyed upon by larvae of *Toxorhynchites rutilus*, a large predatory mosquito. In the northern part of its range, larval *A. triseriatus* do not encounter this predator. Livdahl (1979) conducted functional response experiments in which single first-instar *T. rutilus* were offered 3 to 48 first instar larvae of one of two strains of *A. triseriatus*, one from northern Illinois, where *T. rutilus* is absent, and one from North Carolina, where *T. rutilus* is abundant. Predation proceeded for 24 hours, with prey not replenished. The primary question of interest was number (3) above: Do the functional responses of the predator to the two prey strains differ, and, if so, which parameters differ?

Analyses of these functional response data have already appeared in the literature (Livdahl, 1979; Juliano and Williams, 1985). Both of these analyses were conducted assuming a type II functional response, and using Eq. 8.1, the model appropriate for experiments done at constant prey density (no depletion). However, the experiments were conducted with prey depletion (Livdahl, 1979), hence Eq. 8.4 is in fact the appropriate model. It also appears that explicit tests of the form of the functional responses have never been done (Livdahl, 1979; Juliano and Williams, 1985). Hence, the conclusion from prior analyses (Juliano and Williams, 1985), that the two populations do produce significantly different functional responses, with attack constants (a), but not the handling times (T_h), significantly different, may be based on inadequate analyses.

To test for differences in the parameters of the functional response, nonlinear least squares will again be used. However, indicator variables (Neter and Wasserman, 1974; Juliano and Williams, 1985; Juliano and Williams, 1987) will be employed in order to test the null hypotheses that the parameters for the two groups differ. Such a test could also be implemented by fitting two separate regressions to the data and then comparing parameter estimates (see Glantz and Slinker, 1990, pp. 500-505). The indicator variable approach has the advantage producing an explicit test of the null hypothesis of equal parameters (Neter and Wasserman, 1974). The indicator variable approach is analogous to the computational method employed by SAS in the procedure GLM for linear models with categorical and continuous variables, such as analysis of covariance models.

To compare type II functional responses of two groups, the implicit function with indicator variables is

$$0 = N_0 - N_0 \exp\{[a + D_a(j)]\{[T_h + D_{Th}(j)](N_e) - T\}\} - N_e \qquad (8.11)$$

where j is an indicator variable that takes on the value 0 for population 1 and the value 1 for population 2. The parameters D_a and D_{Th} estimate the differences between the populations in the values of the parameters a and T_h, respectively. If these parameters are significantly different from 0, then the two populations differ significantly in the corresponding parameters. For population 1, $\hat{a}$ and $\hat{T}_h$ are the estimates of the population parameters a_1 and T_{h1}. For population 2, $\hat{a} + \hat{D}_a$ and $\hat{T}_h + \hat{D}_{Th}$ are the estimates of the population parameters a_2 and T_{h2}.

To employ this model, the variable j must be present in the data set. The model is implemented in NLIN in the same way that the previous models for a single population were implemented, using Newton's method. Initial estimates of $D_a = 0$ and $D_{Th} = 0$ usually are adequate. The estimates of a and T_h for the two populations take on the same values that would be obtained from separate regressions on the two populations (Juliano and Williams, 1987).

In analyzing this experiment it is, again, first necessary to determine the shape of the functional response. CATMOD was used to fit Eq. 8.10 to the data for each of the two populations (not shown). For both populations, the cubic coefficient was not significantly different from 0. Fitting a quadratic model to the data resulted in quadratic and linear coefficients that were significantly different from 0 in both populations (not shown). The estimates of the linear terms were significantly less than 0, whereas the estimates of the quadratic terms were significantly greater than 0, indicating a shape similar to that illustrated in Fig. 8.1D, and indicating type II functional responses. Plots of observed mean proportions eaten and predicted values (not shown) support this conclusion. These results indicate that fitting the indicator variable model in Eq. 8.11 is appropriate.

Input for fitting the indicator variable model is given in Appendix 8.2, and partial output is in Table 8.3. In this example, the North Carolina population is arbitrarily designated as population 1 ($j = 0$) and the Illinois population is designated as population 2 ($j = 1$). The populations do not differ significantly in as or T_hs (note asymptotic 95% confidence intervals for D_a and D_{Th} include 0). From this output, t-tests can be conducted for the null hypotheses that D_a and D_{Th} are 0 ($t =$ parameter/asymptotic standard error, degrees freedom $=$ DF RESIDUAL). For the North Carolina population $\hat{a}_{NC} = 0.058$ and for the Illinois population, $\hat{a}_{IL} = 0.244$ ($= \hat{a}_{NC} + D_a = 0.058 + 0.186$). $D_a = 0.186$, with $t_{79} = 1.57$, indicating a nonsignificant difference at $P = 0.10$. Thus the attack constants do not differ significantly. This conclusion differs from that reached by Juliano and Williams (1985), who used an indicator variable model that did not incorporate prey depletion (Eq. 8.1), and concluded that the two populations had significantly different as. The results of the analysis in Table 8.3 actually produce $\hat{a}$s that differ in the same direction ($\hat{a}_{NC} < \hat{a}_{IL}$), and differ by a greater absolute amount ($\hat{a}_{NC} = 0.058$, $\hat{a}_{IL} = 0.244$) than the $\hat{a}$'s from Juliano and Williams (1985)

Table 8.3.   SAS Output from Procedure NLIN for Example 2: Analysis of Toxorhynchites − Aedes *data*[a]

Non-Linear Least Squares Summary Statistics			Dependent Variable NE	
Source	DF	Sum of Squares	Mean Square	
Regression	4	5362.8970908	1340.7242727	
Residual	79	293.1029092	3.7101634	
Uncorrected Total	83	5656.0000000		
(Corrected Total)	82	964.7228916		

Parameter	Estimate	Asymptotic Std. Error	Asymptotic 95 % Confidence Interval	
			Lower	Upper
AHAT	0.057857039	0.01527262547	0.0274575315	0.0882565457
THHAT	2.040486509	0.25143619643	1.5400135252	2.5409594925
DA	0.186433867	0.11875924855	−0.0499513330	0.4228190660
DTH	−0.016849199	0.29005690323	−0.5941950459	0.5604966480

[a]The output from procedure CATMOD is omitted.

analysis ($\hat{a}_{NC}$ = 0.040, $\hat{a}_{IL}$ = 0.068). The conclusions based on the model incorporating prey depletion are more realistic, and the earlier results of Juliano and Williams (1985) appear to have been biased by the use of the model not incorporating prey depletion.

Using this indicator variable approach, it is possible to fit models with some parameters in common between the populations (e.g., by dropping the parameter D_{Th} a model with common T_h would result). In this example, such a model yields the same conclusion regarding the difference between a_{NC} and a_{IL} (not shown).

8.4 Related Techniques and Other Applications

A nonparametric method for estimating parameters for the implicit form of the type II functional response has been described (Juliano and Williams, 1987). This method is based on earlier work describing nonparametric methods for estimating nonlinear enzyme kinetics relationships (Porter and Trager, 1977). This nonparametric method has the advantage of not requiring the assumption of normally distributed error. In simulated data sets, parameter estimates from the nonparametric method are comparable in variability and low bias to those produced by nonlinear least squares (Juliano and Williams, 1987). The nonparametric method has several disadvantages: (1) direct comparisons of parameters between experimental groups is difficult, (2) few if any statistical packages implement this procedure directly, and (3) computational time and programming effort required are considerably greater than those associated with nonlinear least squares.

The general approaches outlined in this chapter have other applications in ecological studies. Logistic regression has been used to test for density dependent survival in field experiments manipulating density (e.g., Juliano and Lawton, 1990), and for comparing survivorship curves under different, naturally occurring conditions (e.g., Juliano, 1988). Nonlinear least squares can be used to fit growth curves (e.g., Sedinger and Flint, 1991; see also Chapter 6), and to estimate parameters for log-normal distributions of species abundances (e.g., Ludwig and Reynolds, 1988).

Acknowledgments

I thank Sam Scheiner and Jessica Gurevitch for the invitation to contribute this chapter, Vickie Borowicz, Sam Scheiner, and Jessica Gurevitch for helpful comments on the manuscript, F. M. Williams, Brian Dennis, and Joel Trexler for valuable discussions of the analysis of functional response experiments, and Aaron Ellison, Jay Ver Hoef, and Noel Cressie for providing drafts of their chapters. Computer resources necessary for this work were provided by grants from NIH (1 R15 AI29629-01), NSF (BSR90-06452), and Illinois State University.

Appendix 8.1

SAS program code for data input, logistic regression, and nonlinear regression analysis of example 1.

A. Data Input Lines

```
DATA NOTONECT;
    INPUT N0 REP FATE NE;            /N0 = initial number of prey, REP = replicate
                                  number, FATE - 0 = prey eaten, 1 = prey alive,
                                       NE = count of prey in each FATE/
    N02 = N0**2;                        /initial number of prey squared/
    N03 = N0**3;                         /initial number of prey cubed/
    CARDS;
5  1 0 0
5  1 1 5
5  2 0 1
...
100 7  1 74
100 8  0 33
100 8  1 67
```

B. Procedure CATMOD for Logistic Regression

```
PROC CATMOD DATA = NOTONECT;
    DIRECT N0 N02 N03;
    MODEL FATE = N0 N02 N03/ML
            NOGLS NOPROFILE;
    POPULATION N0 REP;
    WEIGHT NE;

DATA NOTONEC2;                             /obtaining means and SE's for
    SET NOTONECT;                            observed proportions eaten/
    IF FATE = 0; PROPEAT = NE/N0;
PROC MEANS DATA = NOTONEC2;
    BY N0 NOTSORTED;
    VAR PROPEAT;
    OUTPUT OUT = NOTOMEAN
            MEAN = MEANPROP;

DATA NOTONEC3;                         /generating predicted proportions eaten/
    SET NOTOMEAN;
    K = EXP(-1.3457 + (0.0478*N0)-
        (0.00103*N0**2) +
        (0.000005267*N0**3));
    PRED = K/(1 + K);

PROC PLOT DATA = NOTONEC3;                   /plotting observed means
    PLOT PRED*N0 = 'P'                        and predicted values/
    MEANPROP*N0 = '*'/OVERLAY;
```

C. *Nonlinear Least Squares to Estimate Parameters of the Functional Response Equation—FULL MODEL*

```
PROC NLIN DATA=NOTONEC2;
  PARMS BHAT= 0.001 0.01 0.1                    /initial parameter estimates/
    CHAT= 0.001 0.01 0.1
    DHAT= 0  THHAT=3.0;
  BOUNDS BHAT>0,CHAT> =0,
  THHAT>0;                                      /parameter bounds/
  T=72;                                         /experimental period in H/
  X=NE;                                         /initial predicted value/
  A=(DHAT+BHAT*N0)/
    (1+CHAT*N0);                                /expression for A/

                                                /define the implicit function/
  C1=EXP(-A*T);                            /components of the implicit function/
  C2=A*THHAT;
  H= N0*C1*EXP(C2*X)+X-N0;                       /the implicit function/
  ITER= 0;                                /iterations for Newton's method/

                                          /Newton's method employed to find
  DO WHILE(ABS(H)>0.0001                        predicted number eaten/
          AND ITER<50);                   /stop criteria for Newton's method/
    X=X-H/(N0*C1*C2*
        EXP(C2*X)+1);                            /new predicted value/
    H= N0*C1*EXP(C2*X)+X-N0;              /new value of implicit function/
    ITER=ITER+1;                                 /iteration counter/
  END;
  MODEL NE=X;                             /model for nonlinear least squares/
```

D. *Nonlinear Least Squares to Estimate Parameters of the Functional Response Equation—REDUCED MODEL 1—OMIT CHAT*

```
PROC NLIN DATA=NOTNEC2;
  PARMS BHAT= 0.001 0.01 0.1
    DHAT= 0  THHAT=3.0;                        /initial parameter estimates/
  BOUNDS BHAT>0,THHAT>0;                        /parameter bounds/
  T=72;                                    /experimental period in h/
  X=NE;                                         /initial predicted value/
  A=(DHAT+BHAT*N0);                             /expression for A/
```

*** Remainder of program as in Appendix 8.1C but deleted for brevity ***

E. Nonlinear Least Squares to Estimate Parameters of the Functional Response Equation—REDUCED MODEL 2—OMIT DHAT

```
PROC NLIN DATA=NOTONEC2;
  PARMS BHAT= 0.001 0.01 0.1
      THHAT= 3.0;                              /initial parameter estimates/
  BOUNDS BHAT>0,THH>0;                         /parameter bounds/
  T=72;                                        /experimental period in H/
  X=NE;                                        /initial predicted value/
  A=BHAT*N0;                                   /expression for A/

                                               /define the implicit function/
  C1=EXP(-A*T);                                /components of the implicit function/
  C2=A*THHAT;
  H= N0*C1*EXP(C2*X)+X-N0;                      /the implicit function/
  ITER= 0;                                     /iterations for Newton's method/

                                               /Newton's method employed to find
  DO WHILE(ABS(H)>0.0001                           predicted number eaten/
        AND ITER<50);                          /stop criteria for Newton's method/
    X=X-H/(N0*C1*C2*
        EXP(C2*X)+1);                          /new predicted value/
    H= N0*C1*EXP(C2*X)+X-N0;                    /new value of implicit function/
    ITER=ITER+1;                               /iteration counter/
  END;
  MODEL NE=X;                                  /model for nonlinear least squares/
  OUTPUT OUT=PLOTNOTO
        P=PRED R=RES;                          /output data set for plotting/
PROC PLOT DATA=PLOTNOTO;                        /plotting data/
  PLOT RES*PRED                                /plot residuals vs predicted/
  PLOT PRED*N0='P'
  NE*N0='*'/OVERLAY;          /plot observed and predicted number eaten/
```

Appendix 8.2

SAS program code for a data input and nonlinear regression analysis for example 2. Logistic regression steps omitted, see Appendix 8.1.

```
DATA MOSQUITO;
  INPUT POPN $ N0 REP FATE NE;  /POPN = populations; N0 = initial number of prey,
                                 REP = replicate number, FATE - 0 = prey eaten
                                 1 = prey alive, NE = count of prey in each FATE/
  IF FATE = 1 THEN DELETE;
  IF POPN = 'NC' THEN J=0;                        /create indicator variable/
  ELSE J = 1;
CARDS;
IL 3 1 0 2
IL 3 1 1 1
IL 3 2 0 2
 ....
NC 48 5 1 36
NC 48 6 0 13
NC 48 6 1 35
;

*** Nonlinear least squares to compare parameters of the
    functional response equation - INDICATOR VARIABLE MODEL ***

PROC NLIN DATA = MOSQUITO;
  PARMS AHAT = 0.01 0.1                        /initial parameter estimates/
    THHAT = 0.2 2.0 DA = 0 DTH = 0;
  BOUNDS AHAT > 0, THHAT > 0;                          /parameter bounds/
  T = 24;                                             /experimental period/
  X = NE;                                           /initial predicted value/

                                              /define the implicit function/
  C1 = EXP(-(AHAT + (DA*J))*T);               /components of the
  C2 = (AHAT + (DA*J))*(THHAT +                  implicit function/
       (DTH*J));
  H = N0*C1*EXP(C2*X) + X-N0;                    /the implicit function/

*** Remainder of program as in Appendix 8.1C but deleted for brevity ***
```

9

Multiple Regression: Herbivory

Thomas E. Philippi

9.1 Questions About Herbivory

Herbivory is a paradoxical interspecific interaction. On the one hand, upward of a quarter of all species are herbivorous insects (Strong et al., 1984) and some additional fraction of the noninsect species is also herbivorous, so plant–herbivore interactions are common. On the other hand, the simple observation is that, to a first approximation, the earth is green (Hairston et al., 1960), despite the combined effect of those herbivores. The implication is that most plants, most of the time, are not eaten (Howe and Westley, 1988). Either herbivores are almost universally regulated at population levels well below resource limitation by predators and parasites, or plants are doing something in their own defense against herbivores.

Many different questions can be asked about herbivory. The specific example analyzed in this chapter is taken from the work of Phyllis Coley, who studied herbivory in a lowland tropical forest (Coley, 1983). She measured many plant characteristics that have been suggested to be important defenses against herbivores, and levels of leaf damage, for 46 species of understory plants. The general question addressed by the analyses is: Which of the putative defensive characters are important in explaining variation in rates of loss to herbivory among species of neotropical plants, and what are their relative importances? A more specific subsidiary question is: Do structural defenses reduce the amount of tissue lost to herbivory?

These questions are of the form of a response or dependent variable (tissue loss to herbivory) and one or more explanatory or independent variables. To what extent do the independent variables explain or predict the variation in the dependent variable? An entire family of statistical techniques, general linear models (GLMs), has been developed to address such questions. These techniques differ in both the parametrization of the model addressed and the method of parameter estimation. Multiple regression may be taken as the basic technique, of which all other GLMs are variants. In multiple regression (MR), the independent variables are continuous, and a single "slope" parameter is fitted for each. If the independent variables are categorical, the GLM is ANOVA (Chapter 3), with a "mean" parameter fit for each value of each categorical predictor. In the simplest balanced case, this is equivalent to creating binary dummy variables for each categorical variable (e.g., 0, 0, 0; 1, 0, 0; 0, 1, 0; and 0, 0, 1 for groups A, B, C, and D), and then performing multiple regression on the expanded set of dummy variables. If there are both continuous and categorical independent variables, the GLM is ANCOVA (Chapter 4). Other analyses, such as logistic regression (Chapter 8) and Cox proportional hazards models (Chapter 12), are at heart specialized versions of GLM. The form of parameter estimation is usually ordinary least squares, but other methods, principally maximum likelihood, are also used.

By far the majority of manipulative experiments concerning herbivory would be analyzed by techniques covered elsewhere in this book, especially ANOVA and ANCOVA [e.g., herbivore genotype by host type interactions (Futuyma and Philippi, 1987)]. When independent variables may be freely manipulated, the usual procedure is to use only two extreme values of each variable. This maximizes the statistical power to detect an effect of each independent variable, and if it is a complete factorial design (all possible combinations of high and low for each variable), provides the greatest statistical power to detect generalized interactions as well. However, if the goal is not only to test the existence of an effect, but also to estimate the strength of that effect, or to compare the importance of one factor relative to that of another factor, then matching the naturally occurring amounts of variation in the independent variables is important. The most straightforward way to do this is to use the naturally occurring continuous variation, which requires MR. Another case where MR is called for is when some or all important factors cannot be controlled and held constant, and therefore will confound experiments investigating other factors. For example, while it would be convenient to vary the level of tannins while keeping all other potential plant defenses constant, such a manipulation is rarely possible. Instead, some form of statistical control for the effect of other potential defenses must be employed.

9.2 Multiple Regression

Regression encompasses a large and well-developed body of mathematical techniques, well beyond the scope of a single chapter. Draper and Smith (1981), Neter et al. (1985), and Chatterjee and Price (1991) provide the background to and comprehensive coverage of the usual approaches to multiple regression. Freund and Littell (1986) cover most of the regression procedures built into the

SAS system. Rather than attempt to cover the myriad facets of multiple regression, this chapter will focus on the use of multiple regression in model building, and the problems presented by interactions and collinearity, common problems in ecological research.

In the strict sense, regression is a technique for fitting a line to a set of points. Given a set of (X,Y) pairs, linear regression fits a line of the form:

$$Y = \beta_0 + \beta_1 X + \varepsilon \tag{9.1}$$

If the relationship between X and Y is linear, if the independent variable X is measured without error (at least relative to Y) so the error variation is only in Y, if the error variation is independent of X and the error variation about other points, then ordinary least-squares (OLS) regression produces "best" estimates b_0 and b_1 in the following sense. The estimates b_0 and b_1 produce the smallest mean squared error, and if samples of data were repeatedly drawn from the same population and analyzed, the expectation of the mean value for each b is the corresponding β. Together, these properties are referred to as "best linear unbiased estimator" or BLUE. If the error variation is normally distributed, then the sampling distributions of this process are well defined under the four assumptions, and the statistical significance of the parameter estimates may be tested as either t-tests of $b_i \neq 0$ or as an F ratio of explained mean square over the residual mean square.

Two results of this linear regression may be of interest. The coefficient b_1 is a slope: the estimated effect on Y of a change of 1 unit in X. If the relationship were known to be completely linear, then this slope would hold for all values of X. However, the more usual case is that linearity is established by examination of the residual variation, so the estimated slope should not be extrapolated much beyond the range of X values included in the analysis. A second common interpretation of the result is similar to that in ANOVA: the coefficient of determination (r^2) is the fraction of variation in Y that is explainable by the variation in X. This measure is a function of not only the error variance, but also of both the slope b_1 and the variance in X: for a given slope and error variance, the greater the variance in X, the greater the r^2. Increasing the range of values of X will increase the proportion of variation in Y explained by X.

Multiple regression (MR) is an extension of linear regression to the case where there are more than 1 independent variables:

$$Y = \beta_0 + \beta_1 X_1 + \beta_2 X_2 + \beta_3 X_3 + \ldots + \varepsilon \tag{9.2}$$

If there are two independent variables, MR corresponds to fitting the best plane through (X_1, X_2) such that the squared vertical distances between the (X_1, X_2, Y) points and the plane are minimized. With more independent variables the principle is the same, but a geometric representation is harder to visualize. The coefficients (b_i) are still slopes, and estimate the effect of a 1 unit change in X_i on Y, holding all $X_{j \neq i}$ constant. In multiple regression the coefficient of multiple determination (R^2) measures the fraction of variance in Y explained by the complete model. To assess the importance of a single factor, R^2 for the complete model is compared to R^2 for the model with all except the factor of interest: the difference is the increment associated uniquely with that single factor.

The assumption that the form of the relationship is as specified in the equation is more complicated for multiple regression than for linear regression. For MR, this assumption requires that all of the important independent variables, and only the important variables, are included, and that the effects of the independent variables are independent and additive, as well as the original requirement that they be linear. The additional complexity of MR has two practical repercussions. First, all of the assumptions are rarely if ever met, so at best significance tests provide only approximate P-values. Second, violation of these assumptions means that the model is misspecified: a term should be added or dropped, an independent variable should be transformed to produce linearity, or an interaction term is required. The residuals of a regression model are examined to diagnose such misspecification errors, and to indicate the modifications necessary to correctly specify the regression equation.

9.2.1 Model Building

The strict application of estimating coefficients when the form of the equation is already known is relatively straightforward, but rarely occurs in ecology. A much more common question is, given a dependent variable and a set of potential causes or predictors, which predictors affect the dependent variable, and what are the strengths of those effects? Our herbivory question is of precisely this form: Which plant defenses affect rates of loss to herbivory? Also, note that even if the appropriate set of predictors is known, testing for and correcting misspecification is a form of model building. MR fits a linear model, and not knowing the form of the linear relationship is equivalent to not knowing which of the possible transformations of each predictor produces linearity. Therefore, detection and correction of misspecification errors may be considered a special case of model building, where the set of potential independent variables includes transformations of the predictors as well as interaction terms.

While model building occurs prior to strict multiple regression (parameter estimation), the computations of MR are generally used to evaluate the effects of the independent variables. The idea is to use estimated regression coefficients to determine which variables are important and should be included in the final regression equation, and which have no effect and therefore should be dropped. This is a rather subjective process. While some variables may show strong effects and thus clearly should be included, small coefficients (and thus weak effects) are not as simple to handle. Even if a coefficient equals 0, an appropriate transformation of that variable may still produce strong effects. Conversely, in a finite sample, almost any random variable will explain additional variation of the dependent variable in the sample, even if the true population coefficient is 0. Therefore, model building involves parsimony: trading off a slight reduction in fit or explanatory power for reduction in the complexity of the model.

Choosing a "best" model requires adoption of some rule for the strength of a coefficient that will be omitted or included in the model. The obvious approach would be to test the coefficients of each independent variable, and drop all variables with coefficients equal to 0. Unfortunately, with sample data there is no way to distinguish $\beta = 0$ from $\beta \neq 0$ but small, so any rule for inclusion will occasionally include variables with 0 population coefficients, and occasionally omit variables with nonzero coefficients. The omission of variables that have any effect on the dependent variable, whether by deletion in the regression model or by not measuring them in the experiment, leads to biased estimation of the coefficients for the other independent variables unless the omitted variables are completely orthogonal to the remaining independent variables. [A manifestation of this is the problem of selection on unmeasured traits in multivariate selection models, see Endler (1986).] Inclusion of variables with 0 population regression coefficients does not bias the estimated coefficients for the other variables. However, inclusion of additional variables in the regression model reduces the precision of the estimates of the coefficients of all variables. Deletion of variables usually decreases the variances of the estimates of the coefficients of the remaining variables. This tradeoff between bias and precision is a fundamental property of multiple regression.

Several different criteria are used to evaluate this tradeoff. For a given number of parameters, the "best" model has the largest R^2, although several very different models may have nearly equal R^2. As noted above, adding parameters to the model will always increase R^2, so some other criterion is needed to allow comparison of models of different sizes. Two different criteria are in wide use (Chatterjee and Price, 1991). For a sample size of n and a model with p terms, the residual mean square (RMS) is

$$(RMS)_p = \frac{(SSE)_p}{n - p}$$

Because the sample size n and the total sum of squares (SST) are constant, minimizing $(RMS)_p$ is equivalent to maximizing the adjusted R^2 (available in SAS output):

$$R^2_{ap} = 1 - (n - 1)\frac{(RMS)_p}{SST}$$

The second common criterion is the standardized mean squared error of prediction, estimated as

$$C_p = \frac{(SSE)_p}{\hat{\sigma}^2} + (2p - n)$$

where $\hat{\sigma}^2$ is the estimate of σ^2 obtained from the complete model. C_p is the sum of variance and bias components. The expectation of C_p in the absence of bias is p, so the deviation of C_p from p is a measure of the bias due to dropping variables from the complete model. With this criterion, the smallest model with C_p approaching p is the best model. Values of $C_p > p$ are evidence of biased estimation of parameters, while values of $C_p < p$ result from strong collinearity.

Given these criteria for evaluation, several methods of model selection have been used. Forward selection starts with no variables entered into the equation, and at each step adds the variable that provides the greatest increase in R^2. This process is repeated until no more variables satisfy the conditions for inclusion into the model (usually an F-test of the change in R^2). Backward elimination is the opposite process: starting with the complete model, variables with the least partial R^2 are dropped from the model until no remaining variables satisfy the conditions for removal. Stepwise selection combines both forward and backward selection: starting from no variables, at each step a variable may be added or deleted. These methods share the property of evaluating only a subset of the 2^p possible regression models. However, if there is weak collinearity, none are guaranteed to find the best model. For example, if A is the best single predictor but (B and C) are the best pair of predictors, forward selection will not find the best pair of predictors, and neither will backward selection find the best single predictor.

The alternative is to test all possible subsets of predictors, and identify the models with the best criterion by brute force. This guarantees that the best model of each size, as well as those of similar predictive power, is determined, and thus the overall best models are found. While the number of models to be tested equals 2^p, and the computations involved per model increases with the square of p, CPU time is cheap and getting cheaper. Further, most statistical packages use efficient algorithms to avoid performing all of the calculations for each model, and will print out a list of only the i best models of each size. Therefore, my recommendation is to test all possible subsets of predictors. At worst, this may involve running a low priority batch job on a mainframe or letting a PC run overnight.

9.2.2 Interactions

A common complication is that the effect of one independent variable on the dependent variable is contingent on the value of another independent variable. The presence of such interactions means that the simple model in Eq. 9.2 is not the appropriate model, and something more should be added. In the plant defense example, we a priori might expect interactions: the effect of one type of defense at reducing grazing might be lower in species that are already well defended by other types of defenses. In the framework of ANOVA, this interaction would be measured by comparing the variation explained by the means of each treatment combination, after removing the variation explained by the main effect of each treatment, and tested against the variation within each treatment combination (cell). In multiple regression there is no within cell variation, so there is no direct analog to the interaction term of ANOVA. Instead, the general method is to generate a new variable, the product of the two independent variables, and include it in the multiple regression model. To simplify interpretation, before multiplication these variables should have their means subtracted to center them at 0, but should not have their variances standardized to 1.

At first glance this might seem reassuringly familiar: what is represented as *X1*X2* in ANOVA is a new variable *INT*, equal to *X1*X2*, entered into the multiple regression equation. However, the regression interaction term has only 1 degree of freedom compared to the many degrees of freedom in an ANOVA interaction, and tests only one of the many forms a contingent effect might take. If constructed from centered variables, the regression coefficient of this cross-product estimates the change in slope of either independent variable due to a change of 1 unit in the value of the other independent variable, evaluated at the mean of both independent variables. [If centering is not performed, the interaction term has a similar interpretation, but evaluated at (0,0), which may lie well outside of the possible range of the data involved.] Recently, methods for analyzing more complex interactions have been developed. If interactions are suspected, either Jaccard et al. (1990) or Aiken and West (1991) should be consulted.

9.2.3 Collinearity

A second complication occurs when there are correlations among the independent variables, referred to as collinearity. This is a common occurrence in ecology and evolution: factors such as primary productivity or body size may affect a whole suite of characters of interest, and the very existence of "syndromes" indicates that sets of characters covary nonrandomly. In the presence of this collinearity, the model in Eq. 9.2 may still be the appropriate model, because it only assumes that the *effects* of the independent variables on the dependent variable are independent and additive. The problem lies in estimating the parameters b. In the extreme case, one or more of the independent variables may be written as a linear combination of the other independent variables. Therefore, no unique set of b parameters provide the best fit. For example, if $X_1 = X_2 + X_3$, any set of bs such that $b_1 + b_2 = c_2$ and $b_1 + b_3 = c_3$, where c_2 and c_3 are constants determined by the data, will fit equally well, and there is no unique "best" set of coefficients. As is more often the case, the independent variables may be correlated, but not so tightly that any variable may be written as a linear combination of the others. In this case, the numerical algorithms used to estimate the bs are unstable. Small changes in the data may produce great changes in the parameter estimates, which will be reflected in large confidence intervals around the estimated parameters.

The effect of collinearity (also referred to as multicollinearity) may also be viewed from the perspective of assigning explained sums of squares to the independent variables. Collinearity implies that some portion of the variation in the dependent variable may be attributed to more than one independent variable. Therefore, while the F-test for the significance of the overall model may be highly significant, partial F-tests for each independent variable may indicate that no individual parameter explains a significant amount of the variation. [An equivalent problem occurs in ANOVA, and is manifest as differences between Type I and Type II (or III) sums of squares (Shaw and Mitchell-Olds, 1993).]

Collinearity is relatively simple to detect and measure, and several standard tests are included in SAS and most other computer packages. One, condition number, is the square root of the ratio of the largest eigenvalue of the correlation matrix of the independent variables to the smallest (see Chapters 5 and 11 for explanations of eigenvalues). The larger this number, the greater the collinearity among the independent variables; a crude rule is that values above an arbitrary threshold of 30 require attention (Freund and Littell, 1986). A second common measure is the variance inflation factor (VIF): $1/(1-R_i^2)$, where R_i^2 is the coefficient of determination of the regression when the ith independent variable is regressed against all other independent variables. This measure is the amount that the variance of the ith regression coefficient is inflated due to the collinearity. Once again, there is no hard and fast rule for how big a value is too big, but Chatterjee and Price (1991) suggest that a value greater than 10 "is often taken as a signal that the data have collinearity problems." Freund and Littell (1986) propose a lower, variable threshold of $1/(1-R_{tot}^2)$, with R_{tot}^2 taken from the entire regression model. The interpretation of this threshold is that variables with VIFs greater than this threshold are more related to other independent variables than they are to the dependent variable.

To reiterate, severe collinearity means that there is no unique solution to the regression equation, and no amount of mathematical trickery can solve the problem. However, less severe collinearity may be dealt with in several ways. The obvious solution, dropping one or more variables from the model, is rarely appropriate, and occasionally slightly dishonest. If variables are dropped from the analysis, the estimates of the other regression parameters are more precise in the sense of having lower error variances, but they are biased in unknown ways. If experimental results show that either heat or light could explain a result, dropping one from the analysis and claiming that the analysis demonstrates the importance of the other factor is dishonest. It very well may be that the interpretation of which variable is important is correct, but the interpretation is based on outside biological knowledge, not on the statistical analysis itself.

A more appropriate solution sometimes may be found by examining the rationale for the set of independent variables measured. One common cause of collinearity is that the underlying factor of interest was unmeasurable, so several components or correlates were measured as surrogates. In this case, the surrogates should be highly correlated both with the latent variable and among themselves, so including them all in the regression model produces collinearity problems. However, in the underlying logical model, these measures affect the dependent variable only indirectly through the unmeasurable factor, so they are not expected to have independent effects, and may be combined into a single surrogate for the unmeasurable factor. This combining may be as simple as summing the measures (after standardizing the means and variances to eliminate problems with measurements on different scales). Alternatively, the measures may be combined into a single principal component. Principal components (PCs) are linear combinations of the original variables, with the properties that they are linearly independent of each other (orthogonal), and that the first PC has the largest variance of any linear combination of the original variables, the second has the largest remaining variance, etc. Where a set of measures is chosen as correlates of a single unmeasurable factor, the first PC represents the covariation shared by all measures, and thus may be a better estimate of the unmeasurable factor than a simple sum or average. This form of variable reduction may eliminate collinearity problems without compromising the biological questions to be asked, especially if more than one group of variables may be reduced.

A second and relatively recent approach to working with collinearity is to use biased estimation such as ridge regression or incomplete principal components (IPC) regression (Chatterjee and Price, 1991). As noted in Section 9.2, ordinary least-squares estimation is an unbiased estimator of coefficients when the assumptions of multiple regression hold, so the mean of the parameter estimates of a large number of analyses of independent samples will approach the true populational value. Biased estimation techniques take advantage of the fact that, sometimes, an estimator with a small bias (especially toward 0) may greatly reduce the variance of the resulting estimates, and thus produce more precise estimates (Fig. 9.1).

Ridge regression is one method of biased estimation. The mathematical theory behind ridge regression requires matrix notation of the regression model and is beyond the scope of this book, but may be found in Chatterjee and Price (1991) or Draper and Smith (1981). A small biasing factor k is added during the computations, and serves to stabilize the parameter estimates. The appropriate value of this constant is subjectively determined by plotting the estimated coefficients against the bias factor, usually in the range of 0 to 0.1.

A second approach to biased estimation is to perform principal components analysis on not just specific subsets of the independent variables, but on the complete set. The new set of i principal components contains all of the information of the original i variables. The collinearity inherent in the original data will be expressed as one or more PCs with small eigenvalues. The variances of parameter estimates are proportional to the reciprocals of the eigenvalues, so one or a few

PCs with small eigenvalues contribute most of the imprecision of the estimates of the coefficients for the original variables. Because PCs are orthogonal by definition, dropping one or more from the regression analysis does not affect the estimates involving the other PCs, so the dependent variable may safely be regressed on a reduced set of PCs.

The results of this IPC regression are regression coefficients of the reduced set of PCs, linear combinations of the original variables, and can be back-transformed into terms of the original variables. Because the PCs are orthogonal, this back transformation is straightforward algebra. The back-transformed coefficients for the original variables are biased estimates of the true regression coefficients, because the contributions via the PCs with small eigenvalues to the estimated coefficients are omitted.

In general terms, omitting PCs with small eigenvalues increases the precision of the parameter estimates. If parameters have small loadings on the omitted PCs, they are estimated with little bias. But, if either all PCs have moderate to strong loadings or too many PCs are omitted, the parameter estimates will be so biased as to bear little relationship to the underlying coefficients. Section 9.4 includes an extreme example of omitting too many PCs from the analysis.

In SAS version 6.07 (SAS Institute Inc., 1992) ridge regression and IPC regression are implemented in procedure REG. BMDP includes ridge regression and IPC regression in program 4R. These techniques are by no means a panacea: it is quite possible to have so much collinearity that ridge regression is of no use whatsoever. Rather, biased methods of estimation may help when collinearity is not too severe.

9.2.4 Inference and P-values

Hypothesis testing in multiple regression appears to be straightforward: the mathematics behind t-tests and partial F-tests are well developed, and nearly every computer package provides these computations. Unfortunately, the assumptions necessary for such significance tests to be valid, especially the assumption of independent, identically normally distributed errors in the dependent variable, are rarely met (Mitchell-Olds and Shaw, 1987). Wu (1986) developed a weighted jackknife for multiple regression that reduces some of the necessary assumptions, but even it is not a complete solution for generating valid significance tests (Mitchell-Olds and Shaw, 1987).

Even if the distributional assumptions for significance tests are met, problems remain for generating appropriate tests. Between testing for misspecification and model building, multiple regression is a highly iterative procedure. Because of this multiple analysis of a given set of data, the appropriate null hypothesis to be tested is: Given the *complete* series of analyses, what is the probability of obtaining a fit at least as good as the observed fit at any stage if the data were drawn from a random distribution? Unfortunately, the significance tests included in statistical packages test no such hypothesis, but instead test the probability of obtaining the fit if the final form of the model were known a priori, and thus only the single model tested. There is no general way to compute the appropriate hypothesis tests, and even bootstrapping the sampling distribution would require specifying the universe of all transformations and models that might have been tested in the overall analysis. This presents a dilemma: any initial regression model needs analysis of the residuals to confirm that the assumptions have been met, but a model that has been modified to meet the assumptions no longer has valid significance tests.

The obvious solution to this problem is to not worry about significance levels and hypothesis testing in multiple regression. Iterative model building in multiple regression is perfectly valid as a form of exploratory data analysis. The model obtained from this approach is a descriptor of the sample data, and provides one or more hypotheses to be tested in further experiments (Fig 1.1). If the goal of regression is prediction of additional values, the model provides best estimates.

If one must generate significance tests of hypotheses, and sufficient data are available, it may be possible to generate valid significance tests by cross-validation. Randomly divide the data into two parts, and use only one part for model building by whatever means: anything is fair game in this step. Once a final model has been arrived at, the second, independent subset of the data may be used to validate the model, and generate significance tests. The legitimacy of these significance tests still depends on the assumptions of normality and independence of the error variation in the dependent variable, and so is still likely to be only an approximate test.

9.2.5 Experimental Design and Sample Size

In many cases, the question of experimental design for MR seems moot. If the factors of interest are not manipulable, one does not face the usual question of allocation of experimental units to treatment categories. However, there are at least two important considerations for the design of experiments or surveys that will be analyzed with MR. The first was mentioned briefly in Section 9.1. If the goal is merely to test for the presence or absence of an effect, then extreme values of each factor provide the greatest statistical power. If linearity must be assessed, then three evenly spaced values provide the greatest power (Neter et al., 1985). However, if estimates of the strengths of the effects are needed, then the amount and distribution of natural variation must be known, and if nonlinearities or interactions occur, must be matched in the experimental design. The simplest way to match the variation is with random sampling.

Second, if natural variation produces collinearity among the independent variables, should either experimental manipulations or nonrandom sampling be used to reduce this collinearity? There is no set answer to this question. A few well placed observations may greatly reduce the problems associated with collinearity, which would solve a major problem. However, such observations represent combinations that either do not occur or are rare in nature. If interactions or nonlinearities occur, estimates of the relationships within the narrow band of natural covariation based on what would happen far away from that band may be seriously misleading.

There are no simple methods for calculating sample sizes needed for a given amount of statistical power for a given number of independent variables, and any rule would tend to be misleading. However, interactions and collinearity both greatly increase the sample size required for a given level of statistical power. A pilot survey of naturally occurring variation may give some indication of the sample size required for a complete analysis.

9.3 Worked Examples

Before embarking on the analysis of the data from Coley (1983), a simpler set of artificial data will be analyzed to demonstrate the steps in model building when things work right. The second example analyzes data on plant characters and levels of herbivory, and demonstrates the difficulties inherent in addressing complex processes. These two analyses are a best case and a worst case, most would fall somewhere between.

9.3.1 Artificial Data

Consider the case of three predictors X_1, X_2, and X_3 (say, tannin content, toughness, and nitrogen content) and a single dependent variable Y (say, amount of leaf area lost to a specific herbivore). For this example, a sample of 200 such observations were generated in the form $Y = 4X_1 + X_2^2 + E$, so this is the "true" model that multiple regression will estimate. In the interest of brevity, we will not discuss the form of sampling design that might produce such a set of data, but will proceed with the analysis. All analysis and model building will be performed with just the first 100 observations; the final validation of the model will be performed with the second 100 cases.

The first step in analyzing these data is to inspect the distributions of the dependent and independent variables. While none of the variables need be normally distributed at this point, outliers that may have undue influence on parameter fitting should be identified (Chapter 2), and either corrected if due to errors or flagged for possible omission from later analyses. The second step is to test for collinearity among the three independent variables. SAS performs explicit tests for collinearity with the /COLLIN option of procedure REG. Therefore, we will proceed with testing the first regression model:

$$Y = \beta_1 X_1 + \beta_2 X_2 + \beta_3 X_3$$

and plotting the residuals against both the independent variables and the predicted values (Appendix 9.1A).

Inspection of the variables did not reveal any outliers or distributional anomalies, so these results are not included in the listings. The largest condition number is only 4.2, so multicollinearity is not a problem either (Table 9.1). From the regression ANOVA table, this model was highly statistically significant, explaining 85% of the variation in Y. However, the parameter estimates indicate that X_3 is contributing little to the explanatory power of the model, and that its coefficient does not differ significantly from zero. Therefore, X_3 is a candidate for dropping from the following model. The plots (not shown) of the residuals against X_1, X_3, and the predicted value are all uniform scatter, indicating that neither the independent variables X_1 and X_3 nor the dependent variable need be transformed to produce linearity. Because X_3 contributed little to the explanatory power of the regression model, and does not show a relationship with the residuals, X_3 will be dropped from the next model. However, the residuals appear to be large and positive for extreme values of X_2. This indicates that the relationship between X_2 and Y is curvilinear, with Y increasing faster than linear with changes in X_2. Several functional forms might fit these data, including a polynomial in X_2

Table 9.1. *Initial Analysis of Artificial Data from the Program Listed in Appendix 9.1A*

Analysis of Variance

Source	DF	Sum of Squares	Mean Square	F Value	Prob>F
Model	3	14048812.069	4682937.3565	178.962	0.0001
Error	96	2512058.1501	26167.27240		
C Total	99	16560870.220			
R-square		0.8483			

Parameter Estimates

Variable	DF	Parameter Estimate	Standard Error	T for H0: Parameter=0	Prob>|T|
INTERCEP	1	−291.187689	35.81558497	−8.130	0.0001
X1	1	1.583767	1.68243797	0.941	0.3489
X2	1	38.543848	1.67475156	23.015	0.0001
X3	1	−0.024487	0.79178950	−0.031	0.9754

Collinearity Diagnostics

Number	Eigenvalue	Condition Number	Var Prop INTERCEP	Var Prop X1	Var Prop X2
1	1.90449	1.00000	0.0527	0.0072	0.0528
2	1.09849	1.31671	0.0015	0.4126	0.0010
3	0.88923	1.46346	0.0000	0.5763	0.0010
4	0.10779	4.20343	0.9458	0.0038	0.9452

Table 9.2. Analysis of Adjusted Models from the Program Listed in Appendix 9.1B[a]

A. Model with X_2^2 ($R^2 = 0.94$)

Variable	DF	Parameter Estimate	Standard Error	T for H0: Parameter=0	Prob>|T|
INTERCEP	1	−29.202972	30.29540204	−0.964	0.3375
X1	1	4.222279	1.05023222	4.020	0.0001
X2	1	0.738435	3.18315084	0.232	0.8170
X22	1	1.007016	0.08027424	12.545	0.0001

B. Model with exp (X_2) ($R^2 = 0.16$)

INTERCEP	1	426.392778	38.29101064	11.136	0.0001
X1	1	1.662623	3.92634538	0.423	0.6729
EX2	1	1.309716E−16	0.00000000	4.192	0.0001

[a]Only the relevant output is shown.

Table 9.3. Validation of Final Model Using the Other Half of the Data Using the Program Listed in Appendix 9.1C

Source	DF	Sum of Squares	Mean Square	F Value	Prob>F
Model	2	18613932.080	9306966.0399	1227.180	0.0001
Error	97	735650.34470	7584.02417		
C Total	99	19349582.424			

Root MSE	87.08630	R-square	0.9620	
Dep Mean	525.10485	Adj R-sq	0.9612	
C.V.	16.58456			

Variable	DF	Parameter Estimate	Standard Error	T for H0: Parameter=0	Prob>|T|
INTERCEP	1	7.026293	13.60971755	0.516	0.6068
X1	1	3.513661	0.83252864	4.220	0.0001
X22	1	0.990215	0.01999159	49.532	0.0001

or $\exp(X_2)$. The next step is to evaluate models built upon each of these transforms of X_2 (Appendix 9.1B).

The output from the second run of procedure REG (Table 9.2) is quite informative about the relative merits of the two transformations of X_2. While the X_2^2 term is highly significant, the model with $\exp(X_2)$ explains much less variation than the previous model with X_1, X_2, and X_3 ($R^2 = 0.16$ vs 0.85). Therefore, the exponential model will dropped from consideration. In the model with X_1, X_2, and X_2^2, the coefficient for X_2 is no longer significantly different from 0. Therefore, X_2 will be dropped from the model. The final model is $Y = b_0 + b_1X_1 + b_2X_2^2$.

A final run with the second half of the original dataset validates this model (Table 9.3). The coefficient of determination is 0.96, and the parameter estimates are 0.99 (95% CI = 0.95−1.03) and 3.51 (95% CI = 1.94−5.08). Both confidence intervals include the true values (1 and 4, respectively), but the parameter estimate for the X_2^2 term has much less precision. Inspection of the values of X_2 and X_2^2 explains this decreased precision. X_2 was drawn from a normal distribution, so X_2^2 has few observations with very large values.

9.3.2 Plant Defenses and Loss to Herbivory

Variation in grazing and potential plant defenses were measured for 46 species (representing 23 families) of common canopy and subcanopy trees on Barro Colorado Island, Panama (Coley, 1983). For each species, at least 8 (and usually 10) saplings between 1 and 2 m tall occurring in light gaps were studied. To measure grazing rates, roughly 10 mature leaves per plant were tagged and percent leaf area damaged was measured. Three weeks later, percent area damaged was measured again on the marked leaves. This sampling was performed in the dry season, in the early wet season, and again in the late wet season. Undamaged leaves from 5 to 15 plants of each species were collected for measurement of various putative defensive characters. Due to time and logistical restrictions, the leaf samples were pooled within each species, providing a better estimate of mean values for each species, but masking among-leaf and among-plant variation. Three different measures of tannins [phenol (PHENOL), vanillin (VANIL), and leucoanthocyanin (LEUCO)], four of fiber [neutral detergent fiber (NDF), acid detergent fiber (ADF), lignin (LIGNIN), and cellulose (CELL)], and percent N_2 (N2) were recorded from the pooled samples. Fresh leaves were analyzed for water content (WATER), hairs per square millimeter on the upper and lower surface (UHAIR and LHAIR), and leaf toughness (TOUGH) (measured with a "punchmeter"). Various measures of plant apparency [growth rate (GROWTH), maximum growth rate (GMAX), leaf area (LEAFAREA), species distribution (DIST)] were obtained. Details of the species included, the specific analytical techniques, and the raw data are given in Coley (1983).

9.3.3 Data Analysis

These data present an extreme worst-case for regression analysis. There are 16 independent variables and only 46 cases (41 with complete data). Further, the independent variables are likely to be highly collinear, as some variables represent different methods of measuring the same underlying trait. If there are "defense syndromes," then even variables that measure different traits may be highly correlated. In addition, a priori, interactions are likely: the effect of one defense may be less in species with high levels of other defenses. Therefore, a brute force approach of testing a single large regression model is almost certain to fail. Instead, questions about sets of predictors will be addressed, and then parameters will be estimated in different ways. The approach is philosophically more exploratory data analysis and hypothesis generating than inferential hypothesis testing.

First, the relative importance of chemical defenses, structural defenses, and nutritive value in explaining the interspecific variation in levels of loss to herbivores will be tested with principal components. Second, the effect of one form of interaction will be tested. Next, model building (stepwise) procedures will be used to find a best subset of the independent variables for predicting herbivory within this sample. Finally, ridge regression and incomplete principal components regression will be performed to estimate parameter values for the complete set of independent variables. This set of analyses represents only one of the many potential strategies for analyzing these data.

Transformations, centering, and standardization. The dependent variable is the average percentage of leaf area removed per day for each species. This number is a very small percentage, so the values were transformed to LOG(10000 × grazing rate + 1) as in Coley (1983). As noted above, the independent variables need not be transformed to produce normality. However, as part of the model exploration, we may later need to transform one or more independent variables so that the transformed variable will have a linear effect on the measure of herbivory. We also may need to include interaction terms. Therefore, the independent variables will be centered, but not standardized (Appendix 9.2A).

Tests of collinearity. The design of the study leads to the expectation of strong collinearity among the dependent variables; this collinearity will be measured as a preliminary to the development of regression models (Appendix 9.2A). The largest condition number is over 100, indicating strong collinearity (Table 9.4). There is not, however, an abrupt jump in condition numbers that could indicate the size of a set of noncollinear variables. Instead, the collinearity is more diffuse, with variables correlated with most of the other variables. Therefore, stepwise methods of model building will not identify unique "best" subsets of predictors.

Test of classes of defenses. Part of the cause of the high collinearity in these data is the sets of variables that measure similar traits, and are thus highly correlated. Leaf toughness, both measures of fiber, lignin, and cellulose are a

Table 9.4. *Collinearity Analysis of Coley Data from the Program Listed in Appendix 9.2A*

Number	Eigenvalue	Condition Number	Number	Eigenvalue	Condition number
1	12.66317	1.00000	10	0.04900	16.07519
2	1.64111	2.77781	11	0.03454	19.14631
3	0.80948	3.95521	12	0.02755	21.44008
4	0.57285	4.70167	13	0.01160	33.03426
5	0.37240	5.83132	14	0.00853	38.52595
6	0.34082	6.09550	15	0.00571	47.11167
7	0.20244	7.90910	16	0.00295	65.53466
8	0.13617	9.64330	17	0.00093	116.43160
9	0.12075	10.24063			

Table 9.5. Analysis of the Complete vs Subset Models Using the Program Listed in Appendix 9.2B

Source	DF	Sum of Squares	Mean Square	F Value	Prob>F
Model	16	75.65849	4.72866	3.024	0.0071
Error	24	37.52805	1.56367		
C Total	40	113.18654			
		R-square 0.6684			

Test: CHEM		Numerator:	2.2849	DF: 3	F value: 1.4612
		Denominator:	1.563669	DF: 24	Prob>F: 0.2501
		R-square 0.6079			
Test: STRUCT		Numerator:	4.7836	DF: 5	F value: 3.0592
		Denominator:	1.563669	DF: 24	Prob>F: 0.0282
		R-square 0.4571			
Test: NUTRIENT		Numerator:	0.3942	DF: 2	F value: 0.2521
		Denominator:	1.563669	DF: 24	Prob>F: 0.7792
		R-square 0.6615			

nested series of traits: cellulose and lignin are components of acid-detergent fiber, which contributes to leaf toughness. Leucoanthocyanin and vanillin are two measures of condensed tannins. Therefore, the appropriate independent variables will be grouped into chemical defenses, structural defenses, and nutrition, and the effects of these groups tested in two ways.

First, the coefficients of determination (R^2) will be compared for the complete model versus the model dropping a set of variables. The difference in R^2 reflects that variation in herbivory that can be explained only by the dropped variables. The corresponding F ratio is the additional sum of squares explained by the set of variables, divided by its degrees of freedom, over the mean square error of the entire model. SAS will compute these F ratios automatically with the TEST option (Appendix 9.2B). This is the standard test when data are not collinear and is included as a demonstration of the procedure: it is likely to have little statistical power and perform poorly with these data. The ANOVA tables for the complete and subset models, along with the F-tests of the sets of variables, are in Table 9.5. Dropping the five measures of structural defenses reduced the coefficient of determination by over a quarter (0.67 to 0.46), and produced a significant F ratio. Dropping the other subsets of variables did not greatly reduce the overall R^2.

Given the collinear nature of these data, a better analysis involves reducing each group of variables to two principal components (PC). The first PC represents the major covariance in amount, and the second represents the major variation in type within each group. Initially, two regression models will be fitted: one with just the three first PCs (Appendix 9.3A) and one with both first and second PCs (not shown).

Results of this more appropriate test are in Table 9.6. All first principal components have positive factor loadings, and therefore represent positive covari-

Table 9.6. Initial Analysis of Principal Components Using the Program Listed in Appendix 9.3A ($R^2 = 0.35$)

Variable	DF	Parameter Estimate	Standard Error	T for H0: Parameter=0	Prob>|T|
INTERCEP	1	3.822013	0.21241817	17.993	0.0001
CHEM1	1	0.419245	0.22934854	1.828	0.0747
STRUCT1	1	−0.735336	0.23948595	−3.070	0.0037
NUTRIT1	1	0.557031	0.24391578	2.284	0.0275

ation in the constituent traits. Species with higher levels of one form of phenol tend to have higher levels of all forms. The second PC for chemistry separates phenol (uncondensed tannins) from vanillin and leucoanthocyanin (condensed tannins). The second PC of structural defenses separates cellulose and toughness from lignin and neutral-detergent fiber. The regressions of the first PC of both structural defenses and nutrition against herbivory are significant, and in the expected directions (Table 9.6). Plants with more structural defenses and less nutritive value are eaten less. However, the coefficient for CHEM1 is positive, indicating that if there is a trend at all, plants with higher levels of defensive chemicals are eaten more. Adding the second PCs of each group explains another 12% of the variation in herbivory (not shown). The second PC of structural defenses is strong and negative, indicating that the greater the cellulose and toughness relative to fiber and lignin, the less herbivory. The parameter estimate for CHEM1 has become weaker, and the second chemistry PC indicates that phenols rather than condensed tannins (vanillin and leucoanthocyanin) are associated with increased herbivory, therefore vanillin and leucoanthocyanin may reduce herbivory.

The next stage of this analysis is to test potential causes for the lack of a strong negative effect of chemical defenses. A PC for just vanillin and leucoanthocyanin was generated, to test whether including phenols in the group masked the effect of condensed tannins. The new PC for leucoanthocyanin and vanillin did no better than the PC that included phenols, and still had a positive coefficient (not shown). Another possible cause is that variation in tannins might have less effect on levels of herbivory in plants already well defended by structural defenses. This will be tested in the following section on interactions.

Interactions. Given 16 independent variables, there are 120 potential pairwise interaction terms. These terms test only one of many possible forms for each interaction, yet with 46 observations, an exhaustive examination of interactions is impossible. Instead, interactions among the principal components of the groups of predictors will be investigated.

To test for interactions among the first PCs, three new variables were generated as the pairwise products of the PCs. Note that principal components have a mean of 0 by definition, so no explicit centering was required for generating interaction terms. The new regression model includes the three first PCs and the three pairwise interactions (Appendix 9.3B).

Neither interaction involving the chemical defense PC is significant (Table 9.7), so interactions with the other groups do not explain the weak and positive association between chemical defenses and loss to herbivory. However, there is a strong interaction between structural defenses and nutritional value.

Interaction terms are not straightforward to interpret, so these terms will be explicitly broken down. The regression equation (including just the terms of interest) is

$$\text{Grazing} = 3.9 - 0.72\,\text{Struct1} + 0.53\,\text{Nutrit1} + 0.55\,\text{Struct1} \times \text{Nutrit1}$$

This may be rewritten as:

$$\text{Grazing} = (-0.72 + 0.55\,\text{Nutrit1})\,\text{Struct1} + (0.53\,\text{Nutrit} + 3.9)$$

or

$$\text{Grazing} = (0.53 + 0.55\,\text{Struct1})\,\text{Nutrit1} + (-0.72\,\text{Struct1} + 3.9)$$

At the means of both PCs, a unit increase in nutritional value makes the slope of grazing on structural defenses *less negative*; a unit increase in structural defenses makes the positive effect of nutritional value even stronger. At high levels of defense, nutritional value matters more, not less; and at high nutritional values, structural defenses matter less, not more. This interaction is in exactly the opposite direction than I naively expected.

Regression on principal components and ridge regression. Estimates of the coefficients for each independent model, if attainable, may provide a useful description of the relationships within this sample of plants. The strong collinearity among the independent variables prevents ordinary least squares from providing meaningful estimates of the model parameters. Two biased methods, incom-

Table 9.7. Analysis of Principal Components Including Interactions Using the Program Listed in Appendix 9.3B ($R^2 = 0.44$)

Variable	DF	Parameter Estimate	Standard Error	T for H0: Parameter=0	Prob>|T|
INTERCEP	1	3.933483	0.23129050	17.007	0.0001
CHEM1	1	0.298340	0.24198724	1.233	0.2250
STRUCT1	1	−0.719552	0.23519803	−3.059	0.0040
NUTRIT1	1	0.528031	0.23764734	2.222	0.0322
CXS	1	0.128278	0.24083084	0.533	0.5973
CXN	1	−0.265712	0.29854759	−0.890	0.3789
SXN	1	0.553412	0.23152180	2.390	0.0218

plete principal components (IPC) regression and ridge regression, may provide better parameter estimates (Appendix 9.4B). These methods are not available in all statistical packages, and only with version 6.07 have they been included in the SAS system. Principal components are generated with procedure FACTOR in order to determine the number of PCs to include prior to performing the regression on incomplete principal components (Appendix 9.4A). Two other estimates, the ordinary least-squares estimates from the complete model (Appendix 9.2A) and the slopes for each univariate regression, are generated for comparison with the biased estimates (Appendix 9.4B).

In the principal components analysis, the eigenvalue of the seventh PC is much less than that for the sixth PC (Table 9.8), so parameter estimates from 6 PCs (drop = 10 of the 16 variables) will be used. When IPC regression is effective, only a few PCs are dropped; dropping more than half is far too many for IPC regression to be reliable. For ridge regression, all of the plots of parameter estimates as functions of the bias parameter are required to select a desirable model. In a benign universe or statistics book (e.g., Chatterjee and Price, 1991, p. 219), all parameter estimates would stabilize at the same small value of k. Unfortunately, the collinearity of these data overwhelm even ridge regression. While most parameter estimates stabilize quickly (e.g., cellulose and water), some estimates do not stabilize at all, and even diverge from 0 (e.g., nitrogen) (Fig. 9.1). Therefore, while most parameters that stabilize have done so by $k = 0.06$, I have no more confidence in the overall parameter estimates from ridge regression than those from the other three methods (Table 9.9). For some independent variables such as TOUGH, all methods are in rough agreement, but for others such as ADF and LIGNIN, even the sign of an effect is uncertain.

9.3.4 Interpretation of Results

Because the conditions for valid hypothesis tests and inferences about the population are not met, interpretation of these results differs somewhat from interpreta-

Table 9.8   Principal components analysis showing eigenvalues of the correlation matrix using the program listed in Appendix 9.4A.

Principal component	Eigenvalue	Principal component	Eigenvalue
1	5.6777	9	0.4995
2	2.1030	10	0.3928
3	1.9229	11	0.2121
4	1.3946	12	0.1707
5	1.1271	13	0.1274
6	0.9520	14	0.0511
7	0.6825	15	0.0415
8	0.6300	16	0.0151

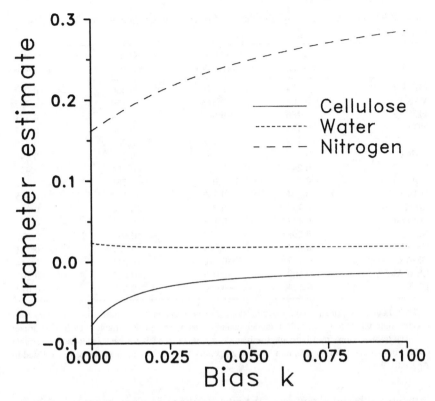

Figure 9.1. Example ridge traces from ridge regression. While the parameter estimates for water and cellulose stabilize as small values of the bias constant k, estimates for the effect of nitrogen content do not stabilize, and in fact enlarge rather than shrink.

tion of most other results in this book. While *P*-values were used during the analysis to help determine which analyses to perform, they do not represent tests of inferences about the entire population of tropical trees, and should not be misrepresented as such. If hypothesis tests are desired, a second set of species would need to be sampled and used to validate these results. However, the lack of *P*-values for rejecting null hypotheses does not mean that these analyses are not informative. The results describe patterns in this particular sample of plant species. Barring additional information, they provide the best guess about relationships in the population of species, sufficient to generate additional hypotheses to test with narrower experiments. Ecology may not progress without rigorous tests of specific hypotheses, but significance tests alone are not enough, and not all worthwhile questions are amenable to manipulative experiments. The dominant result of these analyses is that within this sample, variation in structural defenses, whether measured as physical toughness or as structural components,

Table 9.9. Four Estimates of Standardized Regression Coefficients for Tissue Loss to Herbivory[a]

	IPC (6 PCs)	Ridge (k=0.06)	OLS multivariate	OLS univariate
PHENOL	0.174	0.578	0.436	0.082
VANIL	0.188	−0.313	−0.179	−0.237
LEUCO	0.186	−0.040	−0.114	−0.171
NDF	−0.110	−0.613	−0.480	−0.331
ADF	−0.209	0.448	0.708	−0.415
LIGNIN	−0.079	0.361	0.221	−0.291
CELL	−0.247	−0.110	−0.264	−0.432
TOUGH	−0.389	−1.073	−0.761	−0.642
UHAIR	0.264	0.362	0.238	0.335
LHAIR	0.288	0.243	0.134	0.292
WATER	0.138	0.127	0.102	0.341
N2	0.293	0.124	0.047	0.375
GROWTH	0.072	0.349	0.251	0.331
GMAX	0.117	−0.216	−0.231	0.314
LEAFAREA	−0.109	−0.077	−0.085	0.203
DIST	−0.174	−0.090	−0.017	−0.073

[a]IPC: incomplete principle components; Ridge: ridge regression; OLS Multivariate: ordinary least squares multiple regression with a model including all independent variables; OLS Univariate: ordinary least squares from individual univariate regression models for each independent variable (OLS Multivariate from the program listed in Appendix 9.2A, all others from the program listed in Appendix 9.4B).

explains a substantial portion of the interspecific variation in losses to herbivory. Until Coley (1983), structural defenses were underappreciated in tropical trees, especially relative to chemical defenses.

A rather surprising result (also found in Coley's analysis) is that higher levels of chemical defenses, measured as condensed and uncondensed tannins, are not associated with lower levels of loss to herbivory in this sample. In fact, total phenols were associated with increased levels of herbivory, as were the principal components of the chemical defense measures. These results suggest further investigation of phenols. As suggested by Coley (1983), perhaps phenols are important in the defense of young leaves (which must expand and therefore cannot have high levels of structural defenses), but not in mature leaves. Alternatively, phenols are only cost-effective defenses in species with high herbivore pressure, so that while intraspecifically they reduce the loss to herbivory, they are found only in those species where loss to herbivores (even when defended) is greater than average.

A new result that emerged from this reanalysis is the interaction between structural defenses and nutrient levels. At high levels of nitrogen and water, a given difference in structural defenses has less of an effect on herbivory than at low levels. Low nutritive quality does not act equivalently to a defense. An explanation for this result may hinge on cost-benefit aspects of herbivore diet choice: herbivores may compensate for low nutritional quality by eating more tissue (Slansky and Feeny, 1977).

9.4 Other Applications of Multiple Regression

Many ecological questions involve determining which of a set of potential causative agents is most important, and estimating the strengths of those effects. Such questions parallel the herbivory example presented in this chapter. There is another major application of multiple regression to herbivory, that of measuring selection on several characters [Lande and Arnold (1983); see Simms and Rausher (1992) for an overview of the application to herbivory]. Variation in a fitness measure (e.g., seed set) is regressed on a set of characters (e.g., amounts of various defenses) for different individuals within a species. Quadratic terms for each character are included to distinguish stabilizing selection from directional selection, and all pairwise interaction terms are usually included.

As in the herbivory example, collinearity is a major problem for this application: the goal is to measure direct and indirect selection on a set of correlated characters. A number of evolutionary biologists (e.g., Mitchell-Olds and Shaw, 1987; Endler, 1986; Lande and Arnold, 1983) recommend against using principal components to deal with collinearity. In part, this is due to differences the goals of the analysis: these studies estimate direct and indirect selection on characters by regressing those characters on fitness. However, much of the apparent contradiction is due to different use of principal components, and the omission of too many PCs. The clearest example of this involves Lande and Arnold's (1983) reanalysis of selection on sparrows, which involved nine highly correlated morphological characters. They extracted PCs from the morphological characters, but then used only two of the nine in their regression. This is in accord with the traditional use of PCA, reducing a set of variables to a smaller set of components that contain most of the variation, as the first two principal components included 70% of the variance. The illogical interpretation of their selection coefficients for the second PC is cited against the utility of PCA for dealing with collinearity (e.g., Lande and Arnold, 1983; Mitchell-Olds and Shaw, 1987). The method of IPC regression is subtly different. First, the goal is not to find the smallest number of PCs that includes most of the variation, but rather to eliminate the fewest number of PCs while increasing precision. The technique of Chatterjee and Price (1991) would more likely drop only two or three of the PCs from the regression, and then instead of attempting to interpret the regression coefficients for the PCs, would back-transform them into estimates of the effects of the original variables. If more than two or three PCs need to be dropped, then neither IPC regression nor any other technique will help. To reiterate yet again, if predictors (traits) are so highly collinear (correlated) that most of the PCs must be dropped, then there is simply no mathematical technique to validly estimate the separate contributions (direct selection).

9.5 Related Issues

The difficulties encountered in this analysis of herbivory data were not due to the comparative nature of the example. If feeding trials were performed on foliage from each species of plant, the dependent variable would be different, but the analytical problems would be exactly the same. Neither were the difficulties due to the multivariate nature of the question asked. If the question was simply "does the level of phenols in leaves affect the percent leaf area lost to herbivory," a univariate regression on phenols would be inappropriate. Such an analysis would assume that all other factors were equal—distributed independently of phenols—which they clearly are not. The problems occur because ecological patterns rarely have individual causal factors. Even if only one factor was of interest, multiple regression still would be required to correct for the effects of the covarying factors.

This sobering lesson applies equally well to manipulative experiments analyzed with ANOVA. Not measuring other factors is no different than omitting them from the analysis, and for many ecological questions, the implicit assumption of ANOVA that unmeasured factors are equal is unjustified. Manipulative experiments for ANOVA must be able not only to vary one factor, but also to keep all other factors constant, to provide valid inferences about the population as a whole.

Acknowledgments

Phyllis Coley kindly provided both the data and the inspiration for this chapter.

Appendix 9.1

SAS program code for analysis of artificial data (true model: $Y = 4 * X1 + X2 * X2 + 100 * E$).

A. Initial Analyses

```
PROC UNIVARIATE;                        /inspect the distributions of all variables/
  WHERE (GROUP = 0);                    /moments, medians, and extreme values/
  VAR Y X1 X2 X3;                       /all WHERE statements select
                                            half the data for analysis/

PROC REG;                               /test residuals of simple linear model
  WHERE (GROUP = 0);                        and test for multicollinearity/
  MODEL Y = X1 X2 X3/COLLIN;           /COLLIN requests condition numbers/
  PLOT R*(X1 X2 X3 P)/VPLOTS=2;        /plot residuals (R) against Xs
                                            and predicted values (P)/
```

B. Modified Models with Transformed Variables

```
DATA FAKE2; SET HERB.FAKE;
   X22 = X2 * X2;                                    /new variable X2²/
   EX2 = EXP(X2);                                    /new variable eˣ²/
PROC REG; WHERE (GROUP = 0);
   MODEL Y = X1 X2 X22;                              /polynomial model with X2 and X2²/
   MODEL Y = X1 EX2;                                 /exponential model with eˣ²/
```

C. Final Run for Tests of Significance Using the Other Half of the Data

```
PROC REG; WHERE (GROUP = 1);
   MODEL Y = X1 X22;
```

Appendix 9.2

SAS program code for analysis of Coley data.

A. Data Centering and Initial Collinearity Diagnosis

```
PROC STANDARD OUT=COLEY                             /center data/
   MEAN= 0;
VAR PHENOL VANIL LEUCO NDF
   ADF LIGNIN CELL TOUGH
   UHAIR LHAIR WATER N2
   GROWTH GMAX LEAFAREA
   DIST;

PROC REG DATA=COLEY;                                /test for collinearity/
   MODEL GRAZING = PHENOL
   VANIL LEUCO NDF ADF
   LIGNIN CELL TOUGH UHAIR
   LHAIR WATER N2 GROWTH
   GMAX LEAFAREA DIST                               /STB requests standardized regression
   /COLLIN STB;                                      coefficients for later comparisons/
```

B. Tests of Subsets of Variables

```
PROC REG DATA=COLEY;
   MODEL GRAZING = PHENOL
      VANIL LEUCO NDF ADF
      LIGNIN CELL TOUGH UHAIR
      LHAIR WATER N2 GROWTH
      GMAX LEAFAREA DIST;

   CHEM: TEST PHENOL, VANIL,              /test for sets of parameters = 0/
         LEUCO;
   STRUCT: TEST NDF, ADF, LIGNIN,
           CELL TOUGH;
   NUTRIENT: TEST WATER, N2;
```

*** To obtain R^2 values for reduced models for comparison with total
    R^2 value, analyse new models omitting pertinent variables ***

Appendix 9.3

SAS program code for analysis using grouped predictors produced with principal
components analysis.

A. Initial Analysis Using First and Second Principal Components

```
PROC FACTOR OUT= P1 N=2                       /generate first 2 principal
         DATA=COLEY;                        components for each group/
   VAR PHENOL VANIL LEUCO;
PROC FACTOR OUT= P2 N=2
         DATA=COLEY;
   VAR NDF ADF LIGNIN CELL TOUGH;
PROC FACTOR OUT= P3 N=2
         DATA=COLEY;
   VAR WATER N2;

PROC DATASETS;                              /rename principal components
   MODIFY P1;                                 prior to merging datasets/
   RENAME FACTOR1=CHEM1
          FACTOR2=CHEM2;
   MODIFY P2;
   RENAME FACTOR1=STRUCT1
          FACTOR2=STRUCT2;
   MODIFY P3;
   RENAME FACTOR1=NUTRIT1
          FACTOR2=NUTRIT2;
DATA PRIN;                                  /generate interaction terms for
                                             use in later stage of analysis/
   MERGE P1 P2 P3;                                   /merge datasets/
   CxS = CHEM1*STRUCT1;             /principle components have mean=0 by
   CxN = CHEM1*NUTRIT1;              definition, so interaction terms may be
   SxN = STRUCT1*NUTRIT1;              generated without explicit centering/

PROC REG;
   MODEL GRAZING = CHEM1              /analysis with first components only/
      STRUCT1 NUTRIT1;
```

*** Analysis with second principal components not shown for brevity ***

B. Final Analysis of First Principal Components Including Interaction Terms

```
PROC REG DATA= PRIN;
  MODEL GRAZING = CHEM1
    STRUCT1 NUTRIT1 CxS CxN
    SxN;
```

Appendix 9.4

SAS program code for incomplete principal components regression and ridge regression. The IPC and RIDGE options are available only in SAS version 6.07 and later.

A. Standardize Variables and Generate Principal Components to Determine Number for Inclusion in Regression Analysis

```
PROC STANDARD DATA=COLEY
  MEAN= 0 STD=1;
  VAR PHENOL VANIL LEUCO NDF
    ADF LIGNIN CELL TOUGH
    UHAIR LHAIR WATER N2
    GROWTH GMAX LEAFAREA
    DIST;
PROC FACTOR;
  VAR PHENOL VANIL LEUCO NDF
    ADF LIGNIN CELL TOUGH
    UHAIR LHAIR WATER N2
    GROWTH GMAX LEAFAREA
    DIST;
```

/standardize coefficients for easier/
comparison among methods/

/generate prinicpal components/

B. Incomplete Principal Components, Ridge, and Univariate Ordinary Least-Squares Regression Analyses

```
PROC REG OUTEST=PCEST              /incomplete principal components regression/
    PCOMIT=0 TO 13;                     /PCOMIT drops from from 0 to
    MODEL GRAZING = PHENOL                  13 PCs retaining 16 to 3/
        VANIL LEUCO NDF ADF LIGNIN
        CELL TOUGH UHAIR LHAIR
        WATER N2 GROWTH GMAX
        LEAFAREA DIST;

PROC REG DATA=COLEY                               /ridge regression/
    OUTEST=REST RIDGE=0
    TO .10 BY .002;
    MODEL GRAZING = PHENOL
        VANIL LEUCO NDF ADF LIGNIN
        CELL TOUGH UHAIR LHAIR
        WATER N2 GROWTH GMAX
        LEAFAREA DIST;
PROC PLOT DATA=REST;               /plot parameter estimates against RIDGE/
    PLOT (PHENOL VANIL LEUCO
        NDF ADF LIGNIN CELL
        TOUGH UHAIR LHAIR
        WATER N2 GROWTH GMAX
        LEAFAREA DIST)*
        _RIDGE_/VREF=0;
```

*** Generation of individual regression parameters not shown for brevity ***

10

Path Analysis: Pollination
Randall J. Mitchell

10.1 Ecological Issue

Naturalists have recognized the importance of pollination for plant reproduction for more than 2500 years (Baker, 1983). However, modern attempts to understand the details of this interaction have faced numerous difficulties. The major complication is that plant reproduction involves a number of sequential and relatively distinct stages, and experimental investigation of all the stages simultaneously is not feasible. Consider for example, reproduction of the herbaceous monocarp (i.e., dies after reproducing) Scarlet Gilia (*Ipomopsis aggregata*: Polemoniaceae), found in mountains of Western North America. Reproduction for this self-incompatible plant depends on pollination by broad-tailed and rufous hummingbirds (*Selasphorus platycercus* and *S. rufus*), which probe the red tubular flowers to extract the nectar produced within.

Because of the timing of events, and the basic biology involved, a logical first hypothesis about the factors influencing plant reproduction might be summarized as follows:

Plant Traits → Visitation → Pollination → Reproduction

Each arrow indicates an effect of one class of traits on another. Other interactions are possible (e.g., direct effects of plant traits on reproduction, or of traits such as density or location on visitation), but I will ignore them for this initial example.

One way of determining whether this general hypothesis is correct is to mechanistically (and especially, experimentally) assess the effects of each component in isolation from the others. Assuming that the components can be combined linearly and additively, the expected overall effect of each trait can be calculated using the mean effects from different experiments. For example, three experiments on three different groups of plants might indicate that (1) plants producing more nectar receive more hummingbird visits, (2) plants receiving more visits receive more pollen per flower, and (3) plants receiving more pollen per flower tend to mature more seeds per flower. Multiplication of the numeric estimates for these components predicts the effect of nectar production on seed production. For example, (visits/mg nectar sugar) × (pollen deposited/visit) × (seeds/pollen deposited) = seeds/mg nectar sugar. Combining these estimates is intuitively satisfying, but depends on an untested assumption: each step in the pathway is independent of (uncorrelated with) those preceding it, except for the causal linkages (Welsh et al., 1988). When this assumption is violated, estimated effect sizes can be misleading. Unfortunately, violation may be common; for example, large plants may produce both more nectar and more seeds/fruit (even with equivalent pollen per flower), so that predictions based on the separate estimates would be inaccurate. Such correlations among traits are common (e.g., Campbell et al., 1991), and current understanding of the factors causing the correlations is poor, so the assumption of independence is probably not justified in many cases.

10.2 Statistical Issue

Field biologists generally use two major approaches to deal with intertrait correlations and complicated causal relationships: experimental manipulation and statistical control.

Experimental manipulation of suspected causes is a very effective way to draw strong inferences about causality (e.g., Hairston, 1989), primarily because experiments eliminate correlations among traits through randomization (but see Bollen, 1989, p. 73, for some qualifications). However, experiments are not always logistically possible, ethical, or even a reasonable place to start (see Wright, 1921, p. 557). So, as effective as the experimental approach can be, it is not always appropriate.

Statistical control is another way to deal with correlations among traits (Pedhazur, 1982). A common example is multiple regression, which accounts for correlations among causal variables to estimate the effect of a particular trait with all else statistically "held constant" (see Chapter 9). Statistical control is especially suited for observational data, where one can capitalize on natural variation in the putative causal variables. One disadvantage of multiple regression as a method of statistical control is that it cannot directly deal with complicated causal schemes such as outlined above for pollination, since it deals only with one dependent variable, and does not allow effects of dependent variables on one another. This same problem applies to factor analysis and similar multivariate approaches.

10.2.1 Path Analysis

Path analysis, the focus of this chapter, is a more general form of multiple regression that allows consideration of complicated causal schemes with more than one dependent variable and effects of dependent variables on one another.

Path analysis deals with intertrait correlations in the same way as does multiple regression, thus providing statistical control. Though not a substitute for experimental research, path analysis (and other methods of statistical control) can be especially valuable when used in combination with judiciously chosen experiments. It can be used to analyze experimental data when some variables cannot be randomized (e.g., sex), or when experiments are not possible (Hayduk, 1987, p. xii), provided that treatment levels can be ordered (e.g., amounts of fertilizer), or that there are only two levels per treatment (e.g., pollinated vs not; Schemske and Horvitz, 1988). The main difference is that the experimental method minimizes correlations between the manipulated variables, so no correlation among treatments need be hypothesized (see below).

Path analysis was originally developed by Sewall Wright (1920, 1921, 1934) as a way to partition variation from observational data into causal and non-causal components, according to a particular hypothesis. Along with the required mathematics, Wright introduced the valuable concept of a path diagram, which summarizes an overall hypothesis about how different variables affect one another (as in the simple hypothesis above). Before conducting a path analysis, you must have at least one specific diagram in mind. In proposing the basic causal hypotheses making up the path diagram, use all the information available. As Wright (1921, p. 559) noted: "There are usually a priori or experimental grounds for believing that certain factors are direct causes of variation in others or that other pairs are related as effects of a common cause." The diagram reflects natural history, temporal sequence of events, and intuition and experience with the system. Having clear causal hypotheses in mind before analysis is a critical part of the process. After the fact it is remarkably easy to change your ideas in response to the data, but this compromises significance tests, exactly as with a posteriori comparisons in ANOVA. This is not to say that you should not have several alternative hypotheses in mind. Indeed, this is to be encouraged (Breckler, 1990), and methods for comparing different models will be discussed later.

10.3 Statistical Solution

To demonstrate the implementation of path analysis, I use data from a study of *Ipomopsis aggregata*. The original choice of traits and structure of the path diagram were made a priori, but here I analyze a subset of traits chosen for instructive value. To test the very general hypothesis about pollination described above, I have first customized the hypothesis to more specifically fit the biology of *Ipomopsis*. Flowers of this species are hermaphroditic, so a full accounting of reproductive success would consider both seeds mothered and seeds sired by a plant. Budgetary and logistical considerations prevented estimation of the number of seeds sired, so I only consider a measure of female reproductive success. The customized path diagram is shown in Fig. 10.1. In path diagrams, a one-headed arrow represents a causal effect of one variable on another (e.g., nectar production affects approach rate), a curved arrow represents a correlation, and "U" represents unexplained causes.

In Fig. 10.1 there are three dependent variables (approaches, probes/flower, and proportion fruit set), and below I justify the causal relationships for each. In some cases several alternative causal hypotheses are possible, and a major goal of my analysis is to use the observed data to compare the explanatory value of those alternatives

Approach rate. Hummingbirds may respond to a wide variety of cues when deciding which plants to visit. In this example I consider two plant traits and one indicator of local density of conspecifics. Based on foraging theory, on the work of others, and on observation in the wild (Pyke et al., 1977; Wolf and Hainsworth, 1990; Mitchell personal observation), I hypothesize that birds will preferentially approach and begin to forage at plants with high nectar production rates and many flowers. For similar reasons, birds may more often approach plants in denser clumps. Because little is known about the effects of clumping for *Ipomopsis*, this path is not as well justified as the others; its usefulness will be considered in Section 10.3.4. Thus, the diagram includes causal arrows leading to approach rate from plant traits (flower number and nectar production), and from nearest neighbor distance.

Probes/flower/hour. The proportion of flowers probed might also be affected

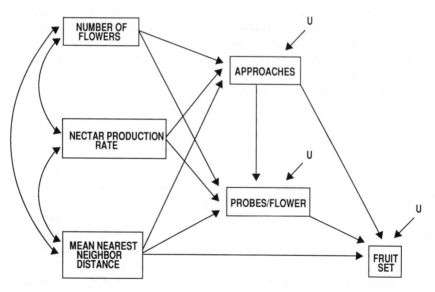

Figure 10.1. Hypothesized relationships determining hummingbird visitation and fruit production for *Ipomopsis aggregata* (Model 1).

by the same characters as approach rate, for similar reasons. Furthermore, before a bird can decide how many flowers to probe on a plant, it must choose to approach a plant, so an effect of approach rate is also hypothesized.

Proportion fruit set. Logically, the chance that a flower matures into a fruit should increase with visitation rate, since each visit deposits additional pollen (Mitchell and Waser, 1992). I use two estimates of visitation rate, each with slightly different properties. The first is approach rate, which may indicate the ability of plants to attract hummingbirds from a distance. Such plants also might consequently receive more cross-pollen, and receive a higher diversity of pollen genotypes, since each approach represents a new opportunity for pollen to arrive from different plants. The second is probes/flower/hour, which represents the chance that a particular flower will be visited in a given time period. More probes/flower/hour should deposit more pollen, which increases fruit (and seed) set for *Ipomopsis* (Kohn and Waser, 1986). Thus, I included both causal pathways in the model.

Pollination is not the only factor affecting fruit set. Among many other potential factors, I have chosen to consider the effects of near neighbors. The proximity of neighbors may indicate the extent of competition among conspecific plants for resources (Mitchell-Olds, 1987), and competition might influence the ability of plants to mature fruits. I have therefore hypothesized effects of neighbor distance on fruit set. Admittedly, little is known of the effects of neighbors for this species, and I will consider these results with caution.

Correlations. Full analysis of a path diagram requires that the pattern of unanalyzed correlations among variables (the curved arrows) be explicitly stated. Traits such as flower number and nectar production may be correlated with one another for both genetic and environmental reasons. With detailed genealogical information, the genetic components of that correlation could be directly analyzed (Wright, 1920; Li, 1975). Likewise, detailed knowledge of the microenvironment experienced by each plant might allow a causal interpretation of this correlation, but that information also is unavailable. Consequently, I have hypothesized that the two traits are correlated for unanalyzed reasons. Nearest-neighbor distance may be one cause of correlation between the plant traits (through effects on resource availability), but since other interpretations are possible I have modeled its relationships with flower number and nectar production as unanalyzed correlations.

10.3.1 Design of Observations

To test the causal hypothesis in Fig. 10.1, I measured floral traits, visitation, and reproduction for a sample of plants in a natural population. Observation of all these variables on the same individuals is required to apply the technique, partly for reasons mentioned in Section 10.1. Measurement methods are described elsewhere (Campbell et al., 1991; Mitchell, 1992). After deciding which variables to consider, my primary consideration was sample size. A good rule of thumb for both multiple regression and path analysis is to have, at minimum, 10–20 times as many observations as variables (Harris, 1985, p. 64; Stevens, 1992, p. 72; SAS Institute Inc., 1989a, p. 140; see also section 10.3.5). For this example there are 6 variables, so a sample size of >60–120 is reasonable. With this in mind, several assistants and I recorded pollinator visitation simultaneously for 5 distinct subpopulations of 23–30 plants each, for a total of 139 plants.

For each plant I counted the number of flowers open on each day of observation, and used the mean across days in the analysis. Mean nectar production rate and mean distance to the three nearest neighboring conspecifics are straightforward to measure and calculate (Campbell et al., 1991). For visitation, I calculated number of approaches/hour of observation, then took the mean across seven observation periods (ca. 70 minutes each). Likewise, for each observation period, I divided the total number of flower probes/hour for each plant by the number of flowers open that day to measure probes/flower/hour, and used the mean across all observation periods. Proportion fruit set was calculated as fruits/total flowers.

Like many parametric techniques, path analysis assumes that the distribution of residuals is normal (see Section 10.3.5). Meeting this assumption is more likely if the variables themselves are normally distributed. To this end, I transformed most variables, and tested the transformed and untransformed variables for deviation from normality using the SAS procedure UNIVARIATE with the NORMAL option (see Appendix 10.1). Transformations were natural logarithms for flower number and nearest-neighbor distance, square root for nectar production, and arcsin-square root for proportion fruit set (Table 10.1).

Path analysis, like other techniques, can be sensitive to missing data. For example, some of my plants were eaten or trampled by animals before I could measure all of the traits mentioned above. There are two ways to deal with missing data. First, do nothing (this is the default method for SAS). The result

Table 10.1. *Observed and Expected Correlations[a]*

	FLR	NPR	NND	APH	PPF	PROPFS
FLR	1.000					**-0.052**
NPR	0.156	1.000				**0.082**
NND	0.148	-0.130	1.000			
APH	0.248	0.271	0.035	1.000		
PPF	-0.145	0.141	-0.064	0.676	1.000	
PROPFS	-0.102	0.213	-0.298	0.182	0.247	1.000

[a]Observed correlations are below the diagonal. The two expected correlations that deviate from observed values are shown above the diagonal, in boldface (see Section 10.3.4). The expected correlations are derived from Model 1 (Fig. 10.3), using the SAS procedure CALIS. FLR, ln(mean number of open flowers each day); NPR, $\sqrt{\text{nectar production rate}}$; NND, ln(mean of distances to three nearest neighbors); APH, mean approaches/plant/hour; PPF, mean probes/open flower/hour; PROPFS, arcsin($\sqrt{\text{proportion fruit set}}$).

is that for some traits there are observations from all plants, and for others the sample size is smaller. However, this method of "pairwise deletion" can occasionally make matrix inversion impossible (SAS Institute Inc., 1989a, p. 287; Hayduk, 1987, p. 327; matrix inversion is a mathematical operation used in solving the regressions). A second method is to completely ignore individuals with incomplete data by deleting plants for which there are no data for any one of the six variables. Most of us are naturally reluctant to discard any data, but such "listwise deletion" is generally the safest and most conservative approach (Hayduk, 1987, p. 326). Unfortunately, there also are drawbacks, especially if the missing individuals are not a random sample of the population (Bollen, 1989, p. 376). Hayduk (1987, p. 327) suggests a practical way to determine if the method of deletion matters: do the analysis once using each method. If the results do not differ appreciably, there is no problem, otherwise listwise deletion is preferable. For the *Ipomopsis* data, I found no substantial difference between results from the two deletion methods, and therefore used listwise deletion for the nine plants with incomplete data.

The observations of plants and pollinators can be summarized in a correlation matrix (Table 10.1). In some respects this answers some questions already— there is a significant positive correlations between nectar production rate and proportion fruit set, in agreement with the idea that nectar production rate is important for reproduction. Unfortunately, study of these simple correlations by themselves does not provide direct information on the causal relationships. That is where path analysis comes in.

10.3.2 Data Analysis

A simple example. Consider the simple path diagram in Fig. 10.2, which represents a simple multiple regression. Recall that one-headed arrows indicate potential direct effects of independent variables (e.g., nectar production) on dependent variables (e.g., approaches), whereas two-headed arrows indicate that indepen-

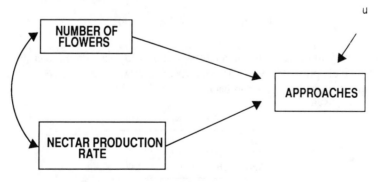

Figure 10.2. Path diagram of a simple multiple regression.

dent variables might be correlated with one another for reasons that are not currently analyzed. The diagram is a symbolic representation of the following equation:

$$\text{Approaches} = p_{\text{APH,FLR}}\,\text{FLR} + p_{\text{APH,NPR}}\,\text{NPR} + p_{\text{APH,U}}\,\text{U}$$

This equation states that all the variation in approach rate is due to causal effects of flower number, nectar production, and all other causes (U). This equation may readily be solved: in SAS, you simply state

PROC REG; MODEL APH = FLR NPR/STB;

where APH = approaches/hour, FLR = ln(mean number of open flowers), and NPR = $\sqrt{\text{nectar production}}$; these and other abbreviations are defined in Table 10.1. SAS implicitly assumes correlations among the independent variables, along with random errors of estimation due to all other causes (U). The option STB requests the standardized regression coefficients in addition to the unstandardized output.

Path coefficients are standardized regression coefficients, and therefore indicate the change in the dependent variable, in standard deviation units, that would be expected following a one standard deviation change in the independent variable. Correlated effects of other independent variables are statistically held constant. For more on the interpretation of standardized coefficients and statistical control see Li (1975, p. 170), Hayduk (1987, p. 39), and Chapter 9. In this simple example, SAS provides estimates of 0.21 for the path from FLR to APH, and 0.24 for $p_{\text{APH,NPR}}$.

The curved arrow in Fig. 10.2 represents the unanalyzed correlation between flower number and nectar production. The magnitude of that path is the simple correlation between flower number and nectar production, and can be found with following simple SAS statements:

PROC CORR; VAR FLR NPR;

A useful way to think about the analysis so far is to focus on the fact that SAS was effectively presented with a 3×3 matrix of correlations between open flowers, nectar production rate, and approach rate:

	FLR	NPR	APH
FLR	1.00		
NPR	0.16	1.00	
APH	0.25	0.27	1.00

The regression analyzed the correlations among the two floral traits and approach rate, but left the correlation between flower number and nectar production unanalyzed.

Consider the correlation between flowers and approaches. Under the hypothesis embodied in Fig. 10.2, $r_{FLR,APH}$ (= 0.25) consists of two components: a direct effect and an indirect effect (Li, 1975 p. 114). The direct effect is the variation in approaches that is uniquely attributable to flowers, and the optimal estimator of this is the path coefficient (partial regression coefficient) from flowers to approaches ($p_{APH,FLR}$; 0.21 in this example). The indirect effect represents the portion of variation in approaches jointly determined by flowers and nectar. The indirect effect occurs because flowers and nectar are correlated ($r_{FLR,NPR} = 0.16$), so that a change in flowers should result in a correlated change in nectar, which itself has a direct effect on approaches ($p_{APH,NPR} = 0.24$). A useful feature of path analysis is that the direct and indirect effects can be read directly from the diagram by tracing the two paths connecting flowers to approaches (the rules for tracing paths are described by Li, 1975, p. 162). Thus, the correlation between flower number and approach rate can be decomposed into a direct effect ($p_{APH,FLR}$) and an indirect effect mediated through nectar production ($r_{FLR,NPR} \times p_{APH,NPR}$).

In equations, this is stated as:

Correlation of FLR and APH = direct effect + indirect effect

Or in symbols:

$$r_{FLR,APH} = p_{APH,FLR} + p_{APH,NPR} \times r_{NPR,FLR}$$

Or, using the SAS estimates:

$$0.25 = 0.21 + 0.24 \times 0.16$$
$$0.25 = 0.25$$

This decomposition indicates that the direct effect of flowers on approaches (0.21) is much stronger than the indirect effect (0.04). A similar decomposition can be applied to the correlation between nectar and approaches.

A more complicated example. More complicated path diagrams are dealt with similarly. Model 1 (Fig. 10.1) is an elaboration of the simple example, modified to include more variables, and effects of dependent variables on one another (e.g., nectar production affects approaches, which in turn affects probes/flower/hour). Just as with the simple example, it is straightforward to write one equation for each dependent variable in the model. Each equation includes a term for every variable affecting that dependent variable, one term for each of the arrows leading to that dependent variable.

$$\text{Approaches} = p_{\text{APH,FLR}} \text{ FLR} + p_{\text{APH,NPR}} \text{ NPR} + p_{\text{APH,NND}} \text{ NND} + p_{\text{APH,U}} \text{ U}$$
$$\text{Probes/flower} = p_{\text{PPF,FLR}} \text{ FLR} + p_{\text{PPF,NPR}} \text{ NPR} + p_{\text{PPF,NND}} \text{ NND}$$
$$+ p_{\text{PPF,APH}} \text{ APH} + p_{\text{PPF,U}} \text{ U}$$
$$\text{Proportion fruit set} = p_{\text{PROPFS,APH}} \text{ APH} + p_{\text{PROPFS,PPF}} \text{ PPF}$$
$$+ p_{\text{PROPFS,NND}} \text{ NND} + p_{\text{PROPFS,U}} \text{ U}$$

These equations and the model in Fig. 10.1 are synonymous, since the correlations among the variables on the far left-hand side are assumed in the mathematics of regression.

One of the beauties of path analysis is that complicated equations can be summarized in fairly simple diagrams. Note that relationships excluded from the diagram (and therefore forced to equal zero) dictate the structure of the equations as much as do those included. For instance, Model 1 specifies that nectar does not directly affect fruit set, but may have indirect effects through approaches and probes/flower; this is reflected by the fact that $p_{\text{PROPFS,NPR}}$ is fixed at zero, and therefore is not in the diagram or equations.

For three of the dependent variables (approaches/hour, probes/flower/hour, and proportion fruit set) there are no unanalyzed correlations (i.e., there are no curved arrows leading to them). The implicit assumption is that any correlations involving those variables are completely accounted for by the hypothesized causal scheme. However, there can sometimes be substantial deviations between the model-implied correlations and the actual correlations, and these deviations can be used to assess the agreement between the model and the observed data (see Section 10.3.4).

10.3.3 Worked Example

Solving the equations for Model 1 is straightforward using the SAS procedure REG as outlined above (Section 10.3.2). The SAS code is in Appendix 10.1, and the results are summarized in Table 10.2 and Fig. 10.3. The path coefficients

Table 10.2. Direct, Indirect and Total Effects for Model 1[a]

Variable	Effect on APH DE	IE	TE	Effect on PPF DE	IE	TE	Effect on PROPFS DE	IE	TE
FLR	0.20	—	0.20	−0.32	0.15	−0.1	—	−0.02	−0.02
NPR	0.24	—	0.24	−0.02	0.18	0.16	—	0.05	0.05
NND	0.04	—	0.04	−0.04	0.02	−0.02	−0.29	−0.00	−0.29
APH	—	—	—	0.76	—	0.76	0.07	0.14	0.21
PPF	—	—	—	—	—	—	0.18	—	0.18
R^2		0.12			0.56			0.14	

[a]Direct effects are path coefficients. DE, direct effect; IE, indirect effect; TE, total effect.

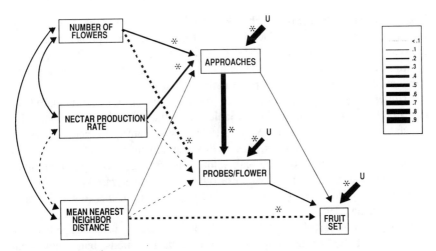

Figure 10.3. Solved path diagram for Model 1 (Fig. 10.1). Solid lines denote positive effects, dashed lines denote negative effects. Width of each line is proportional to the strength of the relationship (see legend), and paths significantly > 0 ($P < 0.05$) are indicated with an asterisk. Actual values for path coefficients are in Table 10.2.

("direct effects") in Table 10.2 can be read directly from the SAS regression output (one regression for each of the three dependent variables) in the column labeled "standardized estimate." Figure 10.3 also presents the solution, and for ease of interpretation, arrow width represents the magnitude of each path. The simple correlations are provided by the SAS procedure CORR. The magnitude of the arrows from "U" (unanalyzed causes) indicate influences on each dependent variable that are unexplained by the causal diagram. Since the total explained variance in a dependent variable is defined to be R^2 (Li, 1975, p. 178), the magnitude of the path from U to any variable ($p_{X,U}$) is calculated as $\sqrt{(1-R^2)}$. As R^2 for PROPFS = 0.14 (Table 10.2), $p_{PROPFS,U} = \sqrt{(1-0.14)} = 0.93$. Calculation of total and indirect effects is laborious but straightforward (Li, 1975, p. 161; Pedhazur, 1982, p. 590; Section 10.3.4).

10.3.4 Interpretation of Results of the Analysis

Once you have performed the path analysis, you must interpret it. What does the diagram as a whole tell you? Which paths are especially important? Was your original hypothesis right? Path analysis has many applications, and each of the questions just raised corresponds to one of the major uses of the technique (Sokal and Rohlf, 1981, p. 649).

Heuristic description. A path diagram is a compact method of presenting abundant information that may not be easily absorbed from regression tables (compare Table 10.2 and Fig. 10.3, which are basically synonymous). Even a brief glance at a path diagram such as Fig. 10.3 can give a substantial understanding of the results: nectar production has a strong effect on approach rate, there are correlations among independent variables, approach rate has little direct effect on fruit set, and so forth. Because path diagrams are easily grasped intuitively, they can be very useful in data presentation (Chapter 2).

Estimation of effect sizes. Sometimes interest is primarily in the magnitude of the direct effect of a particular variable when the effects of other variables are statistically held constant (Wright, 1983). Since path coefficients are really multiple regression coefficients, they fit this requirement nicely. Of course, just as with multiple regression, the path solution is contingent on which variables are included in the model. Inclusion of other independent variables might well change the significance, size, and even the sign of coefficients (e.g., Mitchell-Olds and Shaw, 1987).

Testing hypotheses. Wright (1920; 1921; 1934) originally proposed path analysis as a way to test explicit hypotheses about causal relations, but achievement of this goal was not generally practical until the advent of powerful computers. It now is possible to actually test the hypothesis that a particular path diagram is an adequate description of the causal processes generating the observed correlations, and to compare the descriptive power of alternative models (see below). Aside from directly testing hypotheses, path analysis also can help indicate experiments that might be especially useful. For instance, if path analysis indicates that pollinators respond strongly and directly to nectar production, an experiment to verify that conclusion might be warranted.

Data dredging. This approach is tempting, but should be employed with caution (Kingsolver and Schemske, 1991). Data dredging is the opposite of hypothesis testing—after the data have been collected, one looks for interesting correlations, and assumes that the strongest correlations are causal. There is nothing in the mathematics that prevents one from then solving for a path diagram allowing direct effects between strongly correlated variables, but the search through the data has eliminated any element of hypothesis testing. Just as with a posteriori comparisons in ANOVA, significance levels are not trustworthy for a posteriori data dredging. Furthermore, the resulting causal interpretation will not necessarily be accurate; simulation studies of artificial data reveal that data dredging seldom arrived at the correct causal model (MacCullum, 1986).

Some forms of data dredging are justified, however. For example, pilot studies are often used to generate hypotheses and develop intuition about a system. Further, cross-validation of a dredged model (i.e., attempting to fit the model suggested by the data to an independent data set, or to a previously unanalyzed portion of the original data) also is acceptable and useful (Mitchell-Olds and Shaw, 1987; Bollen, 1989, p. 278; Breckler, 1990).

Interpretation of the *Ipomopsis* model. If the diagram in Fig. 10.3 is correct, it is easy to see that there are significant effects of plant traits on approach rate, that probes/flower is largely determined by approach rate and (negatively) by number of flowers, and that nearest neighbor distance has strong negative effects on fruit set.

Below, I evaluate those interpretations of the individual coefficients. But what about the first proviso: is this diagram correct? This question has been a stumbling block in other fields (e.g., psychology, sociology, economics), as well as in biology, and social scientists have made notable progress toward quantitatively assessing the agreement between the path diagram and the observed data. They do this using the computer-intensive technique of "structural equation modeling," a more general form of path analysis. This is an enormous topic on its own (for introductions see Maddox and Antonovics, 1983; Hayduk, 1987; Loehlin, 1987; Bollen, 1989; Johnson et al., 1991; Mitchell, 1992). For now it suffices to discuss the agreement between model and data. In structural equation modeling, a goodness of fit statistic is used to assess the agreement between the correlations actually observed in the data, and the correlations that would theoretically occur if the path diagram was correct (i.e., if the causal structure in the path diagram was the actual causal structure). Essentially, the goodness of fit statistic is calculated from the difference between the observed and expected correlations (see Table 10.1). The statistic is distributed approximately as χ^2, with degrees of freedom equal to the difference between the number of unique observed correlations and the number of coefficients estimated. In the example, there are 21 observed correlations (see Table 10.1), and 19 coefficients are estimated in Fig. 10.3: 10 direct effects, 3 correlations, and 6 variances. That leaves two degrees of freedom for comparing the observed and expected correlations.

There now are several computer programs that quantify goodness of fit (e.g., EQS, LISREL; see Bentler, 1985; Jöreskog and Sörbom, 1988), including a new procedure in SAS version 6, known as CALIS (Covariance Analysis of Linear Structural Equations; SAS Institute Inc., 1989a). Programming statements to implement CALIS for the example are in Appendix 10.2. The resulting output includes direct, indirect, and total effects, as well as correlations, so it is not necessary to use the REG procedure. However, to use CALIS and structural equation modeling you must learn the rather confusing jargon (see Loehlin, 1987; SAS Institute Inc., 1989a).

The observed and expected correlation matrices for the *Ipomopsis* data are shown in Table 10.1. The observed correlation matrix simply comes from the raw data, and the expected correlation matrix is provided by CALIS in the standard output. There are only two elements that are free to differ between these matrices, corresponding to the two degrees of freedom for comparison. CALIS calculated the χ^2 goodness of fit for Model 1 (Fig. 10.3) to be 3.71, with 2 df, $P = 0.16$, so the lack of fit of model to data is not significant. Nonsignificance in this case is good for the model, not bad; a nonsignificant χ^2 indicates that there is no significant deviation between the observed and expected correlation matrices under this model. This means that the model in Fig. 10.3 has survived an attempted disproof, and remains as a potential explanation of the interaction.

Whether such agreement between model and data also indicates agreement between the model and the actual causal structure is discussed by Bollen (1989, p. 67), but at least the possibility has not been eliminated.

The details of how the goodness of fit statistic is calculated are presented elsewhere (e.g., Hayduk, 1987, p. 132); the important point here is that there now is a way to determine whether the path diagram is reasonable. And for this example, the hypothesis that this is a correct description of the interaction cannot be disproved.

Not all path diagrams can be tested in this manner—the model must be "overidentified." At a minimum, overidentification requires that there be more observed correlations than estimated coefficients (i.e., there must be more knowns than unknowns), but there are other requirements. A full discussion of model identification is beyond the scope of this chapter; see Long (1983), Hayduk (1987), Loehlin (1987), and Bollen (1989).

If the diagram as a whole had a poor fit to the data, it would be premature to place much importance on the values of individual paths, since the model might be of little descriptive value and the magnitude of individual paths might therefore be misleading. But given that the goodness of fit test indicates that Model 1 might be reasonable, inspection of the individual paths is in order. In many cases this may suggest relevant experiments, and results from experiments may suggest improvements or modifications of the path diagram. Such feedback among observations, experiments, and hypotheses is an important part of the strong inference approach to science (Platt, 1964; Chapter 1).

In Model 1, nectar production has significant effect on approach rate by hummingbirds. One interpretation of this is that hummingbirds somehow identify or remember plants with high nectar production rates and preferentially approach them. Because other evidence suggests that identification from a distance is unlikely (Mitchell, 1992), I have experimentally investigated the possibility that hummingbirds actually remember the location of individual plants (Mitchell, 1993), and found that birds do not seem to remember high nectar plants, although they do respond after arriving and sampling some flowers. Since those results disagree with the hypothesis that birds remember individual plants, my current working hypothesis is that birds use spatial location as a cue to identify clumps of plants with high or low nectar production rates. Note that nearest-neighbor information is apparently not used (Fig. 10.3). Although this seems a plausible explanation after the fact, I would not have favored this hypothesis without the path analysis and subsequent experimentation.

Flower number significantly affected both approach rate and probes/flower. Although plants with many flowers were approached more often, probes/flower was negatively affected by flower number. This decrease in probes may represent a proportional cost to plants having many flowers, if it lowers seed production.

Nearest-neighbor distance had no significant effects on visitation behavior, but had strong, negative effects on proportion fruit set. One potential explanation is that short nearest-neighbor distances indicate more intense intraspecific competition (Mitchell-Olds, 1987; see Brody, 1992, for an alternative explanation). As with the examples above, this prediction is amenable to experimental investigation.

The largest paths in all cases involve the unexplained influences on each dependent variable (U), indicating that much of the variance in dependent variables cannot be explained by this model, even though the goodness of fit is acceptable. This may be because many potentially important variables have not been included (e.g., plant resources), but also may mean that chance variation plays a large role.

Comparing alternative path diagrams. Often, several different models are considered feasible a priori. In the past, there was no quantitative way to choose among models, but the goodness of fit test changes this, greatly increasing the usefulness of path analysis. For instance, Model 2 (Fig. 10.4) is an alternative to Model 1 (see Section 10.3), and hypothesizes that nearest-neighbor distance has no direct effect on hummingbird visitation behavior. The observed data do not deviate significantly from what would be expected if Model 2 were correct ($\chi^2 = 4.44$, 4 df, $P = 0.35$), so it also is an adequate description of the interaction. Deciding which of these two acceptable models is a better description of the interaction is facilitated by the fact that they are "nested." Nested models are identical except that one does not include some effects hypothesized in the other. Here, Model 2 is nested within Model 1 because it can be derived from Model 1 by constraining the direct effects of neighbors on visitation to zero (by not estimating those paths). To compare nested models, the difference in goodness of fit is used, employing the fact that the difference between two χ^2 values also

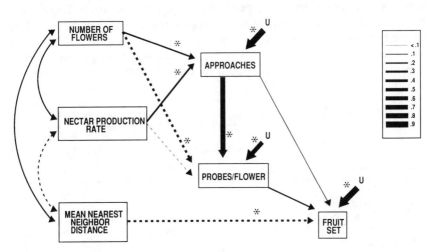

Figure 10.4. Solved path diagram for Model 2, modified from Model 1 to include no effects of nearest neighbor distance on visitation. Conventions follow Fig. 10.3.

is distributed as χ^2, with degrees of freedom equal to the difference in degrees of freedom between the two models (Hayduk, 1987). There is a nonsignificant difference in χ^2 between Model 1 and Model 2 (Table 10.3). Since Model 2 is simpler than Model 1 (it estimates fewer paths), it might be preferred on the principle of parsimony, but given the fairly strong theoretical justification for estimating the effects of neighbors on visitation, I will retain the more general Model 1 for now. Observations or experimental manipulations of nearest-neighbor distance would be useful in further evaluating Model 2. Although these particular acceptable models did not differ significantly, in general there may be strong and significant differences among acceptable models. Indeed, social scientists frequently compare nested models to choose among several otherwise acceptable hypotheses (e.g., Loehlin, 1987, p. 106).

Model 3 is another nested model, going one step further than Model 2 by proposing that nearest-neighbor distance does not affect fruit set (Fig. 10.5). The fit of Model 3 is far inferior to that of Model 1 (Table 10.3), indicating that effects of neighbors on fruit set are very important to an acceptable description of the interaction. As mentioned above, this may indicate intraspecific competition (e.g., for nutrients, space, light, or pollination), among other possibilities. Experimental manipulation of one or more of these potential influences could be pursued, perhaps after some are eliminated by further observational study.

Model 3 illustrates an important difference between structural equation modeling and traditional approaches to path analysis. Without the goodness of fit test there would be no reason to reject Model 3. Just as with Model 1, some paths of biological interest are large, and R^2 values for some dependent variables are reasonable (for Model 3, $R^2 = 0.12$, 0.56, and 0.06 for approaches, probes/flower, and proportion fruit set respectively; compare to Table 10.2). There would be no real clue that a potentially important path is ignored in Model 3, and no quantitative method for deciding which is a more reasonable model. Without a way to compare the fit of models, I probably would only have presented results from one model, and ignored those I did not consider likely. Because of this, the

Table 10.3. Nested Comparison of Alternative Models with Model 1 (Fig. 10.3)

Model Description	Goodness of fit			Nested comparison with Model 1		
	χ^2	df	P	χ^2	df	P
1. See Fig. 10.3	3.71	2	0.16	—	—	—
2. Same as Model 1, but no effects of neighbors on visitation	4.44	4	0.35	0.73	2	0.7
3. Same as Model 2, but no effects of neighbors on fruit set	16.28	5	0.006	12.57	3	0.01

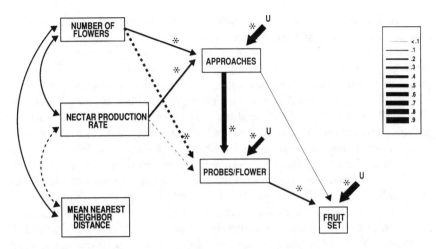

Figure 10.5. Solved path diagram for Model 3, modified from Model 2 to include no effects of nearest neighbor distance on visitation or fruit production. Conventions follow Fig. 10.3.

possibility of alternative models should be carefully considered by both authors and readers, even when goodness of fit is not assessed.

10.3.5 Other Issues

Path analysis relies on a number of assumptions. The assumptions of normal distribution of residuals, additive and linear effects, and inclusion of all important variables are discussed below. Other assumptions commonly made are that residual errors are uncorrelated and that there is no measurement error. The latter assumptions may be relaxed when using structural equation modeling (see Section 10.4).

As with multiple regression, path analysis assumes that the residuals from estimation are normally distributed. Appendix 10.1 demonstrates how to test this assumption using the SAS procedures REG and UNIVARIATE. The same approach can be used to test residuals from GLM and other SAS procedures. For the *Ipomopsis* data, residuals for several variables departed significantly from normality, despite my transformations. Violation of this assumption does not affect the magnitude of the path coefficients (Bollen, 1989, p. 126), but significance tests may be untrustworthy. To deal with this, I have compared the traditional significance levels for individual paths with those from a delete-one jack-knife test that does not assume normality (Mitchell-Olds, 1987; Chapter 13). In no case did significance values differ between the two methods. This robustness gives me confidence in traditionally derived significance levels for these data.

If there had been disagreement between the two approaches, the jackknifed significance levels would probably be preferable.

The goodness of fit test assumes multivariate normality, and further, depends on large sample sizes to utilize the asymptotic χ^2 distribution of the goodness of fit statistic. Surprisingly, the assumption of multivariate normality is not very restrictive, since it is a sufficient but not necessary condition for the goodness of fit test (Cloninger et al., 1983), and the test is robust to skewness and kurtosis (Muthen and Kaplan, 1985). Bollen (1989, p. 79) argues that this assumption is no more restrictive than those made for ANOVA. However, for field ecologists, small sample size will often be a more pressing problem, since observations on 100-200 individuals (sampling units) are usually required to have confidence in the goodness of fit test (Tanaka, 1987). Because of both these limitations, χ^2 values should generally be used more as an index of fit than as a rigorous statistical test. For example, Model 3 does not fit the data well at all, while Model 1 is reasonable; those conclusions are unlikely to be influenced qualitatively by deviations from multivariate normality or by small sample size.

Second, the analysis assumes that the causal relationships are additive and linear. Wright (1983, 1960) claims that this assumption commonly is met. To test for curvilinearity, inspection of raw data plots or residuals can be useful (Cohen and Cohen, 1983, p. 126; Chapter 2; see also Appendix 10.1). Curvilinear relationships often can be modeled using quadratic and higher terms, as in multiple regression (Sokal and Rohlf, 1981, p. 671; Cohen and Cohen, 1983, p. 369; Hayduk, 1987, p. 219).

Third, the analysis assumes that information on all variables having strong causal effects have been included in the analysis. The reasons for and consequences of violating this assumption are covered in Cohen and Cohen (1983, p. 129) and Mitchell-Olds and Shaw (1987). Note that the assumption is not that all traits are included (an impossibility), but instead that all traits with strong causal effects are included (Cohen and Cohen, 1983, p. 354). This assumption should be taken as a warning that the solution may depend on which variables are considered (contrast the significance of the path from probes/flower/hour to fruit set in Figs. 10.3 and 10.5). Just as with experiments, if important variables are omitted or ignored, the answer may be misleading or wrong.

10.4 Related Issues and Techniques

Multiple regression is covered in this book by Philippi (Chapter 9), and elsewhere by Cohen and Cohen (1983) among many others. Good introductions to path analysis and its applications are in Li (1975), Pedhazur (1982), Loehlin (1987), Schemske and Horvitz (1988), Crespi and Bookstein (1989), Crespi (1990), and Kingsolver and Schemske (1991). Currently the only introduction to CALIS is the SAS manual (SAS Institute Inc., 1989a). Consulting the readable introductions to structural equation modeling and LISREL (a program similar to CALIS) of Hayduk (1987) and Loehlin (1987) will be helpful in deciphering the SAS text, which assumes familiarity with the subject. The pros and cons of structural equation modeling are discussed in Karlin et al. (1983), Cloninger et al. (1983), Wright (1983), Bookstein (1986), and Breckler (1990). If variables are strongly intercorrelated (multicolinear), inclusion of latent factors may be useful, and these, as well as correlated residuals and measurement error can be incorporated as an integral part of a structural equation model—see Loehlin (1987), Hayduk (1987), and Bollen (1989), but be sure to read and heed the cautions of Bookstein (1986) and Breckler (1990).

Path analysis has been applied to many other ecological topics, including species interactions (Arnold 1972; Wootton, 1990), gall production (Weis et al. 1989), and determinants of reproductive success (Maddox and Antonovics, 1983; Mitchell-Olds, 1987; Schemske and Horvitz, 1988).

Acknowledgments

Alison Brody, Diane Campbell, Ken Halama, Diana Hews, Karen Mitchell, Mary Price, Hadley Renkin, Rich Ring, Peter Scott, and Nick Waser helped in the field. Valuable discussions and comments on the manuscript came from Alison Brody, Bob Cabin, Ann Evans, Jessica Gurevitch, Ken Halama, Doug Kelt, Diane Marshall, David Reznick, Sam Scheiner, Ruth Shaw, Art Weis, and Nick Waser. Beth Dennis made the illustrations, Tom Mitchell-Olds kindly provided the jackknife program, and David Reznick demonstrated how to test for normality of residuals. The Rocky Mountain Biological Laboratory provided housing, lab space, and a supportive atmosphere during field work. Supported in part by NSF Grant 9001607.

Appendix 10.1

SAS program code for path analysis of Model 1 (Fig. 10.3) using procedure REG.

```
DATA PATHS;
  INPUT FLR NPR NND                    /variables (abbreviations in Table 10.1)/
    APH PPF PROPFS;
  FLR = LOG(FLR);                              /transform variables/
  NPR = SQRT(NPR);
  NND = LOG(NND);
  ASPROPFS = ARSIN(SQRT(PROPFS));
CARDS;                                         /raw data would be here
                                            (6 columns for 139 plants)/

PROC CORR COV;                           /request correlation and covariance
  VAR FLR NPR NND                        matrices, using list-wise deletion/
    APH PPF ASPROPFS;

PROC REG;                                     /regression for approaches/
  MODEL APH = FLR NPR NND/STB;          /"STB" requests standardized output/
  OUTPUT OUT=A P=PRED R=RESID;          /store residual and predicted values/
PROC UNIVARIATE PLOT NORMAL;
  VAR RESID;                                /check normality of residuals/
PROC PLOT; PLOT PRED*RESID;              /check for heteroscedasticity/
PROC REG DATA = PATHS;
  MODEL PPF = FLR NPR                /regression for probes per flower per hour/
    NND APH/STB;
  MODEL ASPROPFS = APH PPF              /regression for proportion fruit set/
    NND/STB;
```

*** Statements similar to those in the regression for APH could be used to check normality and heteroscedasticity, but for brevity are omitted. ***

Appendix 10.2

SAS program code for solving for Model 1 using procedure CALIS. The "ram" statement consists of a description of each of the causal paths to be estimated (other methods of model specification are described in the SAS manual). In each line of the ram statement the first number indicates the number of arrowheads for that effect (1 for a direct effect, 2 for a correlation), the second number indicates the recipient of the effect (the variables are referred to in order), and the third number indicates the causal variable. When a name is provided as the fourth entry, the coefficient is estimated; if a number is provided instead, the coefficient is fixed to that value. Any path or correlation not referred to is fixed at zero, and the correlation of a variable with itself is a variance. For more complicated diagrams it may be necessary to explicitly include a "measurement model" and latent factors—see Hayduk (1987), Loehlin (1987, and Breckler (1990). Although not obvious in this example, implementing path analysis under the assumption of no measurement error actually involves postulating one factor for each measured variable, with factor loadings from each factor to each measured variable of 1.0 (see Hayduk, 1987; Jöreskog and Sörbom, 1988).

```
PROC CALIS CORR                    /invoke procedure CALIS, request total effects/
     DATA=PATHS TOTEFF;
  RAM                              /use "ram" method for specification/
  1 4 1 BFLRA, / FLR -> A /        /specify direct effects, usually
  1 4 2 BNPRA, / NPR -> A /            with a descriptive comment/
  1 4 3 BNNDA, / NND -> A /
  1 5 1 BFLRP, / FLR -> P /
  1 5 2 BNPRP, / NPR -> P /
  1 5 3 BNNDP, / NND -> P /
  1 5 4 BAP, / A -> P /
  1 6 4 BAF, / A -> F /
  1 6 5 BPF, / P -> F /
  1 6 3 BNNDF, / NND -> F /
  2 2 1 CORFLNL,                   /specify correlations/
  2 3 1 CORNDFL,
  2 3 2 CORNDNP,
  2 1 1 VARFLR,                    /specify variances/
  2 2 2 VARNPR,
  2 3 3 VARNND,
  2 4 4 VARAPH,
  2 5 5 VARVPF,
  2 6 6 VARPROP;
```

11

Population Sampling and Bootstrapping in Complex Designs: Demographic Analysis

Mark A. McPeek and Susan Kalisz

11.1 Ecological Issues

For ecologists, the methods of demography have diverse applications: analyzing population stability and structure (Lande, 1987), estimating extinction probabilities (Epling et al., 1960; Goodman, 1987; Grant and Grant, 1992), predicting life history evolution (Lande 1982), predicting outbreaks in pest species (Levins, 1969), examining the dynamics of colonizing or invading species (Lewontin, 1965), and projecting population viability of threatened and endangered species (Crouse et al., 1987; Lande, 1988). The basis of all these issues is an interest in understanding the mechanisms affecting growth, survival, and reproductive schedules in natural populations of plants and animals and how these schedules influence population growth and persistence.

Demographic analyses in general have three goals. The first is to estimate a population's present growth rate and to project future population dynamics. This entails estimating the population's growth rate from measured age and/or stage-specific vital rates (e.g., age-specific survival probabilities and reproductive rates) to address the question: "Would population size increase, remain unchanged, or decrease if conditions were to remain as they were when the data were collected?"

The second goal is to understand and project both population structure and the contributions of different types of individuals to future reproduction. This involves calculating the stable age or stage distribution and reproductive values associated with different classes to address the question: "What would the age or stage structure of the population be and what would the relative contributions of each age or stage class to future reproduction be if conditions did not change?"

Finally, demographic analyses are used to understand the contributions of each age or stage class to determining overall population growth rate and indirectly to determining fitness (Lande, 1982). They are termed sensitivity analysis and elasticity analysis (Caswell et al., 1984; de Kroon et al., 1986; Crouse et al., 1987). Here, the question asked is: "What change in overall population growth rate would occur if the value of a particular survival probability or reproductive rate is changed?"

Two general problems arise in demographic studies. The first is the practical problem of how to sample the population of interest to obtain an accurate representation of demographic conditions. Many features of natural populations and the environment (e.g., genetic structure, spatial and temporal environmental variation) can cause biases in demographic data if sampling programs are not carefully designed. The second problem is how to place some measure of confidence on demographic parameters that are calculated for a population. The object of demographic analysis is to obtain a single number that describes some feature of the population, for example, overall population growth rate. However, an estimate of the confidence one can place in the accuracy of the number is also necessary, especially when comparisons of demographic parameters are made among populations or among different times for the same population.

In this chapter, we discuss how genetic structure and spatial and temporal environmental variation affect estimates of demographic parameters, and how sampling regimes can be designed to reduce biases in demographic analyses because of these features of nature. We also discuss the application of bootstrapping techniques to generate confidence intervals for demographic parameters when complex sampling designs are employed.

11.2 Calculating Demographic Parameters

Almost all populations of plants and animals have structure. The structure of a population can be formalized by categorizing individuals according to their age, size, sex, stage, or other phenotypic trait. The biological techniques for categorizing population structure should be guided by clear age, developmental, morphological, or allocational changes. Often the best depiction of the life cycle of an organism combines two or more of these categories: stage and age (Kalisz and McPeek, 1992), stage and size (Horvitz and Schemske, 1986; van Groenendael and Slim, 1988), and size and sex (Meagher and Antonovics, 1982; Bierzychudek, 1982). Some life stage classes may be defined before data are collected, but reclassification or fine-tuning of the classification scheme may also occur after data has been collected. Statistical techniques for selecting categories using regression and contingency tables with G-tests are given by Caswell (1989). Moloney (1986) presents a generalized method for determining size classes, based on Vandermeer's (1978) earlier algorithm. This method gives a stepwise procedure for optimizing category size by balancing the opposing forces of sampling error (large when category sizes are too narrow) and distribution error (large when categories are too broad). Therefore, the life cycle encapsulated in a demographic model depends on the data collected and subsequent choice of categories and classes within categories.

The sampling regime and the number of individuals included in the field sampling design of a natural population will be a function of the questions being asked and of the life history of the species. In short-lived organisms, the entire life span of each individual can often be easily followed. This type of sampling is called "vertical" or "longitudinal." In organisms with long life spans, with cryptic life stages (e.g., seeds in a seed bank), with life stages having long-distance dispersal, or with rare life stages, it is often impractical to estimate survival and fertility over an individual's lifetime. Therefore, different vital rates will be estimated from data collected on different sets of individuals. This type of sampling is called "horizontal" or "cross-sectional." Censuses of the population to obtain the raw demographic data of vital rates should be taken at regular intervals that correspond to major life history events (e.g., reproduction in synchronously breeding organisms).

Given that the population structure is determined and survival and fertility data are collected, the model life cycle of the population can be depicted with a life cycle graph (Fig. 11.1). Nodes in the graph represent classes that define the structure of the population, while arrows between nodes represent the possible transitions that can occur between any two classes from one census to the next. Each arrow has an associated transition probability or fertility coefficient. These probabilities and coefficients define the number of individuals that will be present in the class to which the arrow points in one census for each individual in the class from which the arrow points in the previous census. In many studies, population censuses are taken each year, because major life history events occur annually.

To calculate most of the interesting demographic parameters of a population, the methods of linear algebra are used to solve the equation $\mathbf{n}(t+1) = \mathbf{A}\mathbf{n}(t)$. [For consistency, we will use Caswell's (1989) notation throughout this chapter.] In this equation, $\mathbf{n}(t+1)$ and $\mathbf{n}(t)$ are vectors of age and/or stage classes (hereafter referred to generically as stage classes) in the population at times $t+1$ and t, respectively. Each element of $\mathbf{n}(t+1)$ and $\mathbf{n}(t)$ is the number of individuals in an age or stage class at that time. $\mathbf{A}$ is the population projection matrix; the elements of $\mathbf{A}$ (a_{ij}) are the transition probabilities or fertility coefficients that give the number of stage i individuals produced (through survival or birth) per individual in stage j over one time interval. The complete life cycle graph is a graphic representation of the elements which comprise $\mathbf{A}$ (Fig. 11.1). $\mathbf{A}$ is called a Leslie matrix (Leslie, 1945, 1948) when only age classes are involved and a Lefkovitch matrix (Lefkovitch, 1965) when stage classes are used. It is also possible to construct models where there are age classes within stages as we discuss below (e.g., Kalisz, 1991; Kalisz and McPeek, 1992), stage classes within ages (Law, 1983), and even more elaborate models [e.g., two-sex models, multiple-population models with dispersal (Caswell, 1989)].

Iterative postmultiplication of **A** by the population vector (**n**) projects the future state of the population. If the initial number of individuals in each stage class is given by **n**(0), the number of individuals in the population at the next census [**n**(1)] will be

$$\mathbf{n}(1) = \mathbf{An}(0) \tag{11.1}$$

If we assume that the elements of the projection matrix do not change, we can project the number of individuals at the second census to be

$$\mathbf{n}(2) = \mathbf{An}(1) = \mathbf{AAn}(0) = \mathbf{A}^2\mathbf{n}(0)$$

Similarly, we can project the population an arbitrary number of time units, t, into the future as

$$\mathbf{n}(t) = \mathbf{A}^t\mathbf{n}(0)$$

Using the methods of linear algebra, this equation can be rewritten as

$$\mathbf{n}(t) = c_1\lambda_1^t\mathbf{w}_1 + c_2\lambda_2^t\mathbf{w}_2 + c_3\lambda_3^t\mathbf{w}_3 + \ldots + c_k\lambda_k^t\mathbf{w}_k \ , \tag{11.2}$$

where the c_is are a set of coefficients determined by **n**(0) and **A** (assuming the number of stage classes is k, $i=1,2,3,\ldots,k$), the λ_is are the eigenvalues of **A**, and the $\mathbf{w}_i$s are the right eigenvectors of **A**, which correspond to each λ_i (Caswell, 1989). [Consult a text on matrix algebra for a more in depth discussion of eigenvalues and eigenvectors (see also Section 5.3.1).] As a convention, the λ_is are numbered in decreasing order of magnitude, with λ_1 being the largest in magnitude and λ_k the smallest. After a sufficiently large number of projections the dynamics of this equation will be dominated by the term having the λ_i of largest magnitude (λ_1), regardless of **n**(0). (λ_1 is often called the "dominant" eigenvalue of **A**.) For most demographic projection matrices, $\lambda_1 > 0$ (Caswell, 1989). When this term comes to dominate, the geometric population growth rate is λ_1. If $\lambda_1 > 1$, the term λ_1^t grows exponentially faster than any other, and the population is projected to increase in size at a geometric rate of λ_1. If $0 < \lambda_1 < 1$, this term decreases exponentially, but the exponential decay of this term is slower than the other terms; in this case the population is projected to decline at a geometric rate of λ_1. If $\lambda_1 = 1$, population size is projected to remain constant.

A number of other demographic parameters can also be derived from the projection matrix, assuming that the term involving λ_1 dominates population dynamics. The fraction of individuals in each stage class will be proportional to the elements of $\mathbf{w}_1$ (Caswell, 1989); $\mathbf{w}_1$ is called the stable stage distribution. Because the proportional representation in each stage class is often of interest, the elements of $\mathbf{w}_1$ are often rescaled so that the elements sum to 1 or 100%. The

A. Life Cycle Graph

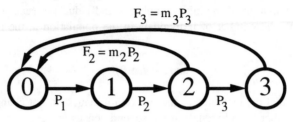

$$F_3 = m_3 P_3$$

$$F_2 = m_2 P_2$$

$$\boxed{0} \xrightarrow{P_1} \boxed{1} \xrightarrow{P_2} \boxed{2} \xrightarrow{P_3} \boxed{3}$$

B. Projection Matrix

Age at t

$$
\text{Age at } t+1 \quad
\begin{array}{c}
0 \\ 1 \\ 2 \\ 3
\end{array}
\begin{bmatrix}
0 & 0 & F_2 & F_3 \\
P_1 & 0 & 0 & 0 \\
0 & P_2 & 0 & 0 \\
0 & 0 & P_3 & 0
\end{bmatrix}
$$

with column headers $0 \quad 1 \quad 2 \quad 3$

C. Matrix Equation

$$
\begin{bmatrix}
n_1 \\ n_2 \\ n_3 \\ n_4
\end{bmatrix}_{t+1}
=
\begin{bmatrix}
0 & 0 & F_2 & F_3 \\
P_1 & 0 & 0 & 0 \\
0 & P_2 & 0 & 0 \\
0 & 0 & P_3 & 0
\end{bmatrix}
\begin{bmatrix}
n_1 \\ n_2 \\ n_3 \\ n_4
\end{bmatrix}_{t}
$$

Figure 11.1. Life cycle graph and associated projection matrix for a population of synchronously breeding individuals with four age classes. The life cycle graph is based on censusing the population immediately after reproduction. Age classes are represented

population will grow with rate λ_1 when the population achieves this stable stage distribution. The reproductive values of the stage classes are given by the left eigenvector $\mathbf{v}_1$ corresponding to λ_1; the left eigenvector satisfies the equation $\mathbf{v}_1\mathbf{A} = \mathbf{v}_1\lambda_1$ (Caswell, 1989). The reproductive value of a stage class is a measure of the potential contribution of an individual in that stage class to future population growth. For comparative purposes the elements of $\mathbf{v}_1$ are usually rescaled so that the youngest stage class has a reproductive value of 1 (each element of $\mathbf{v}_1$ is divided by the original value of the element for the youngest stage class). Therefore, λ_1, $\mathbf{w}_1$, and $\mathbf{v}_1$ are primary descriptors of future population dynamics, structure, and reproductive rates, if the matrix elements do not change. (Hereafter, we will refer to these parameters as simply λ, $\mathbf{w}$ and $\mathbf{v}$. Subscripts with $\mathbf{w}$ and $\mathbf{v}$ will hereafter refer to their elements.) Other important demographic parameters are also calculated from the projection matrix.

The importance of each stage class in determining the population growth rate λ is quantified by two other sets of parameters, namely the sensitivities and elasticities of each matrix element. The sensitivity of λ to a change in the a_{ij}th element of $\mathbf{A}$ when all other elements are held constant is given by

$$\partial\lambda/\partial a_{ij} = v_i w_j / <\mathbf{w},\mathbf{v}>$$

where v_i and w_j are respectively, the ith element of the reproductive value vector and the jth element of the stable stage distribution vector, and $<\mathbf{w},\mathbf{v}>$ is the scalar product of the two vectors (Caswell, 1989). One problem with sensitivities is that they measure changes in λ to perturbations of similar magnitude in all elements of $\mathbf{A}$, but transition probabilities and elements involving fertilities have different scales ($0 \leq$ transition probability ≤ 1; $0 \leq$ element involving fertility $\leq \infty$). Elasticities alleviate this problem by quantifying the change in λ to

---

Figure 11.1.—Continued   by the numbers 0-3: 0 for newborns, 1 for 1 year olds, 2 for 2 year olds, and 3 for 3 year olds. The life cycle graph is given in (**A**). Only age 2 and age 3 individuals reproduce. The probability of surviving from age class i-1 to i is represented by P_i. Each fertility coefficient F_i is a composite function that describes the contribution of offspring to the next census for age i individuals, where m_i is the number of offspring produced by an individual of age i and P_i is as above (Caswell, 1989). The population projection matrix (**A**) for this life cycle graph is given in (**B**). The age classes at times t and $t+1$ are identified above and to the left of the matrix, respectively. The elements of the projection matrix give the number of individuals in age class i at time $t+1$ per individual in age class j at time t; for example, for every age 2 individual at time t, there will be P_3 age 3 individuals at time $t+1$. The matrix equation $\mathbf{n}(t+1)=\mathbf{An}(t)$ for this life cycle graph is given explicitly in (**C**). The vector of the number of individuals in each age class at time $t+1$ is given on the left side of the equation, and the projection matrix and vector of age class sizes at time t are given on the right.

proportional changes in the elements of **A** (Caswell et al., 1984; de Kroon et al., 1986); the formula for calculating the elasticity of a_{ij} is

$$e_{ij} = (a_{ij}/\lambda)(\partial\lambda/\partial a_{ij})$$

11.3 Sampling a Population

The utility of all these demographic parameters depends on three assumptions. The first assumption is that the life cycle model on which **A** is based accurately reflects the life cycle of the population of interest. The second assumption is that the data used to parametrize **A** were collected correctly. The third assumption is that the vital rates of the population do not vary, which is a statement that population growth rate can be described by the term in Eq. 11.2 involving λ_1. Caswell (1989, pp. 19–20) provides a lucid discussion of the issues concerning this third assumption. Caswell points out that the utility of these demographic parameters is in the "projection" of population dynamics into the future and not in the "prediction" of future population dynamics (see also de Kroon et al., 1986). Projection of population dynamics addresses the hypothetical question of "how *would* the population behave *if* the present conditions were to be maintained indefinitely[?]" (Caswell, 1989: italics in original). Prediction attempts to ascribe precise dynamics to future population growth. When interpreted as projections these parameters encapsulate some information about a population and its relationship to the present environment, but they do not necessarily forecast or accurately describe short-term or long-term population dynamics (Caswell 1989). Therefore, interpretations of demographic parameters must always be made with caution and with this critical distinction in mind.

11.3.1 Intrinsic Sources of Variation

Vital rates may vary for a number of reasons and each of these may pose problems for accurately estimating the vital rates of a population. One intrinsic source of variation in vital rates is the inherent variation in survival, growth, and reproductive processes, which is usually termed demographic stochasticity (Goodman, 1987; Menges, 1986). Demographic stochasticity is not caused by a stochastic environment. Even when environmental conditions are constant in both time and space, vital rates have associated variances, because of the probabilistic nature of survival and breeding and the lack of complete uniformity in breeding schedules. With demographic stochasticity the realized values of transition probabilities and fertilities in the population will vary around their expected values from census to census, even when the expected vital rates remain constant. Consequently, the realized transition matrices from census to census will also vary.

A primary cause of demographic stochasticity is small population size, because variances associated with transition probabilities and fertilities are presumed to be inversely proportional to population size. In small populations, projection matrix elements are expected to vary randomly, even when expected vital rates remain constant. Therefore, a population of perpetually small size may never achieve a growth rate of λ and a stable stage distribution. However, estimating the above demographic parameters may not be of primary interest in a small population. Instead, a more useful parameter for a small population may be an estimate of the expected time to extinction (Goodman, 1987; Belovsky, 1987; Shaffer, 1987); e.g., vital rates could be used to parametrize Markovian models of population dynamics for simulation analyses of the expected time to extinction.

The second and possibly more significant source of variation in vital rates is temporal variation in ecological conditions that directly alter vital rates. Temporal variation in vital rates within a breeding population is commonly observed (e.g., Bierzychudek, 1982; Woolfenden and Fitzpatrick, 1984; Kalisz, 1991; Kalisz and McPeek, 1992).

Temporal variation in ecological conditions causes the long-run dynamics of population size to deviate from those predicted by projection matrix analyses. Imagine that we are following the dynamics of a population that experiences a random set of vital rates each year, and each year we can measure the resulting projection matrix. Because we know the projection matrix, we can calculate λ each year. In "classical" demography (i.e., population projection, in which the efficacy of demographic parameters is based on the assumption that the projection matrix were to remain constant through time) λ describes the rate of change in population size over one time interval. Intuitively, it is reasonable to expect that the long-term geometric mean of the λs would describe the average rate of change in population size when vital rates vary temporally. However, this is not the case. For a stochastic environment, two estimators of population growth rate can be defined. One is the average of growth rates (a):

$$a = \lim_{t \to \infty} (1/t)\, E[\ln N(t)]$$

where lim $t \to \infty$ means taking the limit as t approaches infinity, and $E[x]$ is the expected value of x; the other is the growth rate of average population size (ln μ):

$$\ln \mu = \lim_{t \to \infty} (1/t)\, \ln E[N(t)]$$

(Cohen, 1979a; Tuljapurkar, 1982a,b; Tuljapurkar and Orzack, 1980). Cohen (1979a,b) and Tuljapurkar (1982a) have shown that $a \leq \ln \mu$ with temporal variation; in words, the average growth rate of the population is always less than or equal to the growth rate of average population size.

The information derived by calculating λs from projection matrices each year may provide a very imprecise picture of population dynamics (remember the distinction between projection and prediction of future population dynamics). The best descriptor of average population growth rate is calculated directly from population sizes [i.e., average $\ln \lambda = (1/t) (\ln N(t) - \ln N(0))$ where $N(t)$ is total population size (Heyde and Cohen, 1985)), and not an average of the eigenvalues from projection matrices (Cohen, 1979a,b). Tuljapurkar (1989) provides an elegant example to illustrate this; in his example, the eigenvalues calculated from the projection matrices each year are always *less than* 1.0, but the population *increases* in size (i.e., $e^a < 1.0 < e^{\ln \mu}$). The theory and statistics of other demographic parameters in a temporally varying environment are also being developed (see review in Tuljapurkar 1989).

Tuljapurkar (1989) points out that although the long-run behavior of population dynamics may eventually come to a dynamic equilibrium, the population size and structure at any time will reflect recent environmental conditions. Practically speaking, this means that demographic data taken from a population will probably reflect the ecological conditions and population structure of the recent past, more than it will reflect long-term averages. Caution must therefore be exercised in extrapolating from data to the long-term dynamics of the population when substantial year to year variation in vital rates exists (the distinction between "projection" and "prediction").

11.3.2 Genetic and Spatial Variation

Vital rates may also vary as the genetic composition of a population changes or as individuals alter their positions in space. The genetic composition of the population affects the overall vital rates measured in the population, because different genotypes may have different stage-specific transition probabilities and fertilities, even under identical ecological conditions [microbes (Dykhuizen, 1990), zooplankton (Tessier and Consolatti, 1991), aphids (Via, 1991)], and for many sexual plants and animals (Wade, 1991). The vital rates measured for the population as a whole are therefore functions of the transition probabilities and fertilities for all genotypes in the population weighted by their proportions in the population. Changes in the relative abundances of genotypes will alter the expected vital rates for the population as a whole. Consequently, processes such as natural selection (sensu Endler 1986), genetic drift, and gene flow can cause variation in population vital rates over time (Antonovics, 1976).

Vital rates may also vary because of spatial variation in ecological conditions. Spatial variation in vital rates within a breeding population will occur when the population occupies a geographic area with gradients or patches of different abiotic and biotic conditions. Spatial variation in vital rates within a population is also commonly seen (Werner and Caswell, 1977; Kalisz, 1991; Kalisz and McPeek, 1992). With spatial variation, the vital rates measured for the population as a whole are also composite functions, but here they are functions of the vital rates from each unique ecological area occupied by the population weighted by the proportion of individuals in each of these areas. Whether vital rates vary naturally from year to year depends on the proportional representation of individuals among the different ecological areas utilized by members of the population over time. Therefore, issues of optimal patch choice (Fretwell and Lucas, 1970; Pulliam, 1988) or passive dispersal (Bull et al., 1987; McPeek and Holt, 1992) may be critical to the demography of the population (Holt 1985, 1987).

11.3.3 Sampling Design Considerations

Poor sampling design can bias estimates of vital rates when there is either genetic diversity or spatial variability. Ideally, all individuals in a population would be included in censuses to estimate vital rates, which would ensure that no bias due to sampling is introduced into the data set. However, this is usually not feasible. If not, care must then be taken in designing a sampling scheme, so that genotypes are sampled in the proportions they are found in the population and the full range of ecological conditions experienced by the population is sampled with equal intensity. If the researcher has no knowledge of the genetic structure of the population, transects or quadrats should be established at random locations throughout the population to ensure that genotypes in the population are sampled equitably. However, if available, knowledge of the population's genetic structure should be used to design a more elaborate sampling regime that accounts for this structure (e.g., establishing quadrats in blocks that are proportional to the scale of genetic variability within the population). Likewise, sampling the full range of ecological conditions experienced by the population may involve locating sampling units randomly throughout the population, if the extent of spatial variation in vital rates is unknown. However, knowledge of the spatial variation of vital rates should also be used to design sampling programs if available (e.g., establishing quadrats in blocks that are proportional to the scale of spatial variability in the population and their utilization by members of the population) (Chapters 3, 14, and 15). If genotypes are not sampled in proportion to their representation in the population or all areas used by the population are not sampled in proportion to their use by members of the population, estimates of vital rates will be biased toward those of the overrepresented genotypes and/or areas with overrepresented sampling intensities.

Few hard and fast rules exist for the blocking structure or sample sizes needed to account for genetic or spatial structure in sampling programs. If used, the size of blocks should be comparable to the spatial scale of genetic or environmental variation. However, each system will have unique features that will require researchers to tailor their sampling designs to each situation. See Manly (1990b) and Heywood (1991) for more specific guidelines about sampling genetically and spatially structured populations.

11.4 Data Analysis and Statistical Inference

Given that a population has been sampled, the next step is to estimate demographic parameters and their confidence intervals. For simple sampling designs in which all data on survival, growth, and fertility are taken from the same set of individuals, applications of resampling techniques such as bootstrapping and jackknifing are straightforward (see Caswell 1989). We extend that work by considering a more complex case in which all vital rates are not estimated from the same set of individuals. [This analysis was originally suggested to us by Hal Caswell (personal communication).] As an example of how to apply bootstrapping techniques to more complex sampling designs, we will discuss some of our work using blue-eyed Mary, *Collinsia verna* (Kalisz, 1991; Kalisz and McPeek, 1992). *Collinsia verna* is a winter annual plant that grows in the mesic deciduous forests of the eastern United States. Seeds are produced in the spring and germinate in the autumn. Germinated seeds overwinter as small plants, and the following spring plants grow rapidly, flower, produce seeds, and die. Seeds that do not germinate may enter a dormant seed bank.

11.4.1 Sampling Scheme

Because of the presence of a seed bank, *C. verna* has a two-stage, age-structured life cycle in the population we studied (Fig. 11.2). Individuals are "born" when they are produced as seeds on adult plants, not when they germinate. We define an individual in the A stage as an adult plant, and an individual in the S stage as a seed in the seed bank. Numerical subscripts represent the age of individuals. S_1 individuals are seeds that were produced in the previous spring and have entered the seed bank. A_1 individuals are adult plants that germinated from seeds produced in the previous spring. A_2, A_3, and A_4 adults are derived from seeds produced two, three and four springs ago, respectively (these individuals have spent 1, 2, and 3 years in the seed bank, respectively); S_2 and S_3 seeds are derived from seeds produced two and three or more springs ago, respectively. Movement from one stage class to another along an arrow (Fig. 11.2) takes place over 1 year.

An individual in a seed stage (S) can have one of three fates: die, persist in the seed bank (i.e., grow older but remain a seed), or emerge into the adult stage. Seeds on an adult plant can also have one of three fates within a year: die, emerge into the adult stage (A_1), or enter the seed bank (S_1). After emergence a plant is constrained to reproduce and die by its annual habit. Therefore, the age of an individual is determined by its duration in the seed bank. Seeds of *C. verna* can persist for at least 3 years in the seed bank (Kalisz, 1991). Therefore, in our formulation of the life cycle model, age refers to total time alive and not to time spent within a stage. We formulate the life cycle model in this way, because individuals can spend an indefinite amount of time in the seed stage (i.e., S_3 is

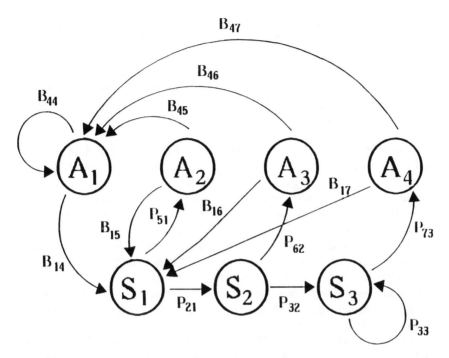

Figure 11.2. Life cycle graph for the two-stage, four age-class life history of *Collinsia verna*. S_1–S_3 represent seeds in the seed bank of ages 1 to 3 years old, respectively. A_1–A_4 represent adult plants of ages 1 to 4 years old, respectively. Arrows connecting two circles (and the italics letters adjacent to them) indicate transition probabilities of individuals at one stage passing into the other stage. The time scale for transition between stages is 1 year.

a carryover loop). Life cycle models for other species with both stage and age classes may be more appropriately formulated where age represents the time spent in particular stage classes and not total lifetime age, especially when definite maximum durations in stages exist.

In any demographic study, it is preferable that all demographic parameters be estimated from the same cohort of individuals. However, following cohorts is often impossible, especially when studying organisms with cryptic life stages, long life spans, stages with long-distance dispersal, or rare life stages. Because seeds are small and cryptic, following the fates of individual seeds and plants that emerged from them was impossible in this study without significantly disturbing the rest of the system and totally altering the demographic rates for the population. Consequently, data were collected separately on the two stages (seeds and adults) in this study.

The data collection scheme was organized around three 75-m transects in the population [see Kalisz (1991) for details]. The locations of these transects were chosen to cover the full expanse of space utilized by the population. Along each transect 25 permanently marked quadrats (40 × 40 cm) were established to quantify vital rates for adult plants. In each of two consecutive years (1982–1983 and 1983–1984) the fates of all seedlings emerging in approximately half of the quadrats on each transect (approximately half in one year and all remaining quadrats in the other year) were followed to estimate their probability of survival to fruit set (designated as s) and the number of seeds produced by surviving plants (designated as f).

An experimental technique was employed to estimate the survival, persistence, and emergence of seeds in the seed bank along the three transects (Kalisz, 1991). Experimental seed banks, constructed of fiberglass window screening and containing autoclaved soil (to kill dormant seeds) and 150 fresh *C. verna* seeds, were established along the transects in the early falls of 1982 and 1983; 12 were established in each year. The number of seeds germinating during the first fall was quantified in all experimental seed banks established that year to estimate the probability that seeds germinate directly without entering the seed bank (designated as $g1$). Half of the experimental seed banks in each transect, chosen at random, were destructively sampled the following spring to determine the viability of the remaining seeds; these data were used to estimate the probability of seeds entering the seed bank (designated as $d1$).

The six remaining experimental seed banks that were established in 1983 were sampled as above in 1984-1985 to estimate the probability of seeds germinating after being in the seed bank for one year ($g2$) and the probability of seeds remaining dormant from age 1 to age 2 ($d2$) (therefore, replicate data are not available for $g2$ and $d2$). Likewise, the six remaining experimental seed banks that were established in 1982 were sampled as above in 1984-1985 to estimate the probability of seeds germinating after being in the seed bank for two years ($g3$) and the probability of seeds remaining dormant from age 2 to age 3 ($d3$) (replicate data is also not available for $g3$ and $d3$).

These estimates of transition probabilities and fertilities were used to construct projection matrices to describe the demography of this population (Table 11.1). Elements of the projection matrix were calculated as follows. P_{21}, P_{32}, and P_{33} are the probabilities of seeds persisting in the seed bank for another year given that the seed survived to that age, where $P_{21}=d2$, $P_{32}=d3$, and $P_{33}=d3$; P_{33} represents a carryover loop whereby seeds can persist for more than 3 years. Therefore, we are assuming that the probability of persistence of seeds in subsequent years is the same as the persistence of seeds from age 2 to 3. P_{51}, P_{62}, and P_{73} are the probabilities of germinating from the seed bank for seeds that have been in the seed bank for 1, 2, and 3 years, respectively, where $P_{51}=g2$, $P_{62}=g3$, and $P_{73}=g3$. The same data were used for both years to derive these matrix elements (P_{21}, P_{32}, P_{33}, P_{51}, P_{62}, and P_{73}). B_{14}, B_{15}, B_{16}, and B_{17} are the joint (probability of a plant surviving from a seedling to fruit set) × (fertility of an adult plant that emerges from seeds 0 to 3 years of age, respectively) × (probability that

Table 11.1. *Projection Matrix for* Collinsia verna *with a 3-Year Seed Bank*[a]

	S_1	S_2	S_3	A_1	A_2	A_3	A_4
S_1	0	0	0	B_{14}	B_{15}	B_{16}	B_{17}
S_2	P_{21}	0	0	0	0	0	0
S_3	0	P_{32}	P_{33}	0	0	0	0
A_1	0	0	0	B_{44}	B_{45}	B_{46}	B_{47}
A_2	P_{51}	0	0	0	0	0	0
A_3	0	P_{62}	0	0	0	0	0
A_4	0	0	P_{73}	0	0	0	0

[a]The numerical subscript on row and column headings indicates the age of the seeds or adult plants in years. Explanations of the matrix elements (transition probabilities and fertilities) are given in the text.

a seed produced by that adult will enter the seed bank). In our analyses we assume that adult plants of different ages all have the same survival probabilities and fertilities, and seeds produced by plants of different ages have the same probability of entering the seed bank in a given year (i.e., $B_{14} = B_{15} = B_{16} = B_{17} = s \times f \times d1$); this assumption is necessary because the ages of adult plants could not be determined in the censuses. Likewise, B_{44}, B_{45}, B_{46}, and B_{47} are the joint (probability of a plant surviving from a seedling to fruit set) $\times$ (fertility of an adult plant that emerges from seeds 0 to 3 years of age, respectively) $\times$ (probability that a seed produced by that adult will germinate directly the next fall). In our analyses we also assume that seeds produced by plants of different ages all have the same probability of germinating in a given year (i.e. $B_{44} = B_{45} = B_{46} = B_{47} = s \times f \times g1$).

11.4.2 Bootstrapping Vital Rates and Demographic Parameters

We applied bootstrapping methods (Chapter 13, see also Efron, 1982; Lenski and Service, 1982; Caswell, 1989) to this data set to estimate means and 95% confidence intervals for the vital rates and demographic parameters of this population. These estimates will be used to compare vital rates and demographic parameters among years (Section 11.4.3). We performed separate analyses for each year. Data for adult survival (s) and fertility (f), and the fraction of age 1 seeds that germinate ($g1$) and remain dormant ($d1$) were available for each year. However, we had only one data set to estimate each of the following parameters: the fraction of age 2 and age 3 seeds germinating ($g2$ and $g3$, respectively) and remaining dormant ($d2$ and $d3$). Therefore, the same data sets were used in analyses of both years for these parameters.

Applying bootstrapping techniques to data taken with such a complicated sampling design requires that the bootstrapped data sets be sampled from the original data set in a fashion identical to that used when the original data set was collected (Caswell, personal communication). For our data this meant that each quadrat on each transect and each experimental seed bank represented a separate set of individuals to be sampled in the bootstrapping design. Data for each year were first organized into separate base data sets, which identified the quadrat or experimental seed bank in which each individual was located, and bootstrapping techniques were applied to each base data set separately.

To construct a bootstrapped data set, individuals are drawn at random with replacement from a given base data set. Specifically, a number of individuals equal to the total number of individuals present in a quadrat (or experimental seed bank) is drawn at random with replacement from the individuals in the base data set for that quadrat (or experimental seed bank), and these sampled individuals constitute a bootstrapped data set for that quadrat (or experimental seed bank). For example, of the 150 seeds initially present in experimental seed bank 1 on transect 3 in year 1 (1982), 77 seeds germinated that fall, 21 were viable the next spring, and 52 died over the winter. A bootstrapped data set for this experimental seed bank consisted of drawing 150 seeds at random with replacement from a set of 77 germinating, 21 dormant, and 52 dead seeds. Because individuals are drawn at random with replacement, some individuals may be drawn multiple times and others may not be drawn. Thus, a single bootstrapped data set might consist of 81 germinating, 20 dormant, and 49 dead seeds; it is even possible (but highly unlikely) that a bootstrapped data set might consist of all dormant seeds or all germinating seeds or all dead seeds. This procedure was performed on each quadrat and experimental seed bank separately.

The bootstrapped data sets for all quadrats and experimental seed banks were then pooled to calculate bootstrap values for dormancy, germination and survival probabilities and fertilities. To continue the above example, seeds in the bootstrapped data sets for all age 1 experimental seed banks on all three transects were pooled to generate a bootstrap value for $g1$ (i.e., the fraction of age 1 germinating seeds in all bootstrap data sets combined) and a bootstrap value for $d1$ (i.e., the fraction of age 1 dormant seeds in all bootstrap data sets combined). This was also done for all age 2 experimental seed banks to generate values for $g2$ and $d2$, all age 3 experimental seed banks to generate values for $g3$ and $d3$, and all quadrats to generate values for s and f. Therefore, after pooling of like bootstrapped data sets, one bootstrap value is calculated for each relevant parameter.

The bootstrapped data set must be drawn from the base data set according to the original sampling design. Again consider constructing the bootstrap sample of seeds from the experimental seed banks to generate bootstrap estimates of $g1$ and $d1$. In each year, 12 experimental seed banks, each originally containing 150 seeds, were sampled to generate the original data to estimate these two parameters. It is incorrect to pool the 1800 individuals in the 12 experimental seed banks first and then draw individuals with replacement. The correct procedure is to first draw a bootstrapped sample of 150 individuals from each experimental seed bank, and then pool the bootstrapped samples to calculate bootstrap values of $g1$ and *d1*. Individuals must be sampled from the base data set exactly as they were sampled in the field.

The bootstrap values of these parameters are then used to construct a bootstrap projection matrix according to the formulas provided above (Table 11.1). This is done by simply inserting the germination, dormancy, survival, and fertility values calculated for the bootstrapped data set into the appropriate elements in the projection matrix. A set of bootstrap values for the demographic parameters (λ, stable stage distribution, reproductive values, sensitivities, elasticities) is then derived by applying the methods of linear algebra discussed in Section 11.2 to this projection matrix. Therefore, for each group of bootstrapped data sets that are drawn from the population, values are calculated for *d1*, *d2*, *d3*, *g1*, *g2*, *g3*, *s*, and *f*, a projection matrix is constructed from these eight values, and a set of demographic parameters is calculated from this projection matrix. All of these derived parameters constitute one set of bootstrap values.

To generate bootstrapped estimates and 95% confidence intervals of each parameter, this whole procedure was repeated 1000 times (1000 is the bootstrap sample size for our study). Bootstrap sample sizes of 500–1000 generally produce accurate results (see Efron, 1982; Caswell, 1989 for more detailed discussions of issues relating to bootstrap sample sizes). In our analysis, we therefore had 1000 bootstrap values each for *d1*, *d2*, *d3*, *g1*, *g2*, *g3*, *s*, and *f*, the projection matrix, and each demographic parameter.

Bootstrapped parameter estimates must be corrected for bias that is introduced by the bootstrapping method. The bias of a bootstrap parameter estimate is calculated by subtracting the value of the parameter calculated for the base population ($\text{PARAMETER}_{\text{base}}$) from the average of the values of the parameter for the 1000 bootstrap values ($\text{PARAMETER}_{\text{boot}}$) (as in Eq. 13.5):

$$\text{BIAS} = \text{PARAMETER}_{\text{boot}} - \text{PARAMETER}_{\text{base}}$$

The bootstrap estimate of the parameter is then calculated by subtracting the bias of the estimate (BIAS) from the value calculated for the base population:

Bootstrapped Parameter Estimate $= \text{PARAMETER}_{\text{base}} - \text{BIAS}$

$$= \text{PARAMETER}_{\text{base}} - (\text{PARAMETER}_{\text{boot}} - \text{PARAMETER}_{\text{base}})$$
$$= (2\,\text{PARAMETER}_{\text{base}}) - \text{PARAMETER}_{\text{boot}})$$

(Caswell 1989). For example, in our study the bias of λ in a year was the average of the 1000 bootstrap values of λ minus λ calculated for the base data set. The bootstrap estimate of λ was then the λ for the base data set minus the bias.

A number of methods exist for calculating confidence intervals of bootstrap estimates (Chapter 13; Efron and Tibshirani, 1986; Caswell, 1989). Here we use the "Percentile Method," a nonparametric method. From the above methods we obtained 1000 bootstrap values of each parameter, which provide an estimate of the distribution of each parameter. To calculate a confidence interval of $(1-\alpha)\times 100\%$ for a given parameter, the 1000 values are ordered from smallest to largest, and the values corresponding to the $\alpha/2$ and $(1-\alpha)/2$ percentiles of the distribution are the lower and upper bounds of the confidence interval, respectively.

To illustrate how incorrect pooling will affect estimates of variances and confidence intervals, consider the following example. Imagine that we had set out only two experimental seed banks (again with 150 seeds each initially) to estimate $d1$ and $g1$. In one experimental seed bank all the individuals germinate, and in the other all individuals remain dormant and enter the seed bank. When bootstrapped data sets are constructed correctly for each, all bootstrapped data sets of the first experimental seed bank have 150 germinated seeds, and all bootstrapped data sets of the second always have 150 dormant seeds. The estimates after pooling are $d1=0.5$ and $g1=0.5$ in every bootstrapped data set because each contains 150 dormant seeds and 150 germinated seeds; there is no variance in the bootstrapped values of $d1$ and $g1$ in this case. However, if the seeds in the experimental seed banks were pooled before constructing the bootstrapped data sets (i.e., the wrong way), the expected values for both $d1$ and $g1$ would still be 0.5, but the values of each would vary among the bootstrapped data sets, which would greatly inflate the estimates of variances and confidence intervals in both. Other examples can be constructed where incorrect pooling will decrease the estimates of variances and confidence intervals. The point here is that incorrect pooling biases estimates of variance and confidence intervals away from their true values.

11.4.3 Comparing Parameters

Most parameters estimated from the population differed significantly between the 2 years (Table 11.2). (Remember that the same data were used to generate bootstrap estimates of $g2$, $g3$, $d2$ and $d3$ for both years.) The fraction of seeds germinating directly into adults ($g1$) was almost 60% larger in year 1 than in year 2, and the fraction of seeds entering the seed bank ($d1$) was almost 30% larger in year 1 (Table 11.2). This means that the fraction of seeds that died in their first year was about 44% higher in year 2 (i.e., 0.45 for year 1 versus 0.65 for year 2). Also, fertilities (f) were more than 300% higher in year 1 than in year 2. In contrast, survival to fruit set was not statistically different in the two years at a 5% type I error rate (i.e., the upper bound of the 95% bootstrapped confidence interval for year 2 and the lower bound of the 95% confidence interval for year 1 were the same). When making such comparisons it may be prudent to adjust the overall type I error rate to account for the multiple comparisons, especially when tests of multiple parameters are not independent as in this case. One method

Table 11.2. Estimated Vital Rates and Bootstrapped Means and Confidence Intervals for Each of the 2 Years of the Study[a]

Vital rate	Population estimate	Bootstrapped estimates	
		Mean	95% confidence interval
Year 1			
f	20.40	20.30	19.40–21.30
s	0.210	0.210	0.200–0.220
$d1$	0.260	0.251	0.245–0.292
$d2$	0.161	0.161	0.149–0.174
$d3$	0.129	0.129	0.111–0.149
$g1$	0.407	0.403	0.398–0.423
$g2$	0.132	0.132	0.116–0.149
$g3$	0.037	0.037	0.025–0.050
Year 2			
f	6.44	6.45	5.88–7.07
s	0.193	0.195	0.181–0.200
$d1$	0.195	0.195	0.178–0.212
$d2$	0.161	0.161	0.148–0.174
$d3$	0.129	0.129	0.109–0.147
$g1$	0.254	0.254	0.243–0.264
$g2$	0.132	0.132	0.114–0.150
$g3$	0.037	0.037	0.025–0.051

[a]The symbols for the vital rates are given in the text. The same data are used in both years for $d2$, $d3$, $g2$ and $g3$.

of adjusting the type I error rate is the sequential Bonferroni correction, which does not require independence of component tests (Holm, 1979; Rice, 1990).

These differences in germination and dormancy probabilities and fertilities are reflected in differences in λ for the 2 years (Kalisz and McPeek, 1992). The bootstrap estimate of λ was 1.80 (95% bootstrapped confidence interval 1.72–1.97) for year 1 and 0.41 (95% bootstrapped confidence interval 0.37–1.44) for year 2 (Kalisz and McPeek, 1992). These estimates suggest that the population was rapidly increasing in year 1 but precipitously declining in year 2. [Bootstrapped means and 95% confidence intervals from this same analysis for the stable stage distribution, reproductive values, and elasticities are presented in Kalisz and McPeek (1992)].

Because of the variability in vital rates, and therefore in **A**, between years, the demographic parameters calculated in this study must be interpreted with caution. The inability to follow seeds in the seed bank and the inability to accurately age adult plants and seeds make it impossible to evaluate whether the population was at or near its predicted stable stage distribution in either year (see Bierzychudek, 1982). Consequently, it is impossible to determine how accurately these parameters describe the demographic conditions of the population during each year (if the population was at its stable stage distribution) or are merely crude estimates (if the population was not near its stable stage distribution). For example, in simulation studies using this data set, we find that it is often possible for λ to be substantially greater than 1.0 but total population size to decrease in a year (M. A. McPeek and S. Kalisz, unpublished data). This occurs when terms in Eq. 11.2 as well as the one involving λ_1 influence population dynamics (i.e., when the population is not at its stable stage distribution). Again, we emphasize that these parameters must be interpreted in terms of population projection (i.e., "what *would* population dynamics be *if* conditions did not change from those when the population was sampled?"), and not as true descriptors of demographic conditions in the population.

11.4.4 Interpreting the Results and Further Studies

The causes of variation in vital rates and projection matrices in this study are not clear. Large numbers of plants and seeds were used to estimate all vital rates (e.g., >5800 adult plants were included in each year's samples, and 1800 seeds were initially present in experimental seed banks each year). Hence bootstrapped error estimates are relatively small and imprecise estimates of parameters are not the cause. Also, it is unlikely that demographic stochasticity generated the observed variation, because the population was large [$>10^6$ individuals in the total population (S. Kalisz, personal observation)].

Variability in genetic composition and temporal variability in ecological conditions both could have contributed to generating the observed variation, but it is impossible to determine from this study alone the contribution of each. The population's genetic composition may have changed over the course of the study. In addition, ecological conditions may have differed among the three transects and between years (Kalisz and McPeek, 1992). For example, two of the three transects flooded in both years of the study, and the amount of canopy cover varied among the transects.

To estimate the contribution of these potential sources of variation, carefully designed experiments and observations of genetic composition would need to have been performed along with the demographic study of the population. For example, electrophoretic studies of allozyme markers done in both years would have provided information on whether significant changes in genotype frequencies occurred from year 1 to year 2. Also, to determine the importance of temporal variation in vital rates due to variation in ecological conditions, clones of genotypes from each transect could have been transplanted into other experimental quadrats on each transect in both years, and experimental seed banks with seeds of known genotypes could have been established along each transect in both years. If vital rates on these experimental plots were similar to those in the natural population, variation in vital rates obtained in this experimental population could be partitioned into genetic and environmental components, and one could potentially assign relative contributions to each of these possible causal factors. Such experiments have rarely been performed (Via, 1991; Wade, 1991), but their results would be invaluable for understanding the driving forces of population dynamics in stage and age structured populations. However, care must be exercised when performing such ancillary studies not to alter the ecological conditions and therefore birth and death rates in the population under study.

This study also illustrates how important the design used to sample the population can be to reducing the bias in estimating vital rates (here we mean bias due to poor sampling design and not bias in bootstrapped estimates). In addition to differences in vital rates between years, vital rates and demographic parameters differed substantially between transects in each year (Kalisz and McPeek, 1992). If only one of the transects had been sampled instead of all three, quite different estimates of the entire population's demographic parameters would have been obtained. Obviously, the best sampling design is to quantify the stage transitions and reproductive schedules of all individuals in the population, but this was not feasible. Great care and forethought must be given to the sampling regime to adequately cover the range of genetic and environmental variability experienced by the population.

It is important to point out that bootstrapping does not remove any bias due to poor sampling design. The bias that is corrected in bootstrapping is a bias generated by the statistical technique and is not related to bias associated with sampling the population. Bootstrapped estimates are only as good as the data used to generate them.

11.5 Related Issues and Techniques

A voluminous literature exists on the methods of matrix algebra as applied to the demography of natural populations. Recent books and reviews of this topic include Caswell (1986, 1989) for the most broad treatment of the subject, Charlesworth (1980) for a treatment of age structure, Manly (1990b) for analysis of stage-structured populations, and Tuljapurkar (1989) for analyses of the consequences of randomly varying environments.

Calculating demographic parameters requires extracting the eigenvalues and associated eigenvectors from matrices of general form (i.e., nonsymmetrical matrices). A number of high-level programming languages are now available for performing such matrix algebra calculations, e.g., GAUSS [Aptech Systems, 23804 SE Kent-Kangley Road, Maple Valley, WA, (206) 432-7855] and MATLAB [MATHWORKS Inc., Cochitulate Place, 24 Prime Park Way, Natick, MA, (508) 653-1415] for personal computers and workstations, and EISPACK (Argonne National Laboratories, Argonne, IL) for mainframe computers. We used GAUSS386i in the analyses presented here. If these languages are not available, the Power Method of extracting eigenvalues and eigenvectors can be used (see Caswell, 1989), which can be coded in BASIC, FORTRAN, PASCAL, or C. The bootstrapping algorithm (and algorithms for other resampling procedures such as jackknifing) is easily coded in all of these programming languages (see also Appendix 13.1). However, researchers will have to develop their own programs for each analysis, because the particular sampling design employed in the study must be incorporated into the algorithm.

With so much potential for variation in the vital rates of a population (especially temporal variation), the application of these mathematical and numerical techniques to demographic data may appear futile. However, with careful planning and the application of data from other types of studies (e.g., electrophoretic studies of genetic variability, experimental studies of the causes of variation, "common garden" experiments of demographic responses) conducted alongside demographic studies in natural populations, many problems can be circumvented, and more information can be brought to bear for understanding population dynamics and structure. Therefore, in closing we echo Caswell's (1989) statement that "It is hard to justify this approach on formal statistical grounds, but harder to justify the loss of potentially valuable insight by not at least trying."

Acknowledgments

We thank J. van Groenendael, H. de Kroon, and E. Menges for many helpful discussions concerning the application of demographic methods to natural populations. James McGraw, Jessica Gurevitch, and Sam Scheiner provided valuable comments on various drafts of the manuscript that greatly clarified our thinking and presentation. This work was supported by NSF Grant BSR-90-00063 to S. Kalisz. This is contribution No. 728 of the W. K. Kellogg Biological Station.

12

Failure-Time Analysis: Emergence, Flowering, Survivorship, and Other Waiting Times

Gordon A. Fox

12.1 Ecological Issues

Do differences among treatments lead to differences in the time until seeds germinate, foragers leave patches, individuals die, or pollinators leave flowers? Do treatment differences lead to differences in the number of events that have occurred by some time, or to differences in the rates at which these events occur? Questions of this kind are at the heart of many studies of life history evolution, demography, behavioral ecology, and pollination biology, as well as other ecological subdisciplines. In all of these cases researchers are concerned either with the time until some event occurs in individual experimental units, or with the related problem of the rate at which these events occur.

Ecologists are interested in these types of data for several different reasons. One is that time itself is sometimes limiting for organisms because their metabolic clocks are constantly working. For example, animals that do not acquire food quickly enough may become malnourished. Time may also be limiting because of external environmental factors. For example, annual plants that do not flower early enough may be killed by frost or drought without successfully setting seed. Time may also be important because of population-level phenomena. For example, in age-structured populations, individuals that reproduce at a late age will leave fewer descendants on average than those that reproduce at an early age, all else being equal. If resource competition is partly age dependent, on the other hand, early reproducers may sometimes be poor competitors and therefore may actually have fewer descendants. Finally, there are many situations in which ecologists are interested in knowing how many events have occurred by a particular time, such as the number of flowers that have been visited by insects by the end of the day, or the number of animals that survive a particular season.

Many ecologists study failure-time data by visually comparing survivorship curves. While this is often a useful part of any exploratory data analysis (Chapter 2), it requires that one make a subjective decision as to whether two survivorship curves are "really" different from one another. I suspect that ecologists do this so frequently because courses in population ecology usually treat survivorship and life tables as fixed features of populations. This makes it easy to forget that failure-time data must be treated statistically.

This chapter discusses data on the timing of events, and special methods for their analyses. These methods originated in several diverse fields in which similar statistical problems arise—e.g., clinical medicine, human demography, and industrial reliability testing—and consequently are quite well developed. To get a feel for the origins and range of applications for these statistical methods, look at the examples in the first chapters of Kalbfleisch and Prentice (1980) and Lawless (1982). In addition to being well developed statistically, this field of statistics has evolved its own jargon. I introduce this jargon in the following section in the context of a discussion of the peculiar nature of this kind of data.

12.2 Statistical Issues

12.2.1 Nature of Failure-Time Data

Ecologists get data on the timing of events by repeatedly observing uniquely identified individuals. As with repeated-measures analysis of variance (ANOVA; see Chapter 6), this repeated-observations structure leads to special statistical methods. These observations may take place continuously (as would be necessary in studying giving-up times of foragers in patches; Chapter 16) or may be in the form of censuses at intervals (as would be appropriate for studying survival of marked plants). At each observation time, the researcher determines whether the event of interest has occurred for each individual. In simple studies, three general outcomes are possible:

1. The event has not yet occurred, and the individual is still in its original state. In this case the individual is said to have *survived*. Note that "survival" in this sense can refer to remaining alive as well as to remaining in a patch, remaining ungerminated, etc.

2. The event has observably occurred. In this case statisticians speak of *failure* and of the timing of the event as the individual's *failure time*. This term originated in industrial reliability testing. In an ecological setting, a failure refers to the individual having been observed leaving the patch, being observably dead, or having observably begun to flower.

3. Finally, the individual may have been lost from the study without a failure being observed. This kind of data point is called a *right-censored* data point, because the researcher can be certain only that the actual

failure time is greater than the last recorded survival time. For example, a failure time (in this case, a "giving up time") cannot be assigned to a forager that was still in a study patch when a predator was observed to eat it, but the researcher knows that the time of abandoning foraging (i.e., the failure time) would have been at least as great as the time of predation (i.e., the censorship time). Similarly, a flowering time or death time cannot be assigned to a plant that has lost its marker, because the researcher is uncertain as to its fate at any time later than the prior census, although it is clear that the failure time was later than the prior census.

Researchers often discard right-censored data points either because they assume that these data are not useful, or because they believe that methods for handling censored data are unnecessarily complex. Both beliefs are wrong. The former belief leads researchers to discard potentially important data, and can contribute to highly biased results. To see this, consider a simple example from clinical medicine. In a study of the effect of a drug on tumor regrowth, some patients are killed by heart attacks or accidents, with no regrowth at time of death. Discarding these cases would not only be an inefficient waste of data, more importantly, it could bias results because those in whom the treatment has been effective are more likely than others to be killed by causes other than cancer.

It is also possible for data points to be *left censored*, if failures have already observably occurred for some individuals when the study begins (Kalbfleisch and Prentice, 1980; Lawless, 1982). An example would be a study of pupal eclosion times for a Lepidopteran population in which there are open chrysalises present at the beginning of the study. Left censorship can frequently be avoided by careful planning (e.g., by beginning the study earlier). In many controlled experiments left censorship is not even possible because failures cannot occur prior to the beginning of the experiment (e.g., because seeds cannot germinate before the researcher plants them). Consequently, this chapter will not discuss left censorship further. Note that in the examples of both left and right censorship discussed so far, censorship is not a planned part of the experiment.

Experimental designs often include censorship: researchers may often plan to end studies before all individuals have failed. This is obviously necessary when individuals have very long lives. Two types of designs can be distinguished. In type I censoring, one can plan to end a study at a particular time, in which case the number of failures is a random variable. With type II censoring, the study is completed after a particular number of failures, in which case the ending time of the study is a random variable (Lawless, 1982). In ecological studies, type I censoring is more common because studies often end when field seasons, graduate careers, or grants end. In most ecological studies some of the censorship is unplanned and random, e.g., some experimental individuals die or are otherwise lost to the study at random times during the course of the study. Consequently, the discussion in this chapter assumes that censoring is type I and/or random. For discussion on data analysis with type II censoring, see Lawless (1982).

This kind of structure means that there are three important elements that require consideration in event time studies. First, measurements are repeated over time. Second, most studies include censored data points. Finally, even if no data are censored, failure times are usually not normally distributed under any standard transformation.

The nonnormality of failure-time data can be partly a consequence of the necessary experimental structure. Normal distributions are symmetric, with infinitely long tails on both sides. But by beginning a controlled experiment at a particular time, the researcher establishes a sharp line of demarcation: no failure can occur prior to the beginning of a controlled study. In many ecological settings, this cut-off line is actually further along in time. For example, in studies of flowering time that begin when the researcher plants seeds, there is usually some minimum time that must elapse before any plant begins to flower. In this sense failure-time data are often intimately related to the particular time at which the experiment is begun, unlike most other kinds of data. In my own studies of the desert annual plant *Eriogonum abertianum*, time to flower in the greenhouse was very far from a normal distribution; plant size at flowering time, plant fecundity, and related traits, fit normal distributions after being log-transformed (Fox, 1990a).

There are also ecological and biological reasons why failure-time data may not be normally distributed. In studies of the survival of marked plants, for example, there is no reason to expect a normal distribution of time to death. In most populations deaths are likely to be concentrated by the timing of major environmental events such as frosts, droughts, or herbivore migrations. Failures that occur when some biological threshold is reached—e.g., onset of reproduction, or deaths due to senescence—are also likely to lead to data sets that are not normally distributed. Finally, there are theoretical reasons why failure times are usually not normally distributed. These involve the nature of the risk of failure, and are discussed in Appendix 12.2.

12.2.2 How Failure-Time Methods Differ from Other Statistical Methods

Failure-time data are frequently of interest to ecologists, but some experimental designs and statistical analyses are better than others. Using what Muenchow (1986) called the "classical" approach, many ecologists have analyzed failure-time data by counting the number of failures among a fixed number of experimental units over a fixed interval. They then compare groups for the mean number of failures, using any of several statistical approaches. In a widely used alternative approach, ecologists conduct experiments designed to measure the failure times of individuals, but then analyze these data with ANOVA.

There are special methods that have been designed to deal specifically with failure-time data. These failure-time methods differ from classical and ANOVA approaches in both experimental design and statistical analysis. Using a failure-time approach, an ecologist measures the time to failure of each uncensored individual. Statistical tests designed for this problem are then used to compare groups over the entire distribution of failure times.

There are several reasons to prefer the failure-time approach. First, the classical approach can compare groups only on a single time scale. This is because the classical approach compares groups for the cumulative number of failures that have occurred by a single point in time—the time for the experiments' end set by the scientist. By contrast, the failure-time approach compares groups' survivorship curves over all failure times. Second, ANOVA assumes that groups' failure times are normally distributed with equal variance, and compares their means: the shapes of the failure-time distributions are assumed to be identical. By contrast, failure-time approaches allow one to compare the shapes of these distributions. Third, neither the classical approach nor the ANOVA approach can account for censored data. Fourth, under the classical approach, it is not clear what to do with multiple failures of an individual—for example, multiple insect visits to a single flower—because these may not be independent. Finally, ANOVA and the *t*-test can be seriously biased methods of analyzing failure-time data, because these tests require approximate normality of the data. This bias can lead to spurious results. Worse yet, the direction of the bias depends on the shapes of the distributions of failure times and on the pattern of data censorship. Consequently one cannot say in general whether ANOVA would be biased for or against a particular hypothesis.

12.3 Example: Time to Emergence and Flowering

12.3.1 Study Species

As part of a larger study on the ecology and evolution of flowering time and other life history traits in wild radish, *Raphanus sativus* (Brassicaceae), I studied the time to emergence and flowering in three populations of wild radish and in two domestic radish cultivars. Wild radish is a common annual (sometimes biennial) weed in many parts of the northern hemisphere. In North America, *R. sativus* is especially common in coastal areas of California and in the Sacramento Valley. There are two principal sources of these old world natives: inadvertent introductions by European settlers and escapees from cultivated radish varieties. The latter are of the same species as wild radish. There are also hybrids with a closely related introduced weed, *R. raphanistrum*.

In the present chapter I will discuss two relatively narrow questions from this research: (1) what are the distributions of emergence and flowering times in these populations, and (2) do these differ among populations? I consider these questions by using two different statistical methods: life table analysis and accelerated failure-time models. These methods are complementary, each having distinct advantages: life tables and associated statistical tests are quite useful for exploratory data analysis, but they lack the statistical power of accelerated failure-time models. On the other hand, accelerated failure-time models require more assumptions than life-table methods. I illustrate the use of a third method—proportional hazards models—with a data set from Muenchow (1986). The power of this method is similar to that of accelerated failure-time models, but it depends on somewhat different assumptions.

12.3.2 Experimental Methods

Seeds were collected from randomly selected individuals in populations around Santa Barbara, California. The Campus Point site is on a bluff overlooking the Pacific Ocean, and the Coal Oil Point site is several hundred meters from the ocean. Both sites are consequently subjected to cool fog during much of the year. The third site, Storke Road, is several kilometers inland and is therefore drier and warmer. The domestic cultivars used were Black Spanish, a late-flowering crop cultivar, and Rapid Cycling, a product of artificial selection for short generation time (Crucifer Genetics Cooperative, Madison, WI.).

In designing failure-time experiments—either in field of laboratory settings—it is important to consider the shapes of failure-time distributions, and the patterns and magnitudes of data censorship. For example, if there is much data censorship, large samples—often involving hundreds of individuals—are necessary to compare treatments. The reason for this is simple: to compare groups, one needs at least a minimum number of actual event times, as opposed to censorship times. If all data points are censored, no between-group comparison is possible. Large samples are also necessary if the tails of the distribution are of particular ecological interest (Fox, 1990b).

As the present study was a pilot experiment, I had no a priori estimate of either the number of censored data points to expect or the shapes of the failure-time distributions. Consequently I used relatively large samples, planting 180 seeds from each of the two domestic cultivars, 140 from both the Storke Road and Campus Point sites, and 130 from the Coal Oil Point site, or a total of 770 seeds.

One seed was planted in each 10-cm pot. Positions on the greenhouse benches were randomized in advance. Because the hypotheses of interest concern differences among populations in the timing of emergence and anthesis, seeds were considered as the experimental units, and individual seeds from the same population as replicates.

Seeds were planted in late October. Plants experienced natural photoperiod in Tucson AZ, in a greenhouse with untinted windows, throughout the experiment. A sterile 1:1 mix of peat and vermiculite was used as a growing medium. Watering before emergence was always frequent enough to keep the soil surface wet. After emergence plants were fed weekly with a solution of 20:20:20 N–P–K.

Temperatures were allowed to vary between approximately 15° and 27°C. Time constraints dictated that the experiment be ended by mid-March 1992.

For each individual seed, I recorded an emergence time (TEMERG) and an anthesis time (TANTH). Each TEMERG and TANTH could be either an actual emergence or anthesis time, respectively, or a censorship time. I also recorded variables EMRGCENS and ANTHCENS, assigning values of 0 for uncensored data points, and 1 for censored data points. Each data record also included a column for population. In general, one could include variables for any other variable of interest, such as treatment factors.

The next section discusses general methods for analysis of failure-time data. These methods are then illustrated by applying them to the radish data.

12.4 Statistical Methods

There is no doubt that the anthesis data require some special handling, because many of these data points are censored. But in the emergence data, the small number of censored data points (Fig. 12.1) might lead one to believe that ANOVA would be a useful way to compare populations. However, there are biological reasons to expect these data to depart from a normal distribution. The germination process began sometime after water was first applied. Since there is no evidence for seed dormancy in this species, one might expect emergence events to be clustered at some point after the start of the experiment, and to trail off after that point. This pattern is suggested by the data in Fig. 12.1, and normal scores plots of the data showed strong departures from normality. No standard transformation succeeded in normalizing these data. Thus even though there is very little censorship in the emergence data (Fig. 12.1), comparison of means using ANOVA would be inappropriate.

Fortunately, there are several well-developed methods for analyzing failure-time data that can handle censored data properly and that do not require normally distributed data. These include life table analysis and two different types of regression models, accelerated failure-time models and proportional hazards models. In the following subsections I describe each of these methods and then apply them in turn.

12.4.1 Life Table Methods: Description

Life tables are a convenient starting point for understanding failure-time statistics. Because life tables preceded the regression models historically, they provide a basis for many of the ideas used in the regression models. Moreover, life tables are a useful way to begin exploratory data analysis.

Formulas for cohort life table estimates are given in Table 12.1. From measurement of the failure rate, it is simple to estimate the proportion of those failing in each interval. There are four other statistics that can be derived from this informa-

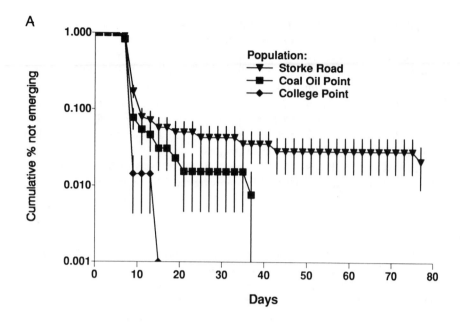

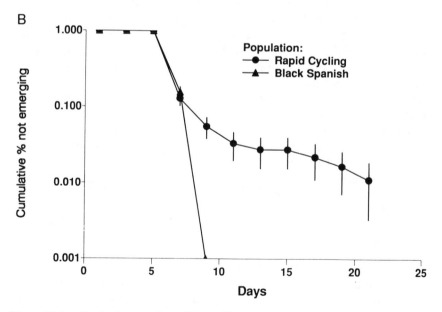

Figure 12.1. Survival curves for radish seedling emergence. Bars are for standard error of the lifetable estimate. Each curve ends when all seeds had emerged (curves intersecting *x*-axis) or when last seedling emerged (all other curves). (**A**) Santa Barbara populations; (**B**) crop cultivars.

Table 12.1.   Definitions of Life Table Statistics

t_i	=	time at beginning of ith interval
b_i	=	width of the ith interval, $t_{i+1} - t_i$
w_i	=	number lost to follow-up
d_i	=	number dying in interval
n_i	=	effective population in interval, $n_{i-1} - w_{i-1} - d_{i-1} - (w_i/2)$
q_i	=	estimated conditional mortality rate, $\dfrac{d_i}{n_i}$
p_i	=	estimated conditional proportion surviving, $1 - q_i$
S_i	=	estimated cumulative survival function $\displaystyle\prod_{j=0}^{i} p_j$
P_i	=	estimated probability density function (unconditional mortality rate), the probability of dying in the ith interval per unit width ($S_i q_i/b_i$)
h_i	=	estimated hazard function (conditional mortality rate, force of mortality, log killing power), the number of deaths per unit time in the ith interval divided by the number of survivors in the interval, $\dfrac{2q_i}{b_i(1 + p_i)}$
e_i	=	estimated median life expectancy, $(t_j - t_i) + \dfrac{b_j[S_j - (S_i/2)]}{S_j - S_{j+1}}$, where the subscript j refers to the interval in which $S_i/2$ occurs.

tion. The estimated cumulative survival function, S_i, is the fraction of the cohort that has not yet failed. The estimated probability density function, P_i, gives the probability that an individual alive at time 0 will fail in the ith interval. Hence P_i is also called the unconditional mortality rate. The estimated hazard function, h_i, is sometimes called the conditional mortality rate—it estimates the chance of failure in the ith interval given that an individual has survived to the beginning of that interval. Finally, the estimated median life expectancy, e_i, gives the median time to failure of an individual that has survived to the beginning of the ith interval.

The ecological significance of the cumulative survival function S_i is obvious. Insight into the probability density and hazard functions can be gained by realizing that if measurements are continuous or nearly so, $P_i = -d(1-S_i)/dt$, and $h_i = -d[\ln(S_i)]/dt$. In other words, the probability density function reflects the rate at which failures accumulate, while the hazard function reflects the per capita risk of failure among those remaining.

Median, rather than mean, life expectancies are calculated because the median is often a more useful and robust estimate of central tendency when the data are not symmetrically distributed. Life tables allow for the use of censored data by treating these data points like any other until the censorship interval, and then discounting them in the censorship interval.

Formulas for calculation of standard errors of life table estimates are shown in Table 12.2. As life table estimates are population-level statistics, standard errors and sample sizes should always be reported so that readers can judge for themselves how much confidence to place in the estimates. Approximate variances for the life table estimates are given by the terms under the square-root signs in Table 12.2.

Life table analyses are always informative in examining failure-time data: their descriptive nature makes them easy to interpret. Consequently I recommend them as a first step in the analysis of most kinds of failure-time data. Hypotheses can also be tested with life table estimates.

There are several ways to compare populations or treatments statistically. Standard errors of the life table estimates can be used for pairwise comparisons. If data are uncensored, standard nonparametric tests such as G-tests or χ^2 tests can be used to test for independence of groups. With censored data, Wilcoxon or log-rank tests (Appendix 12.1) can be used on the failure-time data to test for heterogeneity among groups. Both of these tests compare observed with expected numbers of deaths in each interval.

The log-rank test is more powerful than the Wilcoxon when the hazard functions of the different samples are proportional to one another, and when there is no censoring or random censoring only. The Wilcoxon test is more powerful in many other situations (Lee, 1980). In many cases the tests are likely to give similar results. Lawless (1982) notes that there are circumstances under which neither statistic is likely to be very useful, particularly when distributions are different but the cumulative survivorship or hazard functions cross. Any test is likely to lack power when only a few censuses occur or data are lumped into few intervals, and distributions differ over time but lead to a similar total number of failures (Hutchings et al., 1991). Therefore if the shapes of the curves are likely to be important, censusing must occur frequently enough to detect these differences. Another way of putting this problem is that ties seriously reduce the

Table 12.2. Standard Errors for Life Table Data[a]

$$SE(S_i) \approx \sqrt{S_i^2 \sum_{j=1}^{i-1} \frac{q_j}{n_j p_j}}$$

$$SE(P_i) \approx \sqrt{\frac{(P_i q_i)^2}{b_i} \sum_{j=1}^{i-1} \left(\frac{q_j}{n_j p_j} + \frac{p_i}{n_i q_i} \right)}$$

$$SE(h_i) \approx \sqrt{\frac{h_i^2}{n_i q_i} \left[1 - \left(\frac{h_i b_i}{2} \right)^2 \right]}$$

$$SE(e_i) \approx \sqrt{\frac{S_i^2}{4 n_i (P_j)^2}}$$

[a]From Lee (1980).

power of these tests. Muenchow (1986) observed that in her study of waiting times to insect visits at flowers, she would have had greater statistical power had she recorded times to the nearest second rather than the nearest minute, because she would have had many fewer ties.

These tests, as well as the life table analyses themselves, can be performed with SAS procedure LIFETEST (which uses the χ^2 approximation to the log-rank and Wilcoxon tests described in Appendix 12.1). If these tests reveal heterogeneity and there are more than two groups, one can use the Wilcoxon or log-rank scores to construct Z-statistics for multiple comparisons among groups, as described in Appendix 12.1.4. SAS code for doing this is shown in Appendix 12.3C.

12.4.2 Life Table Methods: Application

Emergence data for the radish study are shown in Fig. 12.1. Cumulative survivorship (i.e., the fraction of plants that had not yet flowered) and its standard error were calculated by SAS procedure LIFETEST, using the code shown in Appendix 12.3. The separation of the survivorship curves implies that the populations may differ in emergence time. This is supported by both the Wilcoxon and log-rank tests (see Appendix 12.1) calculated by procedure LIFETEST: Wilcoxon $\chi^2 = 439.1$, df = 4, P = 0.0001, log-rank $\chi^2 = 335.8$, df = 4, $P = 0.0001$. These tests tell us that the five populations are heterogeneous, but they do not tell us which populations differ from one another. To ask that question, I used the covariance matrix for the Wilcoxon statistic that is automatically generated by procedure LIFETEST, and calculated Z-statistics for each pairwise comparison (see Appendix 12.1 for a description of the Z-statistic, and Appendix 12.3C for SAS code to conduct the multiple comparisons). As noted in Appendix 12.1, these multiple comparisons are not conducted automatically by SAS, and performing them requires a small amount of manual data manipulation. These multiple comparisons, conducted at an overall 0.05 significance level, suggest that all populations differ from one another except possibly Black Spanish and Rapid Cycling (which was a marginally significant comparison) and Coal Oil Point–Campus Point.

Emergence and germination data present a special statistical problem: in general, one does not know whether remaining seeds are capable of germinating, are dormant, have germinated and then died, or were always dead. Inviable seeds should obviously not be considered as part of a study population. In the present case, very few seeds did not emerge, so I analyzed the data by assuming first that these were viable. This means that they were treated as censored data points, with the end of the study as the censorship date. A second analysis assumed that the seeds had always been inviable, so seeds that did not emerge were excluded from the analysis. The results were qualitatively the same; Fig. 12.1 is based on the first analysis. An alternative approach would be to examine the unemergent seeds for viability using a tetrazolium test (Scott and Jones, 1990), and thereby correctly classify each seed. This would be necessary if the two statistical analyses differed qualitatively.

Anthesis data for the radishes are shown in Fig. 12.2. These cumulative survivorship data and their standard errors were also calculated by SAS procedure LIFETEST (Appendix 12.3A). In this case there is considerable censorship, because many plants had not yet flowered by the time the study had to end, and some deaths did occur. Moreover, the censorship affected some populations much more strongly than others. Many more Rapid Cycling plants than others had flowered by the end of the study. Nevertheless, the survivorship curves again imply that the populations differ. This conclusion is supported by the Wilcoxon and log-rank tests calculated by procedure LIFETEST: Wilcoxon $\chi^2 = 582.2$, df $= 4$, $P = 0.0001$, log-rank $\chi^2 = 580.6$, df $= 4$, $P = 0.00001$. Given that these five populations are heterogeneous, which ones are different from one another? To examine this question, I again used the covariance matrix for the Wilcoxon statistic that is automatically generated by procedure LIFETEST, and calculated Z-statistics for each pairwise comparison (see Appendix 12.1 for a description of the statistic, and Appendix 12.3C, for SAS code to conduct the multiple comparisons). These multiple comparisons, conducted at an overall 0.05 significance level, suggest that Rapid Cycling differs from all other populations, and that Black Spanish differs from Coal Oil Point and Storke Road. There were marginally significant differences between Campus Point on the one hand and Coal Oil Point, Storke Road, and Black Spanish on the other.

These life table analyses suggest that *Raphanus* populations differ in time to emergence and flowering. Moreover they have provided a useful description of the populations' responses. There are many cases in which life table analyses and associated significance tests are fully adequate to examine ecological hypotheses (Fox, 1989; Krannitz et al., 1991).

However, two difficulties commonly arise with the use of life tables. First, statistical tests based on life table approaches are sometimes lacking in power. Second, life table methods are difficult to use when there are multiple covariates (Kalbfleisch and Prentice, 1980). For example, emergence time seems likely to affect anthesis time but there is no simple way to account for this effect using life table methods. SAS allows for the calculation of relevant tests (generalizations of the Wilcoxon and log-rank tests), but the algorithms are complex, require relatively large amounts of computation, and the results are not easy to interpret in biological terms [see pp. 1046–1048 in version 6 of the SAS/STAT user's guide (SAS Institute Inc., 1989b)]. The methods discussed in the following section overcome these difficulties, but at the cost of added assumptions.

12.4.3 Regression Models

There are two general types of regression models for censored data: *accelerated failure-time models* and *proportional hazards models*. These models make different assumptions about the affect of covariates such as treatment or initial size.

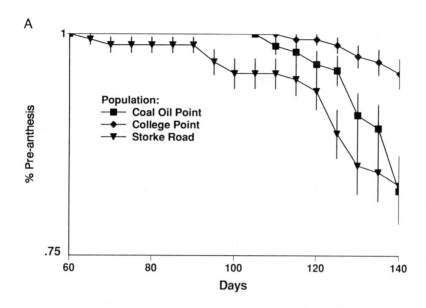

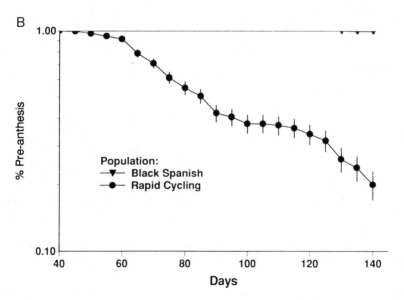

Figure 12.2. Survival curves for radish anthesis. Bars are for standard error of the lifetable estimate. All curves end at censorship date. (**A**) Santa Barbara populations; (**B**) crop cultivars.

Accelerated failure-time models assume that treatments and covariates affect failure time multiplicatively (i.e., the lives of individuals are "accelerated"). An alternative interpretation (Kalbfleisch and Prentice, 1980) is that under accelerated failure-time models, covariates make the clock run faster or slower, so that any period of high hazard will shift in time when the covariates shift. Thus the comparisons of *Raphanus* population time to emergence and anthesis are good candidates for accelerated failure-time models.

Proportional hazards models assume that covariates affect the hazard functions of different groups multiplicatively. Under proportional hazards models, the periods of high hazard stay the same but the chance of an individual falling into one of those periods will vary with the individuals' covariates. A good candidate for a proportional hazards model might be a predation experiment in which the predator density or efficiency changes seasonally: individuals with a "bad" set of covariates do not make the predators arrive sooner, but they are more vulnerable to predation when the predators do arrive. This discussion means that the comparisons of *Raphanus* population time to emergence and anthesis are poor candidates for proportional hazards models.

How does one choose between accelerated failure-time and proportional hazard models? Probably the most useful approach is to consider the ecological hypotheses and ask whether treatments are expected to actually change the timing of periods of high hazard (suggesting the use of accelerated failure-time models) or whether treatments merely change the chance of failure (suggesting the use of proportional hazards models). An additional check on the appropriateness of the proportional hazards model is to plot $\log\{-\log[S(t)]\}$ against time. If the proportional hazards model is appropriate, the curves of different groups should be roughly parallel to one another, for a given level of a covariate (Kalbfleisch and Prentice, 1980). A limitation to this approach is that comparisons must be conducted between treatment groups within levels of a covariate. If all of the predictor variables are covariates, no comparison can be made, regardless of whether the proportional hazards model is appropriate. Consequently it is probably best to rely most heavily on the examination of the ecological hypotheses.

Both kinds of models have been applied in ecological settings; see Muenchow (1986) for proportional hazards models, and Fox (1990a,b) for accelerated failure-time models. These models are closely related; a lucid derivation showing the mathematical relationship between the models is given by Kalbfleisch and Prentice (1980).

Accelerated failure-time models: description. In an accelerated failure-time model, failure times T are modeled as

$$\ln(T) = \mathbf{X}'\boldsymbol{\beta} + \sigma\varepsilon \tag{12.1}$$

where $\mathbf{X}$ is a matrix of covariate values and $\mathbf{X}'$ is its transpose, $\boldsymbol{\beta}$ is a vector of regression parameters, ε is a vector of errors from a specified survival distribution, and σ is a parameter. The covariates in **X** can be dummy variables corresponding to categorical effects, or continuous variables, or both, or their interactions. For example, **X** for the radish emergence data is a 5 × 5 matrix of dummy variables for the five populations of origin, and there are 5 regression coefficients in β. For the flowering data, **X** includes an additional variable, emergence date, and there is an additional regression coefficient for emergence date in β.

Consequently there are two steps to use of these models: (1) choosing a survival distribution and (2) estimating the parameters for the survival distribution and the regression coefficients β. The most commonly used survival distributions are described in Appendix 12.2. All of these are available in SAS procedure LIFEREG. Some additional distributions are discussed by Lawless (1982) and Kalbfleisch and Prentice (1980).

There are three different ways to choose a survival distribution. First, one can sometimes choose a distribution based on a priori ecological or biological grounds. Second, one can take an empirical approach: after the data have been gathered, compare life table estimates of their hazard or survival functions with the hazard or survival functions of various distributions. Several rough empirical methods for doing this are listed in Appendix 12.2. One could also use goodness-of-fit tests, but these require modification if any of the data are censored (Lawless, 1982). A third method for choosing a distribution is suggested by Kalbfleisch and Prentice (1980). These authors propose using a general form (the generalized F distribution) that encompasses all of the distributions in Appendix 12.2 as special cases, to determine numerically which distribution best fits the data.

It may be biologically important in its own right that a data set is, say, gamma- rather than Weibull- distributed. Assessing this importance will usually require some model of the failure process. I know of no such models for any ecological failure process, but Kalbfleisch and Prentice (1980) discuss examples in other fields. On the other hand, knowledge of the survival distribution may sometimes suggest candidates for underlying mechanisms.

Having chosen a survival distribution, the parameters of the distribution and the regression coefficients are determined numerically, using a maximum likelihood approach. Maximum likelihood methods in statistics involve two steps. First, one selects a statistical model, and assumes that it is a correct description of the data. The same assumption is made in regression and ANOVA models. Second, one finds the model parameters that would make it most probable that one would observe the data. This is done by numerically maximizing a "likelihood function," the form of which depends on the model. For more information on construction of likelihood functions for failure-time models, see Kalbfleisch and Prentice (1980), and Lawless (1982). Edwards (1972) provides a comprehensive introduction to likelihood methods, and Shaw (1987) provides an overview of likelihood methods as applied in quantitative genetics.

These methods lead naturally to a set of χ^2 tests for regression coefficients. These test the significance of the covariate's additional contribution to the maximized likelihood function, beyond the contributions of the other covariates already in the model. Thus a nonsignificant value of χ^2 means that the model is adequate without that particular covariate, but does not imply that all covariates with nonsignificant χ^2 values can be deleted. In fact, it is possible for a set of covariates to be jointly significant, even though none of them is individually significant. To examine the possible importance of covariates with nonsignificant χ^2 values, then, one deletes the nonsignificant terms one at a time, reevaluates the model each time, and examines the changes in the other terms.

When there are multiple levels of a categorical covariate like treatment or population, one level is chosen arbitrarily as a reference level with a regression coefficient of 0. The regression coefficients for the other levels therefore provide comparisons with the reference level. Thus a significant χ^2 value for a particular level means only that that level is significantly different from the reference level. It may or may not be significantly different from levels other than the reference level; one needs to perform multiple comparisons to examine this hypothesis. The method for multiple comparisons is analogous to that for life table analyses: one constructs Z-statistics from the estimated regression coefficients and the asymptotic covariance matrix generated by SAS (see Appendix 12.1 for a description of the method, and Appendix 12.3C for SAS code for its implementation).

These χ^2 tests can also be useful when choosing between two distributions, one of which is a special case of the other. The adequacy of the restricted case can be examined by fitting a model to the more general model subject to an appropriate constraint. For example, to choose between the Weibull and exponential distributions, one can fit a Weibull model and subject it to the constraint $p = 1$. The result is, of course (Appendix 12.2), an exponential model, but SAS automatically calculates a test (called a Lagrange multiplier test) that tests the effect of the constraint on maximizing the likelihood function.

Accelerated failure-time models: application. For the radish emergence data, it seemed reasonable to choose either the log-logistic or lognormal distribution: since the species has little or no seed dormancy, one can expect that the rate of germination and subsequent emergence will increase quickly after application of water, reach some maximum rate, and then decline. I chose the log-logistic because, as noted in Appendix 12.2, it has several properties that make it more useful. An a posteriori comparison showed that the log-logistic provided a better fit to the data than the lognormal.

For the radish anthesis data, I thought it likely that the data might best fit a gamma distribution. The gamma hazard monotonically approaches a constant value, and therefore gamma models may often provide good fits for inevitable developmental processes (Appendix 12.2). Such intuition is often wrong, so I tested the gamma model against Weibull and lognormal models. As shown in

Appendix 12.2, the Weibull and lognormal distributions are special cases of the three-parameter gamma. To conduct this comparison, I constrained the gamma shape parameter to yield the other two distributions by assigning a value to "shapel" (0 for the lognormal and 1 for the Weibull distribution; see Appendix 12.2), and then setting the option "noshapel" to instruct procedure LIFEREG not to change this value. SAS code for doing this is shown in Appendix 12.3B. The Lagrange multiplier test for the lognormal distribution was highly significant ($\chi^2 = 22.68$, df $= 1$, $P < 0.0001$). This means that constraining the value of "shapel" to 0 had a significant effect on the likelihood function, and thus the gamma model does provide a significantly better fit than the lognormal.

SAS failed to compute the Lagrange multiplier test for the Weibull distribution: the SAS output merely left blank spaces for the value of the test statistic and its probability. I learned (by calling the SAS Institute when the documentation provided no explanation) that this is a known bug in versions of SAS earlier than 6.07—the only versions available to me. Consequently, I compared these distributions with a likelihood ratio test (see Appendix 12.1). This test is easy to compute by hand, since the test statistic is just twice the logarithm of the ratio of two maximized likelihood functions. Since I already had estimates of the two maximized likelihood functions from my unsuccessful attempt to compute a Lagrange multiplier test, it was simple to calculate

$$R = -2 \ln\left[\frac{L(\boldsymbol{\theta}_{\text{gamma}})}{L(\boldsymbol{\theta}_{\text{Weibull}})}\right] = -2 \ln\left(\frac{-232.48}{-276.97}\right) = 0.35 \qquad (12.2)$$

which, with one degree of freedom, has a large probability of occurring by chance ($P > 0.83$). Thus, the fit of the three-parameter gamma distribution to the data is slightly, but not significantly, better than the fit of the Weibull distribution.

Choosing between these distributions based on the present data set requires a biological, rather than a statistical, rationale. The gamma distribution seems a more biologically reasonable description of the time to anthesis of an annual plant, because the gamma hazard tends toward a constant as time becomes large, while the Weibull hazard goes to either zero or infinity.

Analyses of the radish failure-time models are shown in Tables 12.3 and 12.4, and the SAS code that generated these analyses is shown in Appendix 12.3B. In both of these examples, the regression coefficients for the Storke Road population are zero. The reason for this is that accelerated failure-time models make the clock run slower or faster for some groups. As mentioned above, when there are multiple levels of a class variable (in this case, population), one level (the last one) is taken by SAS as a reference level, and all others are compared to this level. The Storke Road population is thus taken to be the reference population by

Table 12.3. Analysis of Accelerated Failure-Time Model for Raphanus sativus Emergence Time, Using 764 Noncensored Values, 6 Right-Censored Values, and a Log-Logistic Distribution[a]

Variable	df	Estimate	SE	χ^2	P
Intercept	1	1.868	0.014	17442.78	0.0001
Population	4			395.67	0.0001
(Black Spanish)	1	−0.254	0.018	194.03	0.0001
(Coal Oil Point)	1	−0.072	0.020	13.09	0.0003
(Campus Point)	1	−0.084	0.019	19.63	0.0001
(Rapid Cycling)	1	−0.298	0.018	250.84	0.0001
(Storke Road)	0	0	0		
Scale parameter	1	0.100	0.003		

[a]Loglikelihood = 116.3.

virtue of its order. The coefficients for other populations are therefore compared to Storke Road, and the significance tests for each population test whether it differs from Storke Road.

The analysis of the emergence data (Table 12.3) shows that population of origin contributes significantly to the model. The fact that each population's regression coefficient is independently significant shows that each differs from the reference (Storke Road) population. Which other populations differ from one another? To examine this question I conducted multiple comparisons with Z-statistics (see Appendix 12.1.4 for a description of the method, and Appendix 12.3C for SAS code to implement the multiple comparisons). These multiple comparisons are analogous to the ones used in the life table analyses of Section 12.4.2, except that in this case I used regression parameters as the statistics for

Table 12.4. Analysis of Accelerated Failure-Time Model for Raphanus sativus Anthesis Time, Using 210 Noncensored Values, 554 Right-Censored Values, and a Gamma Distribution[a]

Variable	df	Estimate	SE	χ^2	P
Intercept	1	4.974	0.090	3047.02	0.0001
Emergence time	1	0.019	0.010	3.68	0.06
Population	4			495.43	0.0001
(Black Spanish)	1	0.449	0.069	42.90	0.0001
(Coal Oil Point)	1	0.124	0.058	4.58	0.03
(Campus Point)	1	0.324	0.063	26.70	0.0001
(Rapid Cycling)	1	−0.682	0.054	154.35	0.0001
(Storke Road)	0	0	0		
Scale parameter	1	0.394	0.021		
Shape parameter	1	−0.822	0.179		

[a]Loglikelihood = −232.48.

comparison, rather than Wilcoxon rank scores. As with the life table statistics, the multiple comparisons require a small amount of manual data-handling (Appendix 12.3C).

The life table analyses showed significant heterogeneity among populations, and pairwise comparisons suggested that all pairs differ except Coal Oil Point–Campus Point and possibly Black Spanish–Rapid Cycling. The regression coefficients in Table 12.3 and their estimated covariance matrix led to a somewhat different conclusion: the Coal Oil Point–Campus Point comparison still results in a high probability of being from the same population, but the Black Spanish–Rapid Cycling comparison now shows significant differences between these populations. All other comparisons remained qualitatively unchanged. The accelerated failure-time model has thus confirmed the general patterns suggested by the life table analysis, but its greater statistical power has also made it possible to find significant differences among populations that were not revealed by the life table analysis.

To interpret the regression parameters in Table 12.3, recall that they correspond to multiplicative effects of covariates on the probability of seedling emergence. The average time to emergence for a Rapid Cycling plant is $e^{1.868-0.298}p_0(t) = 4.807p_0(t)$, while for a Storke Road plant it is $e^{1.868}p_0(t) = 6.475p_0(t)$. The more negative the coefficient, then, the earlier the average emergence time. The reference distribution $p_0(t)$ is the log-logistic distribution with the scale parameter shown in Table 12.3.

Analysis of the anthesis data (Table 12.4) also show that each population contributes significantly to the model. Late emergence probably tends to delay anthesis, which must be greater than time from planting to emergence. Including emergence time in the model is thus analogous to including a block effect in ANOVA.

The regression coefficients are interpreted in the same manner as for the emergence time data. Only the reference probability distributions differ (emergence is distributed as log-logistic, while anthesis is taken to be gamma-distributed). However, this does not affect the qualitative interpretation of the regression parameters. For example, analysis of the anthesis data shows that Rapid Cycling plants reach anthesis much earlier, and Black Spanish plants much later, than all others. This corresponds well with the results shown in Fig. 12.2.

The significant coefficient for the Coal Oil Point population indicates that it differs significantly from the Storke Road population. By contrast, the pairwise comparisons used following the life table analysis suggested that these populations did not differ. As with the emergence data, multiple comparisons conducted with Z-statistics from this analysis (see Appendix 12.1) point to more among-population differences than were revealed with life table analysis. These comparisons, conducted at an overall 0.05 significance level, suggest that all populations differ from one another, except possibly the Black Spanish–Campus Point pair, the differences among which were marginally significant. Which analysis is correct? The accelerated failure-time model has more statistical power, and therefore it is reasonable to have more confidence in these results than in the life table analysis.

Because the accelerated failure-time model is able to include the effect of emergence time explicitly, we can have more confidence that differences among populations in anthesis time are not simply due to differences in emergence time. Moreover, the accelerated failure-time model—because of its greater statistical power—has revealed several among-population differences that were not identified by life table analysis. Accelerated failure-time models can often provide greater clarity in failure-time studies than life table analysis alone. In the following subsection I illustrate the use of a somewhat different regression model, the proportional hazards model.

Proportional hazards models: description. As mentioned above, in a proportional hazards model, the effect of covariates is to change the chance of falling into a period with high hazard. The covariates act multiplicatively on the hazard function, rather than on the failure time (as in accelerated failure-time models). The hazard function for the ith group is thus

$$h_i(t) = h_0(t) \exp\left[\sum_{j=1}^{r} \beta_j X_{ij} \right] \qquad (12.3)$$

where $h_0(t)$ is a reference hazard function that is changed by the covariates $\mathbf{X}$ and regression coefficients $\boldsymbol{\beta}$.

The Cox proportional hazards model, which is very widely used in epidemiology (see Muenchow, 1986 for an ecological application), estimates the reference hazard function nonparametrically. The regression coefficients $\boldsymbol{\beta}$ are then estimated numerically with a maximum likelihood procedure. The Weibull distribution (Appendix 12.2) can also be used for proportional hazards models comparing two groups, by treating the parameter p as the ratio of two regression coefficients (Kalbfleisch and Prentice, 1980).

Because covariates act on the failure time in accelerated failure-time models, and on the hazard function in proportional hazards models, the regression coefficients can have opposite meanings. A positive coefficient in a proportional hazards model means that the covariate increases the hazard, thereby decreasing the failure time. A positive coefficient in an accelerated failure-time model means that the covariate increases the failure time. The interpretation of the coefficients can also vary with the particular parameterization used by a statistical package: the best advice is to check the documentation to be sure how to interpret the results, and to try a data set with a known outcome.

Proportional hazards models: application. Muenchow (1986) was interested in testing whether male and female flowers of the dioecious plant *Clematis lingusticifolia* are equally attractive to insects, against the alternative hypothesis that males are more attractive. She recognized that this could be treated as a failure-time problem because differences in attractiveness should lead to differences in time to the first insect visit. By treating this as a failure-time problem, Muenchow was able to examine these hypotheses with a creative experimental design: she watched pairs of flowers on the same plant and recorded waiting times until the first insect visit. For each observation, she also recorded the time of day, air temperature, and a categorization of the flower density within ≈ 1 m of the target plant. An important part of the design is that, unlike the radish study discussed above, Muenchow was not following a single cohort through time. Waiting times were recorded to the nearest minute.

This study is one in which the assumptions of the Cox proportional hazards model appear reasonable a priori. If there are differences in attractiveness, they would likely act to increase the "hazard" (i.e., the chance of a visit) of the attractive gender, relative to the hazard for the less attractive gender. Another way of seeing this is that being a member of the attractive gender should have no effect on the number of insects in the area, but does affect the chances of a visit once an insect is within some distance of the plant.

Muencheow's estimate of the survival function is shown in Fig. 12.3. She noted that these data appear to have been drawn from an exponential distribution, but preferred to use a Cox model for her statistical tests.

An important part of Muenchow's study is the way in which she analyzed the four covariates: gender, flower density category, temperature, and time of day. She reported that the starting time and temperature coefficients were not significantly different from zero, i.e., these factors do not appear to influence waiting time to visitation. Because male plants had more flowers, gender and flower density were correlated. Consequently, Muenchow analyzed the data by stratifying: she examined the effect of flower density within each gender, and the effect of gender within each flower density category.

Both gender and flower density independently had significant effects: within a gender, insects visited dense flower groups faster than other groups, and within a density category, insects visited males faster than females. Muenchow concluded that males were more attractive both because they bore flowers more densely and because they had some unknown attractive character.

SAS has recently developed procedure PHREG to analyze this model. Procedure PHREG is incorporated in SAS versions 6.07 and later, and is documented in the SAS/STAT User's Guide for these versions (SAS Institute Inc., 1992). Users of versions 6.06 and 6.04 can obtain procedure PHREG upon request from SAS; users under a site license must direct such requests to their designated SAS representative. Documentation for procedure PHREG as a supplemental program is available at low cost from SAS Institute (SAS Technical Report PP-217, SAS/

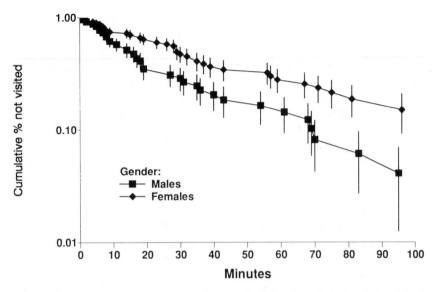

Figure 12.3. Product-limit estimates of the survival function recalculated from Muenchow's (1986) study of waiting times to insect visits at flowers of *Clematis lingusticifolia*. See Appendix 12.1 for discussion of product-limit estimates. Sample sizes are 47 female (8 censored) and 49 male (2 censored) flowers. Nonparametric tests show the genders to be significantly different: log-rank, $\chi^2 = 5.96$, 1 df, $P = 0.01$; Wilcoxon, $\chi^2 = 5.46$, 1 df, $P = 0.01$.

STAT Software: The PHREG Procedure, Version 6). Alternatively, one can analyze this model under procedure NLIN by using a macro provided by SAS on pages 1187–1192 of the SAS/STAT User's Guide, Version 6 (SAS Institute Inc, 1989b, pp. 1187–1192).

Other types of ecological applications. These regression methods are quite versatile, and can accommodate multiple types of failure. An ecological example of this might be the case where a grass either flowers or tillers, or where an animal either switches the food upon which it is foraging or switches patches. It is also possible to model repeated failures. An example might be the time to initiation of nest-building for birds that produce repeated clutches, or time to anthesis of a series of flowers. Such models would require the assumption that the timing of future failures does not depend on the history of past failures. To my knowledge neither type of application has been used in an ecological setting.

12.5 Discussion

12.5.1 Assumption of Independence

The statistical methods discussed here assume that the experimental units are independent, i.e., that the individuals do not interact. Certainly this is a reasonable assumption for the case examined here, time to emergence and flowering of a randomly selected set of seeds grown in individual pots in a greenhouse. However, there are many cases in ecology where this assumption of independence presents problems. Time to flowering in a natural setting will often depend on the number and sizes of neighbors of a plant. Consequently one would have to measure plants that are spaced widely enough that they do not interact. Time to fruit-set can depend on the availability of pollen and pollinators, and can therefore be strongly frequency and density dependent. In this case it may not always be possible to satisfy the assumption of independence. Life table estimates of survival may still be useful in such situations as descriptions. However, the significance tests discussed here will no longer be valid.

When data cannot be treated as independent, it is usually necessary to treat aggregate measurements of treatment groups (e.g., mean sizes of individuals) as single data points. With uncensored failure-time data, one might take the median failure time of each treatment group as a datum. This approach cannot be used if some data points are censored before 50% have failed, as then only an estimate of the median and its variance are available. There are currently no methods for appropriately analyzing failure times in this situation. Rather than comparing failure times, it may be more useful to compare groups for the probability of failing within a specified time.

12.5.2 Presentation of Results

How can results be communicated so that other biologists can interpret them? With standard normal statistics, the most useful information is usually the mean and its standard error or specified confidence limits. Values for test statistics (t or F) and associated probabilities may also be useful. One can describe the data very well by also specifying the sample variance or standard deviation and sample size.

With failure-time data, description of results is more complicated as they are not adequately described by their mean and variance: the mean failure time is a poor estimate of central tendency, and the distribution of failure times is not symmetrical. Censorship can further complicate the picture, especially if groups differ in patterns of censorship. In most cases, graphic representation of data is quite useful (see Chapter 2 and Figs. 12.1 and 12.2).

The most useful information is usually a set of survivorship curves estimated with any of the methods described here. Error bars or confidence regions should be included in such presentations. For life table estimates, standard errors for the cumulative survivorship function should be shown for each interval. The median is a more robust measurement of central tendency than the mean. The methods described in Tables 12.1 and 12.2 allow one to calculate the median life expectancy and its standard error for an individual at the outset of the experiment. If groups differ in patterns of censorship, it may also be useful to describe the patterns of censorship for different groups. For example, in my study of flowering times in *E. abertianum*, increasing the moisture available to plants decreased the chance that they would die before flowering. This kind of information can affect the interpretation of results.

12.5.3 Concluding Remarks

The kinds of studies discussed in this chapter—time to emergence and flowering in a cohort of experimental plants, and time to insect visits for male and female flowers—illustrate some of the kinds of questions that can be addressed with failure-time approaches. Research in areas like life history evolution can clearly benefit from using these approaches. However, Muenchow's (1986) insect visitation study also shows how failure-time methods can be used to bring a lot of statistical power to bear on problems that may not initially seem to be failure-time problems.

This said, it is important to note that failure-time methods are not as well developed as ANOVA. One cannot use failure-time methods to estimate variance components. Erratic patterns of censorship and the nonnormality of most distributions of failure times makes such advances seem quite unlikely.

On the other hand, there are many types of ecological experiments for which failure-time methods are the best approach, both in terms of experimental design and statistical analysis. As shown in this chapter, the designs are generally quite simple once one realizes that censored data points are still data points. Statistical analyses are accessible to ecologists, especially inasmuch as the analyses can generally be related to the study of life tables. In addition to the SAS implementation discussed in this chapter, most of the methods are available in SPSS, BMDP, and other statistical packages. FORTRAN programs for performing many of these analyses are provided by Kalbfleisch and Prentice (1980) and Lee (1980). These methods should be more widely used by ecologists.

Acknowledgments

I thank Susan Mazer and Lorne Wolfe for assistance with *Raphanus* seed collection and advice on radish cultivation, and Abreeza Zeeger and Jeff Seminoff for assistance in the greenhouse. Neil Willits provided a thorough and constructive reading of the manuscript, and suggested several of the illustrative examples. I am also grateful to the editors for their support, patience, and criticism, and am especially grateful to Kathy Whitley for nearly endless patience. Facilities were generously provided by the Department of Ecology and Evolutionary Biology at the University of Arizona. I was supported by NSF Grant BSR-900230.

Appendix 12.1 Tests for Equality of Survival Distributions

A large number of different tests have been suggested by various authors. Most of these are closely related forms of log-rank, Wilcoxon, and χ^2 tests, and can be regarded as generalizations of χ^2 approaches suggested by Mantel and Haenszel (1959). The following may be the most generally used versions.

12.1.1 Logrank Test for Two Samples

These tests are described by Peto and Peto (1972). Pool the data for the samples. For each failure time t_i, find the number D_i of failures in the interval between t_i and t_{i-1}, and the total number of individuals that had not failed or been censored at the last census. The latter is the number "at risk" during the interval, r_i. For each of these failure times, estimate the log of the cumulative survival function (under the null hypothesis that the samples are equivalent) as

$$e(t_i) = \sum_{j \leq i} \frac{D_j}{r_j} \qquad (12.4)$$

Assign to each of the observed failure times the scores

$$w_i = 1 - e(t_i) \text{ for uncensored observations} \qquad (12.5)$$
$$w_i = -e(t_i) \text{ for censored observations}$$

Let S be the sum of the scores in one of the two groups. The variance of S is given by

$$V = n_1 n_2 \sum_{i}^{n_1+n_2} \frac{w_i^2}{(n_1 + n_2)(n_1 + n_2 - 1)} \qquad (12.6)$$

where the ns are the numbers of individuals in each group. The test statistic $L = S/\sqrt{V}$ is asymptotically normally distributed, and the critical region for significance testing is therefore $L < -Z_\alpha$.

Peto and Peto (1972) and Peto and Pike (1973) suggested an intuitively simple χ^2 approximation to this test, that has been discussed in an ecological context by Pyke and Thompson (1986) and Hutchings et al. (1991). A more general version of this χ^2 approximation is discussed under Section 12.1.3.

12.1.2 Wilcoxon Test for Two Samples

These tests are described by Peto and Peto (1972). Estimate the cumulative survival function S' using the product-limit (also called Kaplan-Meier) method. Under this method the survival function is recalculated after every individual observation. In contrast the life-table methods in Table 12.1 involve calculations at specified intervals. These estimates are generally quite similar. Score each observation as

$$u_i = S'(t_i) + S'(t_{i-1}) - 1 \text{ for uncensored observations} \qquad (12.7)$$
$$u_i = S'(t_i) - 1 \text{ for censored observations}$$

The rest of the calculation follows Section 1.1 of this appendix.

12.1.3 Chi-Square Approximations to Log-Rank and Wilcoxon Tests for N Samples

These tests are described by Kalbfleisch and Prentice (1980) and Lawless (1982). Although the above tests look quite different from one another and from χ^2 tests, they can be approximated by tests similar in form to χ^2 tests (Peto and Peto, 1972), and generalized for comparison of an arbitrary number of samples. If a Strata statement is used, SAS procedure LIFETEST automatically performs both of the following N-sample tests.

If there are r samples, form a vector $\mathbf{v}$ of length r, with

$$v_j = \sum_{i,j}^{k} \frac{w_i(d_{ij} - n_{ij}d_i)}{n_i} \qquad (12.8)$$

where d_{ij} is the number of failures in the jth sample in the ith interval, d_i is the sum of failures over all samples in the ith interval, n_{ij} is the number of individuals at risk in the jth sample during the ith interval (the number that had not failed or been censored as of the beginning of that interval), and n_i is the sum of individuals at risk over all samples in the ith interval. The w_i term is a weighting term. For the logrank test, the w_is are all 1, and for the Wilcoxon test they are n_i. In summary, v_j is a weighted sum of differences between observed and expected failures.

Form the $(r - 1)$ by $(r - 1)$ estimated covariance matrix $\mathbf{V}$, the elements of which are given by

$$V_{jl} = \sum_{i}^{k} w_j^2 \frac{(n_i n_{il} \delta_{jl} - n_{ij} n_{il}) d_i s_i}{n_i^2(n_i - 1)} \qquad (12.9)$$

where the index i identifies specific failure times, $s_i = n_i - d_i$, and $\delta_{jl} = 1$ if $j = l$, and 0 otherwise. The test statistic is $\mathbf{v}'\mathbf{V}^{-1}\mathbf{v}$, where $\mathbf{v}'$ is the transpose of $\mathbf{v}$. This has a χ^2 distribution with $r - 1$ degrees of freedom.

12.1.4 Post hoc Pairwise Comparisons among Groups

It is possible to construct post hoc pairwise comparisons among groups from information generated by tests for heterogeneity among groups. For example, a Wilcoxon test or an accelerated failure-time model may show that treatment groups differ, but if there are more than two groups these tests do not provide information on which groups differ from each other. For any two groups, the comparison statistic is

$$Z = \frac{W_i - W_j}{\sqrt{V_{ij} + V_i - 2V_{jj}}} \qquad (12.10)$$

where W_i is the previously estimated statistic for the ith group, and V_{ij} is the $(i,j$th element in the covariance matrix. To test for significance, one must take account of the fact that there are k populations being subject to pairwise comparison, or $p = k(k - 1)/2$ comparisons being made. Thus the significance test for pairwise comparisons of k populations at level α is whether $Z > Z_\gamma$, where Z_γ is the upper 100_γ percentage point of the normal distribution, and

$$\gamma = 1 - \left(\frac{1}{2}\right)_{1/p} (2 - \alpha)_{1/p} \qquad (12.11)$$

Values for Z-statistics are provided as appendices to most statistics texts, as well as by SAS.

12.1.5 Likelihood Ratio Tests

Likelihood ratio tests are parametric tests that are often used with life table analysis to consider whether two data sets come from the same (specified) parametric distribution (see Appendix 12.2). Lawless (1982) describes likelihood ratio tests for a number of different distributions.

When a Strata statement is used, SAS procedure LIFETEST automatically performs a likelihood ratio test that assumes the data to be exponentially distributed, and tests whether the scale parameters are equal (see Appendix 12.2). This test is not very powerful (Lee, 1980). Worse yet, statistics based on the exponential distribution are not robust with other distributions (Appendix 12.2), so results of this test should be treated with considerable caution unless data are known to be exponentially distributed. It is sometimes possible to transform data to an exponential distribution and then use the likelihood ratio test performed by procedure LIFETEST. However, it is not clear that one gains much advantage by doing so as compared with using an accelerated failure-time model.

These tests can, however, be useful in evaluating particular hypotheses about parametric distributions, such as when one wants to compare two distributions, one of which is a special case of the other (see Section 12.4.3). If the parameter vectors for the two distributions are θ_1 and θ_2, and the maximized likelihood functions are $L(\theta_1)$ and $L(\theta_2)$ respectively, the test statistic is (Kalbfleisch and Prentice, 1980)

$$R = -2 \ln\left[\frac{L(\theta_1)}{L(\theta_2)}\right] \qquad (12.12)$$

This statistic is distributed as χ^2, with degrees of freedom equal to the difference in the number of parameters of the two distributions being compared.

Appendix 12.2 Distributions of Failure Times

Failure-time distributions can be compared in many ways. For ecologists the most useful are probably comparisons of the hazard functions associated with each distribution, because the assumptions about the failure process are made explicit. Comparisons of the cumulative survivorship function and the probability density function for lifetimes can also be useful, especially because some distributions do not have closed form expressions for the hazard but do for one or both of these functions.

A number of failure-time distributions have been used in empirical studies of failure times. While in principle any distribution can be used with an accelerated failure-time model, some are more likely than others to be ecologically relevant, and most statistical packages offer a limited choice of distributions. Some of the most important distributions are described below. Note that the SAS manual uses considerably more complex, but equivalent, notation. These differences should be taken into account in interpreting SAS output. Other distributions are discussed by Lawless (1982).

Analytical expressions for the hazard, survival, and probability density functions are shown in Table 12.5. Kalbfleisch and Prentice (1980) provide readable derivations for these expressions. Representative cases are illustrated for each distribution. The Mathematica program used to evaluate these functions and generate the graphs is available on request from the author.

12.2.1 Exponential Distribution

Under the exponential distribution, the hazard function is constant (Table 12.5). Fitting data to the exponential thus requires estimation of only the single parameter λ, which must be > 0. This generates the cumulative survivorship and probability density functions shown in Fig. 12.4.

A constant hazard generates a "type II" survivorship curve. There are probably

Table 12.5. Functions of Failure-Time Distributions[a]

Distribution	Parameters	$h(t)$	$P(t)$	$S(t)$
Exponential	λ	λ	$\lambda e^{-\lambda t}$	$e^{-\lambda t}$
Weibull	p, λ	$\lambda p(\lambda t)^{p-1}$	$\lambda p(\lambda t)^{p-1} e^{-(\lambda t)^p}$	$e^{-(\lambda t)^p}$
Two-parameter gamma[b]	k, λ	$\dfrac{P(t)}{S(t)}$	$\dfrac{\lambda(\lambda t)^{k-1} e^{-\lambda t}}{\Gamma(k)}$	$1 - \Gamma(k, \lambda t)$
Three-parameter gamma[b]	k, λ, β	$\dfrac{P(t)}{S(t)}$	$\dfrac{\lambda \beta(\lambda t)^{k\beta-1} e^{-(\lambda t)^\beta}}{\Gamma(k)}$	$1 - \Gamma\left[(k, \lambda t)^\beta\right]$
Lognormal[c]	p, λ	$\dfrac{P(t)}{S(t)}$	$\dfrac{\sqrt{2\pi}}{pt} e^{\frac{-p^2 \log(\lambda t)^2}{2}}$	$1 - \Phi[p \log [\lambda t]]$
Log-logistic	p, λ	$\dfrac{\lambda p(\lambda t)^{p-1}}{1+(\lambda t)^p}$	$\dfrac{\lambda p(\lambda t)^{p-1}}{[1+(\lambda t)^p]^2}$	$\dfrac{1}{1+(\lambda t)^p}$

[a]h, P, and S are the hazard, probability density, and survival functions, respectively. Note that some SAS parameterizations differ from these.

[b]$\Gamma(x)$ is the gamma function $\int_0^\infty u^{x-1} e^{-u} du$;

$\Gamma(k, x)$ is the incomplete gamma function $[1/\Gamma(k)] \int_0^x u^{k-1} e^{-u} du$.

[c]$\Phi(x)$ is the cumulative normal distribution function $\Phi(w) = (1/\sqrt{2\pi}) \int_{-\infty}^w e^{-(u^2/2)} du$.

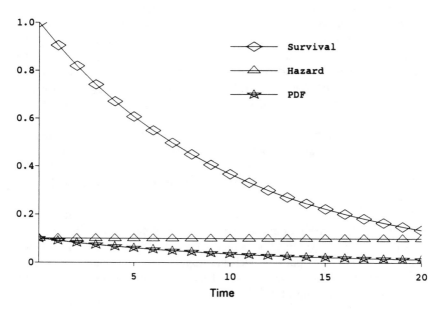

Figure 12.4. Hazard, survival, and probability distribution functions for the exponential distribution for $\lambda = 0.1$.

few ecological situations in which hazards are truly constant for very long periods of time. However, it may be realistic to treat some hazards as approximately constant. A well-known example is survivorship among many species of adult Passerine birds.

The exponential distribution is probably of greatest use as a starting point for understanding failure-time distributions. The constant hazard means that the exponential distribution plays a role in failure-time models that is somewhat similar to that of the normal distribution in linear statistical modelling: the exponential provides a simple null model and a point of departure for more complex failure processes.

A simple empirical check for the exponential distribution is to examine whether a life table estimate for the hazard function is approximately constant. Equivalently, if the exponential distribution adequately describes the data, a plot of $\log[S(t)]$ vs t should be approximately linear, through the origin. SAS procedure LIFETEST produces this plot if you specify Plots=LS. Such empirical tests are quite important, because significance tests assuming exponentially distributed data are less robust than tests assuming other distributions (Lee, 1980).

12.2.2 Weibull Distribution

The Weibull distribution is a generalization of the exponential. In the Weibull distribution the hazard function is a power of time (Table 12.5). The two parameters p and λ are > 0. Note that if $p = 1$, the Weibull reduces to the exponential distribution. Otherwise, the hazard is monotonically increasing for $p > 1$ and decreasing for $p < 1$. Considering its effect on the hazard function, it should not be surprising that the shape of P depends on p, and p is often referred to as a shape parameter. Figure 12.5 shows examples of the hazard, survival, and density functions.

The monotonic trend in the Weibull model means that it can be realistic in many systems. For example, epidemiological models use a Weibull distribution with $p > 1$ for time of onset of fullblown AIDS in HIV-infected patients. Mortality in humans and many other mammals can sometimes be approximated with a Weibull distribution, if only adults are considered.

A simple empirical check for the applicability of the Weibull distribution to a data set can be derived from the Weibull survival function: using life table estimates for the survival function, a plot of $\log\{-\log[S(t)]\}$ against $\log(t)$ should be approximately a straight line, with the slope being an estimate for p and the $\log(t)$ intercept an estimate for $- \log(\lambda)$. This plot is produced by SAS procedure LIFETEST by specifying Plots=LLS.

12.2.3 Gamma Distribution

The gamma distribution is another generalization of the exponential. In contrast to the Weibull distribution, the hazard function in the gamma either increases or

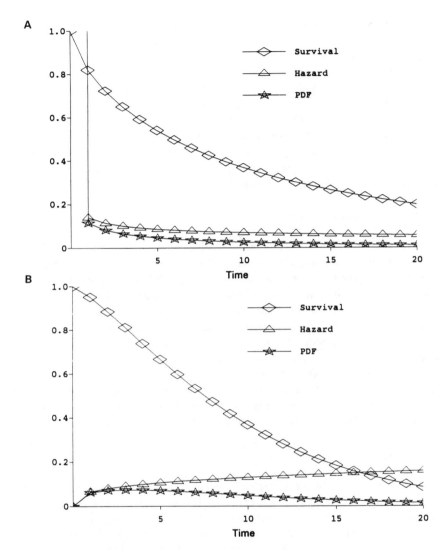

Figure 12.5. Hazard, survival, and probability distribution functions for the Weibull distribution. (**A**) $\lambda = 0.1$, $p = 0.7$; (**B**) $\lambda = 0.1$, $p = 1.3$.

decreases monotonically towards a constant as *t* goes to infinity. The monotonic nature of the gamma hazard makes it particularly useful for some developmental processes. In addition to its use in this chapter, I have found flowering time (Fox, 1990a) and senescent mortality (Fox, 1990b) in other plants to be well described by gamma distributions.

The two parameters k and λ (Table 12.5) are both > 0; k determines the shape of the distribution and λ its scale. The complicated SAS parameterization allows the shape parameter to be negative. If $k > 1$, the hazard function is 0 at $t = 0$, and increases monotonically toward λ. If $k < 1$, the hazard function is infinite at $t = 0$, and decreases monotonically toward λ with time. If $k = 1$, the gamma reduces to the exponential distribution. Examples are shown in Fig. 12.6.

The gamma distribution can be made even more flexible by generalizing to a three-parameter model (Table 12.5). This three-parameter gamma includes all of the preceding distributions as special cases: the exponential ($\beta = k = 1$), Weibull ($k = 1$), and two-parameter gamma ($\beta = 1$). The lognormal distribution (below) is the limiting case when k goes to infinity. Clearly a very wide range of survival data can potentially be fit with this distribution.

12.2.4 Lognormal Distribution

Analytical expressions for the lognormal involve indefinite integrals (Table 12.5), but the salient feature of the lognormal is that the hazard function is 0 at time 0, increases to a maximum, and then decreases, asymptotically approaching 0 as t goes to infinity. The latter property means that the lognormal is probably not useful for studies involving long lifetimes (although these are quite rare in ecology) because a hazard of 0 is implausible. The lognormal is a two-parameter distribution in which both p and λ are assumed > 0. See Fig. 12.7 for representative examples of the hazard, survival, and density functions.

As one might guess from the name, under the lognormal distribution the log failure times are normally distributed. This suggests that simple empirical checks for the lognormal distribution can be done by log-transforming the data and using any of the standard methods of testing for normality.

12.2.5 Log-Logistic Distribution

The log-logistic distribution is roughly similar in shape to the lognormal, but it is often easier to use the log-logistic. The major reason for this is that the functions of interest have closed-form expressions (Table 12.5). Especially when there are censored data, the lognormal can involve considerable computation because of the indefinite integrals in the hazard and survival functions. The log-logistic can also be somewhat more flexible than the lognormal, as can be seen by considering the hazard, survivorship, and probability distribution functions (Table 12.5, Fig. 12.8). If $p < 1$, the hazard function is infinite at 0 but decreases monotonically with time. If $p > 1$, it decreases monotonically from λ. When $p > 1$, the hazard is similar to the lognormal hazard: it increases from zero to a maximum at $t = (p - 1)^{1/p}/\lambda$, and then decreases toward zero.

The humped nature of the log-logistic and lognormal hazards may make them especially useful for describing organisms' responses to environmental factors. The seedling emergence data analyzed in this chapter are one example. On the

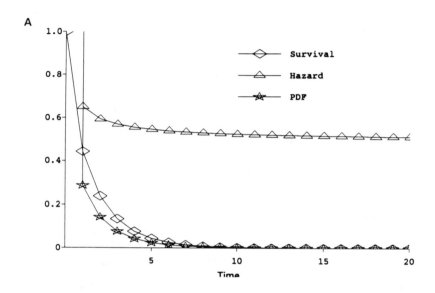

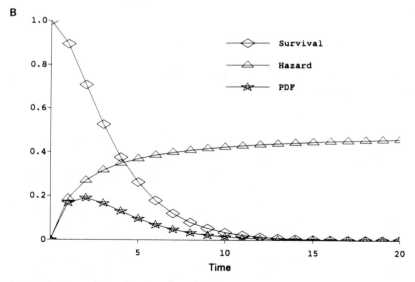

Figure 12.6. Hazard, survival, and probability distribution functions for the two-parameter gamma distribution. (**A**) $\lambda = 0.5$, $p = 0.7$; (**B**) $\lambda = 0.5$, $p = 1.9$.

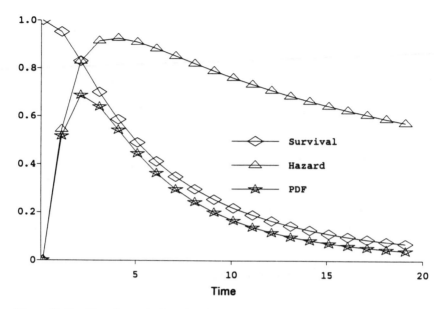

Figure 12.7. Hazard, survival, and probability distribution functions for the lognormal distribution for $\lambda = 0.2$, $p = 1.1$.

other hand, few mortality hazards are likely to begin as very small, reach a maximum, and then decline.

Appendix 12.3

SAS program code for life table analyses, failure time analyses, and multiple comparison procedures.

A. Life Table Analysis

These statements generate the data in Fig. 12.1, the Wilcoxon and log-rank tests for heterogeneity among populations, and the covariance matrix for Wilcoxon and log-rank statistics, to be used below for multiple comparisons. A similar procedure LIFETEST statement, but with WIDTH=5 and TIME TAN-TH*ANTHCENS(1), generated the data in Fig. 12.2 and the relevant test statistics.

```
PROC LIFETEST METHOD=LIFE          /choose life table analysis/
    WIDTH=2;                       /interval width = 2 days/
    STRATA POP;                    /do separate analysis for each population and
                                   test for heterogenetity among populations/
    TIME TEMERG*GERMCENS(1);       /TEMERG = time variable,
                                   GERMCENS = censorship variable, a value
                                   of 1 meaning a right-censored point/
```

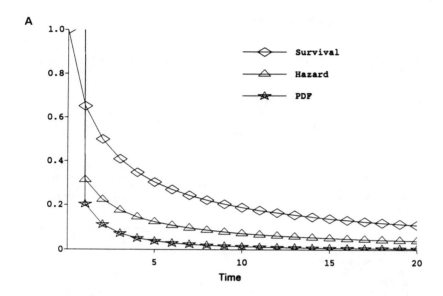

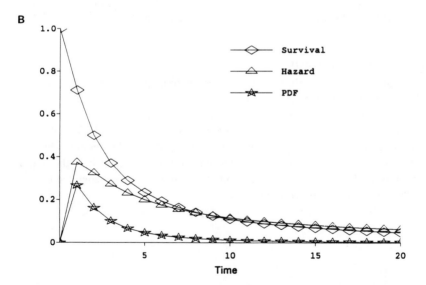

Figure 12.8. Hazard, survival, and probability distribution functions for the log-logistic distribution. (**A**) $\lambda = 0.5$, $p = 0.9$; (**B**) $\lambda = 0.5$, $p = 1.3$.

B. Accelerated Failure Time Models

Note that procedure LIFEREG allows multiple MODEL statements.

```
PROC LIFEREG; CLASS POP;                    /identify classification variables/
                                            /analysis of emergence time (Table 12.3) and
                                            estimated covariance matrix for the regression
                                    parameters for multiple comparisons (section 12.4.2)/
MODEL TEMERG*GERMCENS(1)
          = POP                             /POP is the independent variable/
          /DIST=LLOGISTIC                   /fit to a log-logistic distribution/
          COVB;                   /request covariance matrix for parameter estimates/


                                            /Analysis of anthesis time/
MODEL TANTH*ANTHCENS(1) =                    (Table 12.4) (section 12.4.2)/
          TEMERG POP/              /emergence time is treated here as a covariate/
          DIST=GAMMA                         /fit to a gamma distribribution/
          COVB;


                                   /comparison of the gamma distribution with the
                                            Weibull and lognormal distributions/
MODEL TANTH*ANTHCENS(1) =          /Lagrange multiplier test: gamma vs lognormal/
          TEMERG POP/
          DIST=GAMMA                         /specify general distribution/
          SHAPE1=0                  /lognormal is the gamma with SHAPE1=0/
          NOSHAPE1;                           /force SHAPE1 to stay at 0/

MODEL TANTH*ANTHCENS(1) =          /Lagrange multiplier test: gamma vs Weibull/
          TEMERG POP/                        /Note: this test does not work in
          DIST=GAMMA                         SAS versions lower than 6.07/
          SHAPE1=1 NOSHAPE1;        /Weibull is the gamma with SHAPE1=1/
```

C. Multiple Comparisons

Make one disk file containing the names of the treatment groups. Make a second disk file containing the test statistics for all of the groups (e.g., the Wilcoxon rank scores for each groups generated by procedure LIFETEST, or the regression coefficients for each group generated by procedure LIFEREG). Make a third disk file containing the covariance matrix for the test statistics. Procedure LIFETEST automatically generates the covariance matrix and will print it if the /COVB option is used in the MODEL statement. These files could be generated by using a text editor on the SAS output.

```
DATA;                                              /declare array names for
                                                processing in the loop below/
    ARRAY NAMES[5] $;                                  /population names/
    ARRAY STATS[5];                    /statistics for test: in this case, Wilcoxon scores/
    ARRAY COVAR[5,5];                      /covariance matrix from Wilcoxon test/
    ARRAY ZSCORE[5,5];                                    /Z-scores/
    ARRAY PROB[5,5];                                    /probabilities/

    INFILE 'NAMES.DAT';                         /read in the data from the 3 files/
        INPUT NAMES[*] $;
    INFILE 'STATS.DAT';
        INPUT STATS[*];
    INFILE 'COVAR.DAT';
        INPUT COVAR[*];

    FILE 'ANTHCOMP.OUT';                        /specify name of output file/
    PUT 'MULTIPLE COMPARISONS                     /write the file headings/
        OF RAPHANUS ANTHESIS';
    PUT '   BASED ON WILCOXON
        RANK STATISTICS';
    PUT;

    DO I = 1 TO 5;                              /calculate the Z-scores and their
        DO J = (I + 1) TO 5;                     single-comparison probabilities/
            ZSCORE[I,J] = ABS(STATS[I]                /calculate Z-scores/
            - STATS[J])/
            SQRT(COVARI[I,I] +
            COVAR[J,J] - 2*COVAR[I,J]);
            PROB[I,J] = 1 -                    /probabilities for single comparison/
            PROBNORM(ZSCORE[I,J]);
            PUT 'COMPARISON:                            /write output to file/
            ' NAMES[I]'& ' NAMES[J]':
            ' 'Z = ' ZSCORE[I,J] ',
            PR{FROM SAME SAMPLE}
            = ' PROB[I,J];
            PUT;
        END;
    END;
    PUT;
    PUT 'THESE GIVE PROBABILITIES
        FOR SINGLE COMPARISONS ONLY.';
    PUT 'FOR MULTIPLE COMPARISONS,
        USE THE FOLLOWING:';

    ALPHA = .05;                              /for multiple comparisons, among NPOP
    NPOP = 5;                             populations and a significance level of alpha,
    NCOMPARE = NPOP*(NPOP-1)/2;              calculate the corrected cutoff point for
    INVCOMP = 1/NCOMPARE;                             the significance test/
    ADJUST = 1 - ((2 - ALPHA)**INVCOMP)
            *(1/2)**INVCOMP;
    ZADJUST = PROBIT(ADJUST);
    PUT 'TO ACCEPT A DIFFERENCE AS
        SIGNIFICANT AT THE ', ALPHA ,' LEVEL,';
    PUT 'USE ONLY Z-VALUES > ',
        ADJUST;
```

13

The Bootstrap and the Jackknife: Describing the Precision of Ecological Indices

Philip M. Dixon

13.1 Introduction

Many different indices and coefficients are used in quantitative ecology. Some examples include diversity indices (Magurran, 1988), similarity indices (Goodall, 1978), competition coefficients (Schoener, 1983), population growth rates (Lenski and Service, 1982), and measures of size hierarchy (Wiener and Solbrig, 1984). All of these indices are statistics, calculated from a sample of data from some population and used to make inferences about that population (Chapter 1). To help a reader interpret these results, good statistical practice is to report both the value of a statistic and an estimate of its accuracy. Although it is easy to calculate the value of many ecological statistics, it is often hard to estimate their accuracy. This chapter discusses two techniques, the bootstrap and jackknife, that can be used to estimate the accuracy of ecological indices.

13.1.1 Precision, Bias, and Confidence Intervals

To understand how the bootstrap and jackknife work, we must first quickly review some concepts of statistical estimation. Consider how to estimate the mean weight of a seed in a large collection of seeds. If every seed were weighed, then the true mean weight would be known exactly. But, such a complete sample is usually impractical. Instead, a random sample of seeds is selected and weighed. The average weight of the sample is an estimate of a parameter, the mean weight of a population of the entire collection of seeds. How can the accuracy of the statistic be described? Because each seed probably has a slightly different weight, different samples of seeds will have different sample means. Some sample means will give too low an estimate of the population mean, while other estimates will give too high an estimate (Fig. 13.1). The set of average seed weights from all possible samples of seeds is the sampling distribution of the sample mean; characteristics of this distribution describe the accuracy of the statistic.

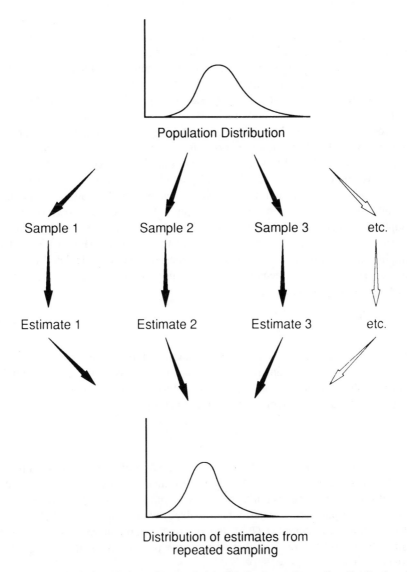

Figure 13.1. Relationship between population distribution and sampling distribution. If the values in the population were known, then the sampling distribution of the estimate can be obtained by repeatedly drawing samples from the population and calculating the statistic from each sample.

The accuracy of a statistic has two components: bias and precision (Zar, 1984, pp. 4–5). Bias is a measure of whether a statistic is consistently too low or too high. It is defined as the difference between the population value and the average of the sampling distribution. If the sampling distribution is centered at the population parameter, then the statistic is unbiased. Precision is a measure of the variation in the sampling distribution. At a minimum, one should estimate the variance or standard error of a statistic, but it is better to estimate a confidence interval or test some statistical hypothesis (Young and Young, 1991; Matloff, 1991). Remember that bias and precision are separate components of accuracy. A precise statistic may be biased if its sampling distribution is concentrated around some value that is not the population value. However, many commonly used statistics, such as the sample mean and sample variance, are unbiased.

The distribution of estimates from all possible samples is a nice theoretical concept, but an experimenter has only one sample of data. For sample means and a small number of other statistics, the sampling distribution is known from one sample of data because there are known mathematical relationships between the properties of the sample and the properties of the population (Fig. 13.2). If the sampling distribution is known, bias and precision may be calculated exactly. For example, the sample mean from random samples is unbiased and its standard error, $S_{\bar{x}}$ can be estimated by

$$S_{\bar{x}} = S_x / \sqrt{n} \tag{13.1}$$

where S_x is the sample standard deviation and n is the sample size.

Confidence intervals can also be calculated if the sampling distribution is known. Under certain general conditions, the sample mean is normally distributed, so that a 95% confidence interval is given by

$$(\bar{x} - t_{n-1}S_{\bar{x}}, \bar{x} + t_{n-1}S_{\bar{x}}) \tag{13.2}$$

where t_{n-1} is the critical value for a two-sided 95% confidence interval from a t distribution with $n-1$ degrees of freedom. Confidence intervals are commonly misunderstood. A confidence interval is a random interval with the property that it includes the population mean, which is fixed, with a specified probability. If a 95% confidence interval is (1,2), this does not mean that 95% of the sample values are between 1 or 2 or that 95% of the possible population means are between 1 and 2.

13.1.2 Precision and Bias of Ecological Indices

In general, it is difficult to calculate the precision or bias of most ecological indices. The mathematical techniques used with the sample mean do not work for many ecological indices because there is no exact formula for the variance of

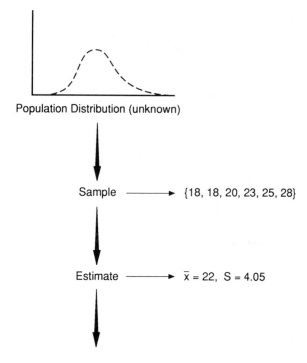

Population Distribution (unknown)

Sample   ⟶   {18, 18, 20, 23, 25, 28}

Estimate   ⟶ $\bar{x} = 22$, $S = 4.05$

Statements about precision

- Standard error of mean $S_{\bar{x}} = \dfrac{S}{\sqrt{n}} = \dfrac{4.05}{\sqrt{6}} = 1.65$

- 95% confidence interval for μ $(\bar{X} \pm t_{n-1} \cdot S_{\bar{x}}) = (22 \pm 2.54 \cdot 1.65) = (17.8, 26.2)$

Figure 13.2. Use of data and statistical theory to infer a sampling distribution from one sample of data. The sampling distribution of certain statistics is known theoretically. Statements about the precision of the mean can be made using sample information (e.g., sample mean and standard deviation).

the statistic and the sampling distribution is unknown. However, it is more important to choose an ecologically useful coefficient than a statistically tractable one. Bootstrapping and jackknifing are important techniques for ecologists because they can be used to estimate the accuracy of ecologically useful coefficients, even ones that are otherwise statistically difficult. The jackknife can be used to estimate the bias and variance of a statistic, while the bootstrap can be used to estimate bias, variance, and a confidence interval. Both techniques are computer-intensive methods that replace analytical equations with numerical computation.

The focus in this chapter is statistical estimation, especially calculating confidence intervals, because estimation is generally more useful than hypothesis testing (Gardner and Altman, 1986; Salsburg, 1985). However, the bootstrap can be used to construct hypothesis tests by estimating the sampling distribution of a test statistic under some null hypothesis (Hall and Wilson, 1991). A related computer-intensive technique, Monte-Carlo randomization (see Chapter 15), can also be used to test statistical hypotheses.

In the last few years, the jackknife and bootstrap have been used to estimate the bias or precision in a variety of ecological applications: population growth rates (Meyer et al., 1986), population sizes (Buckland and Garthwaite, 1991), ratios of variables (Buonaccorsi and Liebhold, 1988; Dorfman et al., 1990), factor loadings from principal components analysis (Gibson et al., 1984), genetic distances (Mueller, 1979), diversity indices (Heltshe, 1988; Heltshe and Forrester, 1985), diet similarity (Smith, 1985), and niche overlap indices (Manly, 1990a). General introductions to the bootstrap and jackknife can be found in Lanyon (1987), Manly (1991), Efron and Tibshirani (1991), and Potvin and Roff (1993). In this chapter, I will describe the bootstrap and jackknife techniques and illustrate their use with the Gini coefficient of size hierarchy and the Jaccard index of community similarity.

13.1.3 Gini Coefficients and Similarity Indices

The Gini coefficient (Weiner and Solbrig, 1984) is used in plant population biology to summarize a distribution of plant sizes. In a typical experiment, a collection of plants is grown under specified conditions. The Gini coefficient, G, measures the inequality in the observed plant sizes. It ranges from 0, representing exact equality, when all plants have the same size, to a theoretical limit of 1, representing extreme inequality, when one plant is extremely large and all other plants are extremely small. G can be calculated from a set of data by

$$G = \frac{\sum_{i=1}^{n} (2i-n-1) X_i}{(n-1) \sum_{i=1}^{n} X_i} \tag{13.3}$$

where n is the number of individual plants and X_i is the size of the ith plant, when plants are sorted from smallest to largest, $X_1 \leq X_2 \leq \ldots X_n$. After calculating G, we often want to describe the precision of G, which can be done using the bootstrap (Dixon et al., 1987).

A similarity index estimates the similarity between two communities. Many similarity indices have been proposed (see Goodall, 1978 for a review); the Jaccard (1901) index measures differences in species presence or absence. Typically, the species in two communities are sampled, and a similarity index is calculated from the two lists of species. The Jaccard coefficient is:

$$J = \frac{a}{t} \tag{13.4}$$

where a is the number of species found in both communities and t is the total number of species found in either community. The coefficient is ecologically useful but it is difficult to calculate its precision. Again, the bootstrap and jackknife are methods for describing the precision of similarity indices (Smith et al., 1986).

13.2 The Bootstrap

13.2.1 Ecological Example: Gini Coefficients to Measure Size Hierarchy in Ailanthus Seedings

Consider the following study of the development of a size hierarchy in the pioneer tree, *Ailanthus altissima* (Evans, 1983). As one part of the study, six seeds were randomly chosen from a large collection of seeds and planted in individual pots. Another 100 seeds were planted in a common flat so that they could compete with each other. After 5 months, each surviving plant was measured (Table 13.1). The Gini coefficient for the individually grown plants ($G = 0.112$) was smaller

Table 13.1. Number of Leaf Nodes for 5-Month-Old Ailanthus altissima *Grown under Two Conditions: in Individual Pots and in a Common Flat*

6 plants grown individually

| 18 | 18 | 20 | 23 | 25 | 28 |

75 surviving plants grown together in a common flat

8						
10						
11	11	11				
12	12	12				
13	13	13	13	13	13	13
14	14	14	14	14		
15	15	15	15	15	15	15
16	16					
17	17	17				
18	18	18	18	18		
19	19	19	19			
20	20	20	20	20	20	
21	21	21				
22	22	22	22	22		
23	23	23	23	23	23	23
24	24	24				
25	25	25	25			
26						
27	27	27				
30						

than that for the competitively grown plants ($G = 0.155$), consistent with the hypothesis that competition increases the inequality in the size distribution (Weiner and Solbrig, 1984). However, because the sample sizes were small (6 individual plants and 75 surviving competitively grown plants), the sample coefficients may not be very precise estimates of the population coefficients. The bootstrap procedure can be used to quantify the precision and calculation a confidence interval for both Gini coefficients.

The six seeds that were grown individually are a random sample from all the seeds that were collected. If, by chance, a different set of six seeds had been used, the individual plant sizes would probably not be the same as those in the observed data, so the estimated Gini coefficient would probably be different. The sampling distribution of Gini coefficients (Fig. 13.1) could be calculated if one could take multiple samples from the same population. However, only one sample of six plants is available. The techniques used to make conclusions from sample means (Fig. 13.2) cannot be used for Gini coefficients because the sampling distribution is unknown and there is no formula for the variance of a small sample.

The bootstrap technique estimates the precision of a statistic by approximating the unknown sampling distribution in a two step procedure. First, the unknown distribution of values in the population is approximated by a discrete distribution (Fig. 13.3). Then, many bootstrap samples are drawn from this distribution. The unknown sampling distribution is approximated by the distribution of estimates from many bootstrap samples (Fig. 13.3). This bootstrap distribution is then used to estimate the bias, standard error, and confidence interval.

Because the bootstrap distribution is calculated by random sampling, two analyses of the same data will probably produce two different bootstrap distributions. Hence, estimates based on the bootstrap distribution may not be the same. This variability is minimized by taking a sufficiently large number of bootstrap samples. In general, more bootstrap samples are needed to estimate a confidence interval than are needed to estimate bias or standard error.

13.2.2 Calculating the Bootstrap Distribution

The sizes of the six individually grown *Ailanthus* plants were 18, 18, 20, 23, 25, and 28 leaf nodes. The Gini coefficient is $G = 0.112$, using Eq. 13.3. The estimated population distribution is a discrete distribution in which the value 18 occurs with probability 2/6 (because 2 plants had 18 leaves) and the values 20, 23, 25, and 28 each has probability 1/6. To draw a bootstrap sample from this distribution, one could write the size of each plant on the sides of a six-sided die. The number 18 will be on two sides, while the numbers 20, 23, 25, and 28 will each be on one side. The die is tossed, and the number that shows is written down. This is repeated five more times to generate a random sample of six sides. To generate the bootstrap sample on a computer, draw a simple random sample of six observations with replacement. Then, calculate the Gini coefficient for the bootstrap sample. For example, a bootstrap sample might be the values 20, 23,

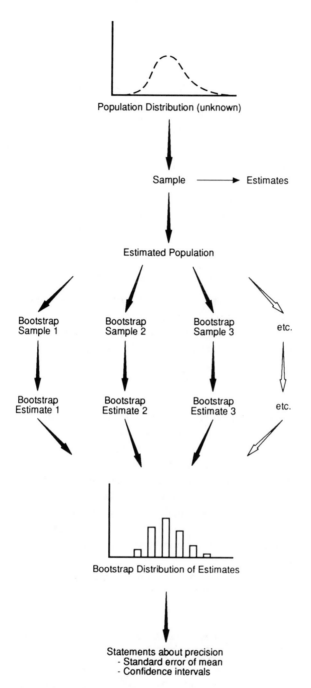

Figure 13.3. Use of data and the bootstrap distribution to infer a sampling distribution. The bootstrap procedure estimates the sampling distribution of a statistic in two steps. The unknown distribution of population values is estimated from the sample data, then the estimated population is repeatedly sampled, as in Fig. 13.1, to estimate the sampling distribution of the statistic.

18, 23, 28, and 18, for which, $G = 0.105$. Repeat this process for many bootstrap samples. Typically, 50 to 100 bootstrap samples are used to estimate a standard error; 1000 or more bootstrap samples are recommended to calculate confidence intervals (Efron, 1987). The choice of number of bootstrap samples will be further discussed in the practical details section.

Five bootstrap samples from this set of data are shown in Table 13.2. There are two important features of the bootstrap samples: they are not identical, and each bootstrap sample omits some observed values and repeats others. Both features result because the observed data are sampled with replacement. Notice that some of the bootstrap samples have Gini coefficients larger than the observed value, $G = 0.112$, while other samples have smaller coefficients.

13.2.3 Estimating Bias and Standard Error from the Bootstrap Distribution

Bias and standard error for a statistic such as the Gini coefficient can be estimated from the bootstrap distribution. The bootstrap estimate of bias is just the difference between the average of the bootstrap statistic and the statistic calculated from the original sample. Using Gini coefficients as an example,

$$\overline{bias} = \overline{G} - G \tag{13.5}$$

where G is the sample Gini coefficient and $\overline{G}$ is the mean of the bootstrap values. For the data from the individually grown *Ailanthus* plants, the sample Gini coefficient was 0.1121 and the average Gini value for the five bootstrap samples of Table 13.2 was 0.1099. Hence, the bootstrap estimate of bias was $0.1099 - 0.1121 = 0.0021$. In practice, one should use 50 to 100 bootstrap samples to estimate bias (Efron, 1987). Using 1000 bootstrap samples (data not shown), the biases for the individually and competitively grown *Ailanthus* are estimated to be -0.018 and -0.0020. In both cases, the bias is small. If the bias were larger, the estimate of bias could be subtracted from the observed value to produce a less biased estimate.

The standard error of the sample Gini coefficient is estimated by the standard deviation of the bootstrap distribution. That is

Table 13.2. *Five Bootstrap Samples for the Data from Individually Grown Plants*

Bootstrap sample						Gini coefficient
18	18	23	25	28	28	0.117
18	18	18	25	25	25	0.098
18	18	23	23	28	28	0.116
18	18	18	20	23	28	0.107
18	18	20	23	25	28	0.112

$$s_G = \sqrt{\frac{\sum_i (G_i - \overline{G})^2}{B-1}} \qquad (13.6)$$

Where G_i is the Gini coefficient calculated from the ith bootstrap sample, $\overline{G}$ is the average Gini coefficient in all the bootstrap samples, and B is the number of bootstrap samples. For the five bootstrap samples shown in Table 13.2, the estimated standard error is 0.0079. Again, five bootstrap samples are used only to demonstrate the calculations. Using 1000 bootstrap samples (data not shown), the estimated standard errors for the Gini coefficient of individually and competitively grown plants were 0.022 and 0.0097.

A standard error is a single measure of precision. It can be turned into a confidence interval if we assume a particular distribution, such as the normal. However, it is possible to use other features of the bootstrap distribution to directly estimate a confidence interval without assuming normality.

13.3 Calculating confidence intervals from the bootstrap distribution

A bootstrap confidence interval can be calculated from the bootstrap distribution in three different ways: the percentile bootstrap, the bias-corrected bootstrap, or the accelerated bootstrap (Efron, 1987). The three methods differ in their complexity and generality. The percentile bootstrap is the simplest, but least general. The bias-corrected bootstrap adjusts the percentile bootstrap for bias, while the accelerated bootstrap adjusts the percentile bootstrap for bias and skewness (Efron, 1987).

13.3.1 Percentile Bootstrap

The percentile bootstrap is the simplest and most commonly used method to construct bootstrap confidence intervals. In this method, the 2.5 and 97.5 percentiles of the bootstrap distribution are used as the limits of a 95% confidence interval. To calculate the 2.5th percentile from 1000 bootstrap replicates, sort the estimates from the bootstrap samples in order from smallest to largest and average the 25th and 26th largest values. The two values are averaged because 1000 times 2.5% is an exact integer (25). Similarly, the 97.5 percentile is the average of the 975th and 976th largest values. A histogram of 1000 bootstrap Gini coefficients for competitively grown *Ailanthus* is shown in Fig. 13.4. The 25th, 26th, 975th, and 976th largest values were 0.133, 0.133, 0.171, and 0.171, so (0.133,0.171) was the 95% confidence interval using the percentile method.

Percentile bootstrap confidence intervals can be calculated for any statistic, but they do not always work very well (Schenker, 1985). A confidence interval can be evaluated by simulating many data sets from a population with some known

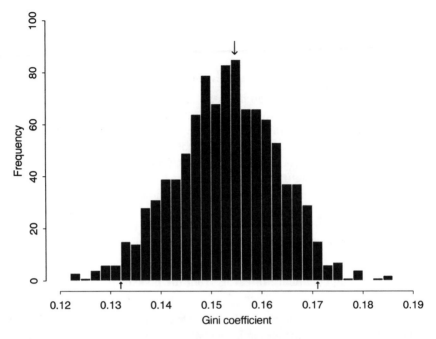

Figure 13.4. Histogram of 1000 bootstrap values for competitively grown *Ailanthus*. Arrows below the histogram mark the 25th and 976th largest values. Arrow above the histogram marks the Gini coefficient for the observed data ($G = 0.155$).

parameter, calculating a confidence interval from each data set, and counting how many confidence intervals bracket the true parameter. A 95% confidence interval should include the true parameter in 95% of the simulated data sets. When sample sizes are small (i.e., less than 50), the percentile confidence intervals for the variance (Schenker, 1985) and Gini coefficient (Dixon et al., 1987) are too narrow. For example, with samples of 20 points, 90% confidence intervals for the variance include the true variance only 78% of the time and 95% confidence intervals for a Gini coefficient of a lognormal population cover the true value of 0.30 only 85% of the time (Schenker, 1985; Dixon et al., 1987).

The problem with the percentile method is that it assumes that the bootstrap distribution is centered around the observed value. The percentile bootstrap produces correct confidence intervals when the observed value is the median of the bootstrap distribution (Efron, 1982). This is not the case for the Gini coefficient in this example, where 56% of the bootstrap values are smaller than the observed value (Fig. 13.4). For individually grown *Ailanthus*, 74% of the bootstrap values are smaller than the observed value. In both cases, the observed values are larger than the median of the bootstrap distribution, so the upper and lower confidence bounds are too low.

13.3.2 Bias-Corrected Bootstrap

The bias-corrected percentile bootstrap adjusts for a bootstrap distribution that is biased, i.e., systematically too low or too high. The bounds of bias-corrected intervals are found by calculating the bootstrap distribution, as before, then determining F, the fraction of bootstrap replicates that are smaller than the observed value. A second value, z_0, is calculated as the probit transform of F, that is

$$z_0 = \Phi^{-1}(F) \tag{13.7}$$

where Φ^{-1} is the inverse cumulative normal distribution. The appropriate percentiles for a 95% confidence interval are calculated by

$$P_1 = \Phi(2z_0 - 1.96) \tag{13.8}$$
$$P_u = \Phi(2z_0 + 1.96) \tag{13.9}$$

where Φ is the normal cumulative distribution function. The normal and inverse normal cumulative distribution functions are available in many statistical packages, or they can be computed from tables of areas of the normal curve (e.g., Zar, 1984). The value 1.96 is the normal critical value for a 95% confidence interval; it can be changed to generate other confidence intervals. Finally, the bounds of the bias-corrected confidence interval are given by the values in the bootstrap distribution that match the calculated percentiles, P_1 and P_u.

For the bootstrap distribution shown in Fig. 13.2, 56.4% of the values were smaller than the observed Gini coefficient of 0.1548, so $F = 0.564$. Hence, $z_0 = 0.166$, $P_1 = \Phi(-1.628) = 5.181\%$, and $P_u = \Phi(2.29) = 98.91\%$. The 5.181th and 98.91th percentiles of the bootstrap distribution are estimated by 52nd and 990th largest bootstrap estimates. The 52nd and 990th values were used because the number of bootstraps (1000) times the percentiles (5.181% or 98.91%) were not exact integers, so the observations corresponding to the next largest integer were used. The bounds of the bias-corrected 95% confidence interval were 0.136 and 0.176, which are both larger than the bounds of the percentile bootstrap, which were 0.133 and 0.171. The bias-correction procedure shifted the confidence interval to adjust for bias.

13.3.3 Accelerated Bootstrap: A Further Refinement

The accelerated bootstrap makes a second correction that is helpful when bootstrapping statistics like the variance or the Gini coefficient, where a few extreme points will have a large influence on the observed value. This technique is so named because it "accelerates" the bias correction (Efron, 1987). These confidence intervals depend on two constants, the z_0 used to correct for bias, and the acceleration factor, a, which corrects for skewness. The a coefficient is nonzero when a few points have a large influence on the observed estimate.

A nonparametric estimate of a can be computed by calculating the skewness of the empirical influence function. This function measures the sensitivity of the observed Gini coefficient to the presence of each point in the original sample. It can be estimated for each point using n perturbed sets of data, where n is the number of data points. The first perturbed data set has $t+1$ copies of the first data point and t copies of the other points. In general, the ith perturbed data set has $t+1$ copies of the ith data point and t copies of the rest. I have found $t = 5$ to give adequate estimates for Gini coefficients. The influence of the ith point on the Gini coefficient is calculated as

$$I_i = (G_i - G)/(nt+1) \tag{13.10}$$

where G_i is the Gini coefficient from the ith data set, G is the Gini coefficient from the original data set, and t is defined above. The coefficient a is proportional to the skewness of the n influence values. That is

$$a = \frac{\sum_{i=1}^{n} I_i^3}{6(\sum_{i=1}^{n} I_i^2)^{(3/2)}} \tag{13.11}$$

The percentiles for the upper and lower 95% confidence bounds are calculated as

$$P_1 = \Phi\left(z_0 + \frac{z_0 - 1.96}{1 - a(z_0 - 1.96)}\right) \tag{13.12}$$

$$P_u = \Phi\left(z_0 + \frac{z_0 + 1.96}{1 - a(z_0 + 1.96)}\right) \tag{13.13}$$

where Φ is the normal cumulative distribution function, a is defined by Eq. 13.11, and z_0 is defined by (13.7). As with the bias-corrected confidence intervals, the value 1.96 is the normal critical value for a 95% confidence interval. A different critical value could be used to generate other confidence intervals. The bounds of the accelerated confidence interval are given by the values in the bootstrap distribution that match the calculated percentiles, P_1 and P_u. If the distribution is symmetric, $a = 0$, and the accelerated bootstrap confidence intervals are the same as the bias-corrected intervals (Eqs. 13.8 and 13.9). When both a and z_0 are zero, the accelerated bootstrap confidence intervals are the same as the simple percentile bootstrap intervals.

Using the sizes of competitively grown *Ailanthus* (Table 13.1), the constant a was estimated as 0.0193. For the bootstrap distribution in Fig. 13.4, $z_0 = 0.166$, so $P_1 = \Phi(-1.568) = 5.85\%$, and $P_u = \Phi(2.383) = 99.14\%$. Hence, the bounds of the accelerated confidence interval were the 59th and 992nd largest values in the set of 1000 bootstrap replicates, i.e. (0.137, 0.176). For these data, both the bias-corrected and accelerated bootstrap procedures adjust the confidence intervals upward to account for the skewed and biased sampling distribution.

The three variations of the bootstrap described here are nonparametric techniques. They make no assumptions about the population (e.g., that population values follow a normal distribution) and use only information from the sample to estimate the precision of a statistic. All three techniques do assume independence. This assumption is critical; if it is violated, the bootstrap confidence intervals will be wrong. Other types of bootstrap procedures have been suggested, including the parametric bootstrap, which incorporates distributional assumptions (Efron, 1982), the percentile-t bootstrap, which uses the distribution of t-statistics rather than the distribution of estimates (Efron, 1992), and the Bayesian bootstrap, which samples the data points with unequal probabilities (Rubin, 1981). Although each may have theoretical advantages over nonparametric techniques (Efron, 1992), none has been frequently used in ecological studies.

Bootstrap confidence intervals for Gini coefficients can be calculated using the SAS program in Appendix 13.1. There are no SAS procedures to construct a bootstrap distribution or calculate a Gini coefficient, but one can do both using SAS data steps (see Appendix 13.1 for details). The SAS program in Appendix 13.1 takes about 8 seconds to run on a SUN Sparkstation I or 3 minutes to run on a 20 MHz 386 PC. FORTRAN subroutines to estimate bootstrap confidence intervals are available (Buckland, 1985). Programming languages like Fortran, Pascal, and C are more appropriate for bootstrapping because they are faster, but the advantage of SAS is that the program in Appendix 13.1 can be easily adapted to bootstrap any statistic that SAS can store in a data set. For example, to bootstrap variances or medians, replace the data steps that calculate Gini coefficients (DATA BOOTS; and DATA SKEWS;) with SAS procedures that store the desired statistics in an output data set.

13.4 The jackknife

The jackknife procedure (Miller, 1974) is a second general technique to answer the question, "How precise is my estimate?" It can estimate bias or standard error, but not a confidence interval (Efron, 1982). Although the bootstrap has theoretical advantages, the jackknife is often used in ecological applications because it requires considerably less computation and includes no random component. The basic idea behind the jackknife is that the bias and standard error of a statistic can be estimated by recalculating the statistic on a subset of the data.

13.4.1 Jackknifing the Gini Coefficient

The jackknife estimates of bias and standard error (Miller, 1974) are best described using an example such as the Gini coefficient for individually grown plants (Table 13.1). First, calculate the Gini coefficient based on all the data points ($G = 0.112$). Then, remove the first data point and calculate G_{-1} based on the remaining five data points (Table 13.3). Repeat the calculations, removing each point in turn, to calculate six G_{-i} values. Then, calculate p_i, called a pseudovalue, for each data point.

$$p_i = G + (n-1)(G - G_{-1}) \qquad (13.14)$$

where n is the number of data points, and G and G_{-i} are defined above. Table 13.3 lists the six jackknife samples, G_{-i} values, and pseudovalues for individually grown *Ailanthus*.

The jackknife estimate of bias (Efron, 1982) is

$$\overline{bias} = G - \bar{p} \qquad (13.15)$$

where G is the sample Gini coefficient and $\bar{p}$ is the mean of the jackknife pseudovalues. For the data from the individually grown *Ailanthus* plants, the sample Gini coefficient was 0.1121 and the average jackknife pseudovalue was 0.1128, so the jackknife estimate of bias is $0.1121 - 0.1128 = -0.0007$. The bias was small, but if it were larger, it could be subtracted from the observed value to produce a less biased estimate. The jackknife is especially effective at correcting an statistic for first-order bias, i.e. a bias that is linearly proportional to the sample size (Miller, 1974).

The jackknife estimate of the standard error is just the standard error of the pseudovalues:

$$s_G = \sqrt{\frac{\sum(p_i - \bar{p})^2}{n(n-1)}} \qquad (13.16)$$

where n is the sample size and $\bar{p}$ is the mean of the pseudovalues. From the data in Table 13.3, the estimated standard error of our Gini coefficient was 0.024, which was close to the bootstrap estimate of 0.022. The similarity is not surpris-

Table 13.3. Jackknife Samples with Gini Coefficients for Individually Grown *Ailanthus*

Jackknife sample					Gini coefficient	Pseudovalue
18	20	23	25	28	0.110	0.124
18	20	23	25	28	0.110	0.124
18	18	23	25	28	0.120	0.070
18	18	20	25	28	0.124	0.053
18	18	20	23	28	0.117	0.089
18	18	20	23	25	0.091	0.216

ing; theoretical results show that the jackknife is a linear approximation to the bootstrap procedure and that they converge at large sample sizes (Efron, 1982). For convenience and clarity, these calculations have been demonstrated using a small sample of data, but estimates of bias and standard error from small samples must be treated with some caution. As with the bootstrap, estimates from larger samples are more precise.

A SAS program to calculate the jackknife estimate of standard error is listed in Appendix 13.2. There is no built-in SAS procedure to jackknife a statistic, but the necessary operations can be programmed using a SAS data step or any other programming language. As with the bootstrap program, an advantage of SAS is that the program in Appendix 13.2 can be modified easily to jackknife any statistic that can be stored in a SAS data set.

In general, the jackknife cannot be extended to calculate confidence intervals (Efron, 1982). Unlike the bootstrap, it is not possible to calculate a nonparametric confidence interval directly from the jackknife replicates. There are too few of them, because the number of jackknife replicates is limited to the number of data points in the sample. Some attempts have been made to construct confidence intervals by assuming a normal distribution (Smith et al., 1986; Meyer et al., 1986). Such confidence intervals would have the form $(G \pm t_k s_G)$, where t_k is a critical value from a t distribution with k degrees of freedom. The problem with this approach is that the appropriate number of degrees of freedom is unknown, in spite of a lot of theoretical research (Efron and LePage, 1992). Using $n-1$ degrees of freedom, where n is the original sample size, has worked well in some cases (Meyer et al., 1986).

13.4.2 Ecological Example: Similarity of Zooplankton Communities before and after Disturbance

The jackknife has been used to estimate the precision of various similarity measures, including measures of diet similarity (Smith, 1985), community similarity (Smith et al., 1979; Smith et al., 1986; Heltshe, 1988), and niche overlap (Manly, 1990a). As part of a study on zooplankton community dynamics in a cooling water reservoir, Taylor and Mahoney (1988) counted the zooplankton present in four replicate samples during hot water discharge and in four replicate samples after discharge (Table 13.4). Eight species were found in both "during" and "after" samples, out of a total of 19 species. Hence, the Jaccard similarity was 0.42. The next step was to estimate the precision of the estimate.

In the zooplankton data, there were two levels of replication, samples and animals, so there were two possible ways to jackknife (or bootstrap) the data. One approach is to assume that animals are independently caught; the other is to assume that that samples are independently collected. If animal captures are independent, then the jackknife estimates of accuracy would be computed by sequentially removing each of the 1313 animals from the sample and recomputing

Table 13.4. Species Composition of Zooplankton Samples during and after Discharge of Hot Water[a]

Species	During				After			
Alona spp.	0	0	0	0	0	0	0	1
Bosmina sp.	0	4	2	1	5	0	13	6
Ceriodaphnia lacustris	1	0	0	0	0	0	0	0
Cyclopoid spp.	0	0	0	1	1	0	7	1
Chydorus spp.	0	0	0	0	0	0	1	1
Diaphanosoma brachyurum	0	0	1	0	0	0	0	0
Eurytema affinis	0	0	0	0	0	0	0	1
Euchalanis spp.	1	0	0	0	0	0	0	0
Ilyocryptus spp.	0	0	0	1	0	0	0	0
Keratella spp.	1	1	0	0	10	28	0	0
Kellicottia bostoniensis	0	0	0	0	0	0	1	0
Cyclopoid + *Diaptomus nauplii*	2	3	6	2	19	16	58	39
Oligochaeta	2	0	0	0	0	1	0	0
Polyarthra spp.	1	0	0	0	12	11	8	8
Tricocerca similis	0	0	0	0	3	1	0	3
Monostylus sp.	1	2	0	1	2	2	3	3
Rotifer, unknown spp.	1	1	1	2	3	4	3	1
Trichocerca spp.	0	0	0	0	8	15	3	2
Synchaeta stylata	0	0	0	0	362	374	66	168

[a]Data are counts per sample. Four samples were collected at each time.

the Jaccard coefficient. Bootstrap estimates would be computed by drawing a "during" bootstrap sample of 39 animals by resampling from the population of 39 animals collected during discharge and an "after" bootstrap sample of 1274 animals by resampling from the 1274 animals collected after discharge.

If samples are independent, jackknife estimates are computed by sequentially removing each of the eight samples. Bootstrap estimates are computed by resampling from the four samples during discharge and the four samples after discharge. The choice between these assumptions depends on how the samples were collected. In most ecological problems, treating samples as independent is more appropriate because of the sampling technique and clustering of individuals (Heltshe, 1988).

The Jaccard coefficient is estimated from two samples of data. Its precision can be estimated by bootstrapping (Smith et al., 1986), or by a two-sample jackknife (Smith et al., 1979), which is slightly different from the one-sample jackknife described previously. To use the two-sample jackknife, estimate jackknife values by sequentially removing each sample and recomputing the Jaccard coefficient. Let J_{ij} by the Jaccard coefficient after removing the jth sample from the ith group. The subscripts D and A will identify values from the "during" and "after" discharge groups. Removing the first sample from the "during" discharge samples changed the components of the Jaccard formula to six species found at both times, out of 17 species total, so $J_{D1} = 0.353$. Jackknife values for the other "during" samples and all the "after" samples are estimated by each sample in turn. Calculate a pseudo-value from each J_{ij} by

$$p_{ij} = (n_i - 0.5)J - (n_i - 1)J_{ij} \qquad (13.17)$$

where n_i is the number of samples in the ith group (here 4), and J is the Jaccard coefficient from the complete data. Table 13.5 presents the jackknife Jaccard coefficients and pseudovalues for each sample. Then, calculate the variance of the pseudovalues in each group:

$$s_i^2 = \frac{\sum_{j=1}^{n_i}(p_{ij} - \bar{p}_i)^2}{n_i - 1} \qquad (13.18)$$

where $\bar{p}_i$ is the mean of the pseudovalues of the ith group.

The jackknife estimate of the standard error of J is calculated by pooling the two variances. The estimate is

$$s_J = \sqrt{\frac{s_A^2}{n_A} + \frac{s_D^2}{n_D}} \qquad (13.19)$$

where n_A and n_D are sample sizes for the "after" and "during" groups, and s_A^2 and s_D^2 are the sample variances of the pseudovalues in the "after" and "during" groups. The variances of the pseudovalues in Table 13.5 were 0.0142 and 0.0170, so the jackknife estimate of the standard error of J was $\sqrt{0.0142/4 + 0.0170/4} = 0.088$. An SAS program to calculate the Jaccard similarity and the two-sample jackknife estimate of standard error is listed in Appendix 13.3.

Table 13.5. Jackknife Values and Pseudovalues for Zooplankton Samples.

Group	Sample omitted	Jaccard value	Pseudovalue
During	1	0.353	0.415
	2	0.421	0.211
	3	0.444	0.140
	4	0.389	0.307
After	1	0.421	0.211
	2	0.368	0.368
	3	0.444	0.140
	4	0.471	0.062

[a]Raw data are given in Table 13.4.

13.5 Practical details

13.5.1 Choice of Technique

The bootstrap and the jackknife are two techniques to answer the same question, "How precise is a particular statistic?" The choice of technique depends on the measure of precision. Only the bootstrap estimates a confidence interval, but either technique estimates the bias or standard error. However, the two techniques will not usually give the same answer. Which is better? Some theoretical results show that in extremely large samples, the techniques will give the same answers (see Efron, 1982; Efron and Tibshirani, 1986 for details). Often both techniques give similar answers, but if the differences are important, the performance of the two techniques has to be evaluated on a case-by-case basis.

To estimate a confidence interval, statistical theory suggests that one should use a bootstrap (Efron, 1992). Confidence intervals based on the jackknife standard error assume that the statistic has a normal distribution. If the distribution is skewed or heavy-tailed, then the jackknife confidence intervals are likely to be inaccurate.

The distribution of the statistic influences which bootstrap confidence interval technique (percentile, bias-corrected, or accelerated) is appropriate. The accelerated interval is the most general of the three; it reduces to the percentile confidence interval if the two constants, z_0 and a, are zero. However, there are two costs associated with the accelerated interval: it requires more complicated computations and it requires more bootstrap replicates than does the percentile method, because the two constants are estimated. Theoretical arguments suggest that the accelerated confidence interval is generally better than the other two techniques (Efron, 1987). It uses features of the data to make adjustments to the confidence interval.

Simulation studies of the coverage of the three bootstrap confidence intervals suggest that there are few differences between the three intervals for Gini coefficients (Dixon et al., 1987) or ratios (Dorfman et al., 1990), and few differences between the percentile and bias-corrected intervals for population growth rates (Meyer et al., 1986). All three of the above cited studies examined the coverage of a confidence interval by repeatedly simulating samples from some population, calculating confidence intervals, and counting the number of intervals that included the true population parameter. However, none of these studies reported the fraction of intervals that was too high or low. A 95% equal-tailed confidence interval should be too low 2.5% of the time and too high 2.5% of the time, for an overall coverage of 95%. Simulations with confidence intervals for the Gini coefficient suggest that the accelerated interval is more likely to produce equal-tailed confidence intervals (Dixon, unpublished results).

13.5.2 What Should I Bootstrap?

The key assumption behind the bootstrap is independence. The data are assumed to be independent samples from some population. This assumption determines what should be bootstrapped. In the zooplankton example, there were two levels of sampling: samples and individuals. If there were any patchiness, individuals were not independent, so bootstrapping individuals would not be appropriate. If the sampling scheme was appropriate, the four samples were independent samples from areas of the reservoir, so bootstrapping samples would be appropriate.

The choice of what to bootstrap becomes very important in more complicated problems. For example, consider bootstrapping a regression or correlation coefficient. Bootstrapping a correlation coefficient between two traits on individuals is relatively easy (Efron, 1982). If individuals were randomly sampled from some population, then the bootstrap replicates are formed by randomly choosing individuals with replacement from the sample, just as was done for Gini coefficients in my example. Bootstrapping a regression is more complicated because it can be done in two different ways (Efron and Tibshirani, 1986, Section 5). One can fit a regression, estimate the residuals and predicted values, and generate bootstrap replicates by adding randomly sampled residuals to the predicted values. This procedure estimates the precision of the regression coefficients, assuming that the regression model is correct (Efron and Tibshirani, 1986). The other procedure is to randomly select individuals, just like bootstrapping a correlation coefficient. This procedure estimates the precision of the regression coefficients, even if the regression model is not correct (Efron and Tibshirani, 1986). This bootstrap can also be used to estimate the precision of model selection procedures like stepwise selection and nonparametric regressions (Efron and Tibshirani, 1991). It is also more appropriate when the independent variables are random, not fixed in advance by the investigator.

13.5.3 How Many Bootstrap Replicates Should I Use?

Increasing the number of bootstrap replicates increases both the precision of the estimated standard error or confidence interval and the cost of computing it. Some recommended numbers of bootstrap replicates are approximately 100 to estimate a variance or standard error and 1000 or more to estimate a confidence interval (Efron, 1987, section 9). Variances can be estimated more precisely because the variance is an average of squared deviations, while the endpoints of the confidence interval are individual data points. Around 1500 bootstrap replicates should be used to compute the bias-corrected or accelerated confidence intervals (Efron, 1987, Section 9). The extra replicates are needed because the estimation of the z_0 and a constants introduces extra variability into the endpoints of the confidence intervals.

Although generating more bootstrap replicates increases the precision of the estimated standard error or confidence interval, there is a limit. Bootstrap estimates have two sources of variability, one due to variability among the bootstrap replicates and one due to sampling variability in obtaining the original data. Jackknife estimates have only one source of variability, that from the original data. The accuracy of inferences made by either bootstrapping or jackknifing is limited by the amount of information in the original data. Collecting more data will increase the accuracy of either technique. Both techniques solve an otherwise extremely difficult problem, the estimation of precision, but neither technique is a panacea for small sample size or extraneous variability.

Acknowledgments

Preparation of this manuscript was supported by Contract DE-AC09-76-SR00-819 between the U.S. Department of Energy and the University of Georgia. Alexandra Webber drafted Figs. 13.1–13.3. Karen Garrett, Bob Keeland, Milda Vaitkus, Gene Schupp, Gene Rhodes, Andy Schnabel, and an anonymous reviewer provided helpful comments on drafts of this manuscript.

Appendix 13.1

SAS program code for bootstrapping Gini coefficients.

```
DATA BOOT SKEW;                    /read the data and create two data sets:
                                   BOOT contains the original data and all the bootstrap
                                   samples, SKEW contains the original data and
                                   samples needed to estimate the coefficient in
                                   the accelerated bootstrap/
ARRAY DATA{200} D1-D200;           /array to store the original data, set up for a
                                   maximum of 200 data points, increase here and in
                                   next statement if necessary/
RETAIN D1-D200                     /tell SAS not to clear the array/
       NPT 0;                      /NPT counts the number of data points/
KEEP IBOOT                         /IBOOT is the number of the bootstrap
                                   sample, 0 = original data/
     ISKEW                         /ISKEW is the number of sample in the
                                   skew data set, 0 = original data/
     X;                            /X is an observed value/
INFILE CARDS EOF=WRITE;            /goto the section labeled write
                                   when done reading the data/
                                   /USER specified parameters/
NBOOT = 1000;                      /number of bootstrap samples to draw/
ALPHA = 0.05;                      /1 - size of confidence interval/
                                   /read in original data and copy to boot data set/
IBOOT = 0;                         /iboot = 0 = > working with original data/
INPUT X @@;                        /read a data value, possibly more than one on a line/
NPT = NPT + 1;                     /update the number of data values/
DATA{NPT} = X;                     /store this value in the data array/
OUTPUT BOOT;                       /output to the boot data set/
RETURN;                            /and go get another input value/
                                   /after reading all the data, create
                                   the boot and skew data sets/
WRITE:                             /generate nboot bootstrap samples and output them/
  DO IBOOT = 1 TO NBOOT;           /loop executed once for each sample/
    DO I = 1 TO NPT;               /loop executed once for each point/
      PT = 1+FLOOR(NPT*            /randomly choose an integer
           UNIFORM(0));            from 1 to npt, with replacement/
      X = DATA{PT};                /find that data value in the array/
      OUTPUT BOOT;                 /and store it on the data set/
    END;
  END;

                                   /use SAS macro variables to store useful
                                   pieces of information between data steps/
CALL SYMPUT("NPT",NPT);            /the number of data points/
CALL SYMPUT("NBOOT",NBOOT);        /the number of bootstrap samples/
CALL SYMPUT("ALPHA",ALPHA);        /and the alpha level/
```

*** The accelated bootstrap needs the coefficient a, which is the skewness of the influence function. Estimate this by the skewness of the empirical influence estimates. For each data point, compute the empirical influence by comparing the statistic calculated from the original data set repeated 5 times with the statistic calculated from 6 repeats of the target data point and five repeats of rest data set. ***

```
T = 5;                              /number of copies of each point/
CALL SYMPUT("T",T);                 /store in a macro variable for future use/
DO ISKEW = 0 TO NPT;                /loop to repeat once for each data point,
                                     ISKEW = 0 indicates just multiplying
                                     original data, not adding any points/
   IF ISKEW > 0 THEN DO;            /if need to add a point, do it/
      X = DATA{ISKEW};             /by finding that data value in the array/
      OUTPUT SKEW;                  /and storing it on the skew data set/
   END;
   DO J = 1 TO NPT;                 /loop to add t copies of every data value/
      X = DATA{J};                 /find the jth value from the array/
      DO K = 1 TO T;                /and add t copies to the skew data set/
         OUTPUT SKEW;
      END;
   END;
END;
```

*** USER adds raw data here. Current input command is set up to read values of one variable. These may be stored 1 value per line or more than one value per line. An unlimited number of lines of data is permitted. If more than 200 data values, users needs to increase the size of the data array and increase the number of D variables used (see top of this data step). ***

```
CARDS;
18 18 20 23 25 28
;
```

*** Calculate Gini coefficient for the original data and each bootstrap sample. To bootstrap other statistics, replace the following PROC SORT and DATA step with a procedure that calculates a statistic BY IBOOT and stores it in a variable named BOOTX on an output data set named BOOTS. ***

```
PROC SORT DATA=BOOT;                /sort each bootstrap sample so that the
   BY IBOOT X;                            values are in increasing order/
DATA BOOTS; SET BOOT;
   FILE 'BOOT6.DAT';
   RETAIN GINISUM                   /accumulates (2i-n-1)*X/
          SUM                       /accumulates X, so we can calculate mean/
          IPT                       /index of data value, 1=smallest, NPT=largest/
          NPT;                      /number of values in this sample/
   BY IBOOT;                        /need to do these calculations separately
                                          for each bootstrap sample/
   IF FIRST.IBOOT THEN DO;          /if this is the first value in the sample
      GINISUM = 0;                        initialize sums and counters/
      IPT = 0;
      SUM = 0;
      NPT = SYMGET("NPT");
   END;
   IPT = IPT + 1;                   /then add 1 to the number of data points/
   SUM = SUM + X;                   /add the value to the sum/
   GINISUM = GINISUM +             /and add (2i-n-1)X to its sum/
             (2*IPT - (NPT + 1))*X;
   IF LAST.IBOOT THEN DO;           /if just did the last value in the sample/
      BOOTX = GINISUM/((NPT-1)*SUM); /calculate Gini coefficient for sample/
      OUTPUT;                       /store on boots data set/
      IF IBOOT = 0 THEN CALL        /if this was the Gini coeff for the observation data/
         SYMPUT("SAMPLE",BOOTX);    /store it it a macro variable/
```

```
    PUT BOOTX;
    END;
    KEEP BOOTX IBOOT;                    /BOOTS data set only needs the estimate from the
                                          bootstrap sample (BOOTX) and the number of the
                                          bootstrap sample (IBOOT), needed to keep the
                                          original sample (IBOOT=0) separate from the
                                          bootstrap replicates (IBOOT > 0)/

*** Calculate the Gini coefficient for each set of data in SKEW. Each unique set identifer
    by a unique value of ISKEW. N.B. This is needed only to compute the a coefficient
    for the accelerated bootstrap CI. The algorithm identical to that used for
    the bootstrap replicates above, except that the number of points in each skew
    data is n*t or n*t+1 ***

PROC SORT DATA=SKEW;
    BY ISKEW X;
DATA SKEWS; SET SKEW;
    RETAIN GINISUM IPT SUM NPT;
    BY ISKEW;
    IF FIRST.ISKEW THEN DO;
      GINISUM = 0;
      IPT = 0;
      SUM = 0;
      NPT = SYMGET("NPT") *              /number of points in unperturbed skew data/
          SYMGET("T");
      IF ISKEW = 0 THEN                  /one extra point in each perturbed/
          NPT = NPT + 1;
    END;
    IPT = IPT + 1;
    SUM = SUM + x;
    GINISUM = GINISUM +
          (2*IPT - (NPT + 1))*X;

    IF LAST.ISKEW THEN DO;
      BOOTX = GINISUM/((NPT-1)*SUM);
      OUTPUT;
    END;
    KEEP BOOTX ISKEW;
                                         /calculate influence function for each data point;
                                          again, only needed to get the a coefficient for
                                          the accelerated bootstrap/
DATA _NULL_;                             /will not create an output data set/
    RETAIN S0                            /S0 is the Gini coefficient in the original data/
           U3 0                          /U3 accumulates third moments of I/
           U2 0;                         /U2 accumulates second moments of I/
    SET SKEWS;                           /work with the Gini coeff calculated
                                          for each perturbed data set/
    NPT = SYMGET("NPT");                 /NPT = number of points in original data,
                                         used to figure out number of skew data sets/
    IF ISKEW = 0 THEN S0 = BOOTX;        /remember the original Gini value/
      ELSE DO;                           /if value from a perturbed data set/
        INFL = (BOOTX - S0)/             /calculate influence/
            (SYMGET("T") + 1);
        U3 = U3 + INFL**3;               /accumulate third moments/
        U2 = U2 + INFL**2;               /accumulate moments/
      END;
    IF ISKEW = NPT THEN DO;              /if this is the last perturbed data set/
```

```
A = U3/(6*U2**(3/2));          /calculate skewness of influence/
CALL SYMPUT("A", A);           /and store it in a macro variable/
END;

                               /calculate the percentiles for the bias-corrected
                                        and accelerated bootstraps/
DATA  NULL_; SET BOOTS;
RETAIN M0                       /M0: Gini coefficient in original data set/
       P0;                      /P0 will count number of bootstrap Gini coefficients
                                    less than M0. This is used to get the Z0 coefficient
                                                   for bias correction/
IF IBOOT = 0 THEN M0 = BOOTX;   /store the original Gini coefficient/
  ELSE DO;                      /if a bootstrap replicate/
    IF BOOTX < M0 THEN          /see if less than original value/
      P0 = P0 + 1;              /if so, add 1 to P0/
END;

                               /calculations for the percentiles/
IF IBOOT = SYMGET("NBOOT")     /after looking at all bootstrap replicates/
    THEN DO;
  ALPHA = SYMGET("ALPHA");      /remember what the alpha level was/
  A = SYMGET("A");              /calculated for A coefficient/
  Z0 = PROBIT(P0/              /calculate Z0 coefficient/
       SYMGET("NBOOT"));
  ZA2 = PROBIT(1-ALPHA/2);      /use alpha to calculate critical value/
  PL = 100*ALPHA/2;             /percentiles for percentile confidence interval/
  PU = 100*(1-ALPHA/2);         /these depend just on alpha/
  PLBC = 100*PROBNORM           /percentiles for bias-corrected CI/
         (2*Z0-ZA2);
  PUBC = 100*PROBNORM           /these depend on Z0/
         (2*Z0+ZA2);
  PLBCA = 100*PROBNORM(Z0 +     /percentiles for accelerated method/
         (Z0-ZA2)/(1-A*(Z0-ZA2)));
  PUBCA = 100*PROBNORM(Z0 +     /these depend on Z0 and A/
         (Z0+ZA2)/(1-A*(Z0+ZA2)));
  CALL SYMPUT("PL",PL);         /store percentiles in macro variables
  CALL SYMPUT("PU",PU);           to be used by PROC UNIVARIATE/
  CALL SYMPUT("PLBC",PLBC);
  CALL SYMPUT("PUBC",PUBC);
  CALL SYMPUT("PLBCA",PLBCA);
  CALL SYMPUT("PUBCA",PUBCA);
  CALL SYMPUT("CI",100*         /also store 1-alpha value
         (1-ALPHA));
  CALL SYMPUT("Z0",Z0);           and the Z0 value to use in a title/
END;
PROC UNIVARIATE DATA=BOOTS;     /find the appropriate percentiles
                                    of the bootstrap distribution/
WHERE IBOOT > 0;               /do not include sample value/
VAR BOOTX;
OUTPUT OUT=PERC                /store percentiles in a data set/
    PCTLPTS = &PL &PU &PLBC    /specifies desired percentiles/
      &PUBC &PLBCA &PUBCA
    PCTLNAME=CI_L CI_U BC_L    /and their variable names/
      BC_U A_L A_U
    PCTLPRE =BOOT;             /needed, but ignored by SAS ver 6.07/
```

```
PRINT TGLSV[FORMAT=6.2];
RES=Y-X*WTS;                                          /residuals/
FILE 'RESID.OUT';                                   /system file/
I=YD[,1];                                      /vector of row positions/
J=YD[,2];                                  /vector of column positions/
DO ITER = 1 TO 25 BY 1;             /output residuals to system file/
    II=I[ITER,1];
    JJ=J[ITER,1];
    RS=RES[ITER,1];
    PUT @1 II 1.0 @10 JJ 1.0 @ 20 RS 15.7;
END;
```

Appendix 13.2

SAS program code for jackknifing Gini coefficients.

```
DATA JACK;                              /read the data and create data
                                        set with the jackknife samples/
    ARRAY DATA/200/D1-D200;            array to store the original data, set up for a
                                       maximum of 200 data points; increase here and
                                       in next statement if necessary/
    RETAIN D1-D200                     /tell SAS not to clear the array/
        NPT 0;                         /NPT counts the number of data points/
    KEEP IJACK                         /IJACK is the number of the jackknife sample/
        X I;                           /X is an observed value/
    INFILE CARDS EOF=WRITE;            /go to the section labeled write
                                       when done reading the data/
                                       /read in original data and copy to jackknife data set/
    IJACK = 0;                         /IJACK = 0 => working with original data/
    INPUT X @@;                        /read a data value, possibly more than one on a line/
    NPT = NPT + 1;                     /update the number of data values/
    DATA/NPT/ = X;                     /store this value in the data array/
    OUTPUT JACK;                       /output to the jack data set/
    RETURN;                            /and go get another input value/
                                       /after reading all the data, create the jack data sets/
    WRITE:                             /remove each point in turn from the data set/
        DO IJACK = 1 TO NPT;           /loop executed once for each sample/
            DO I = 1 TO NPT;           /look at each point in turn/
                IF I = IJACK THEN DO;  /if it is in this jackknife sample/
                    X = DATA/I/;       /get its value from the data array/
                    OUTPUT JACK;       /and store it on the data set/
                END;
            END;
        END;
                                       /use SAS macro variables to store useful
                                       pieces of information between data steps/
    CALL SYMPUT("NPT",NPT);            /the number of data points/
```

*** USER adds raw data here. Current input command is set up to read values of one
variable. These may be stored 1 value per line or more than one value per line. An
unlimited number of lines of data is permitted. If more than 200 data values, users
needs to increase the size of the data array and increase the number of D variables
used (see top of this data step). ***

```
CARDS;
18 18 20 23 25 28
```

*** Calculate Gini coefficient for the original data and each jackknife sample. To jackknife other statistics, replace the following PROC SORT and DATA step with a procedure that calculates a statistic BY IJACK and stores it in a variable named JACKX on an output data set named JACKS. ***

```
PROC SORT DATA=JACK;            /sort each jackknife sample so that the
  BY IJACK X;                            values are in increasing order/
DATA JACKS; SET JACK;
  RETAIN GINISUM              /accumulates (2i-n-1)*X/
     SUM               /accumulates X, for calculating mean/
     IPT         /index of data value, 1=smallest, NPT=largest/
     NPT                     /number of values in this sample/
     N                   /number of values in original sample/
     SAMPLE;          /value of Gini coefficient from original data/
  BY IJACK;                /need to do these calculations separately
                                    for each jackknife sample/
  IF FIRST.IJACK THEN DO;             /if this is the first value in the
     GINISUM = 0;              sample initialize sums and counters/
     IPT = 0;
     SUM = 0;
     N = SYMGET("NPT");      /find number of points in original data/
     IF IJACK = 0 THEN NPT = N;     /number of points in this sample/
              ELSE NPT = N-1;  /there is one less point in jackknife samples/
  END;
  IPT = IPT + 1;             /then add 1 to the number of data points/
  SUM = SUM + X;                       /add the value to the sum/
  GINISUM = GINISUM +               /and add (2i-n-1)X to its sum/
        (2*IPT - (NPT+1))*X;
  IF LAST.IJACK THEN DO;        /if just did the last value in the sample/
     JACKX = GINISUM/((NPT-1)*SUM);  /calculate Gini coefficient for sample/
     PSUEDOX = SAMPLE + (N-1)*             /and pseudo value/
           (SAMPLE-JACKX);
     OUTPUT;                        /store on JACKS data set/
     IF IJACK = 0 THEN DO;   /if this was the Gini coefficient for the observation data/
        SAMPLE = JACKX;       /remember the value in a variable/
        CALL SYMPUT("SAMPLE",JACKX);   /and store it it a macro variable/
     END;
  END;
  KEEP IJACK PSUEDOX;         /JACKS data set only needs the estimate of the
                             pseudovalue (pseudox) and the number of the
                             jackknife sample (IJACK) needed to keep the
                             original sample (IJACK=0) separate from
                             the jackknifed values (IJACK > 0)/

PROC UNIVARIATE DATA=JACKS;
  WHERE IJACK > 0; VAR PSUEDOX;
```

Appendix 13.3

SAS program code for computing a jackknife estimate of the standard error for the Jaccard coefficient of similarity using the two sample jackknife of Smith et al. (1986).

```
DATA SPP;                        /read species name and absence/presence or
                                 abundance data.  Program is currently setup for four
                                 replicates in each group; if different, change A4 and
                                 B4 in the next three lines to indicate the number of
                                 replicates in each group./
  INPUT SPNAME $ A1-A4 B1-B4;           /read data for one species/
  ARRAY ALLA A1-A4;              /ALLA stores replicates from one group/
  ARRAY ALLB B1-B4;              /ALLB stores replicates from the other group/
  IJACK = 0;                             /IJACK = 0 = > full data set/
  A = SUM(OF ALLA{*});            /A = 0 if species absent in group 1/
  B = SUM(OF ALLB{*});            /B = 0 if species absent in group 2/
  C01 = (A = 0)*(B > 0);         /calculate the bits of the Jaccard
  C10 = (A > 0)*(B = 0);           coefficient; C01 = in 2 only,
  C11 = (A > 0)*(B > 0);         C10 = in 1 only, C11 = in both/
  OUTPUT;                                /store on data set/
  DO OVER ALLA; GROUP = 1;       /then jacknife the first group of replicates/
    IJACK = IJACK + 1;           /these are numbered IJACK = 1 to NA/
    JA = A - ALLA;               /remove data from the ith sample/
    C01 = (JA = 0)*(B > 0);      /recalculate bits of the jaccard/
    C10 = (JA > 0)*(B = 0);
    C11 = (JA > 0)*(B > 0);
    OUTPUT;                                /store on data set/
  END;
  NA = IJACK;                    /NA = number of samples in 1st group/
  DO OVER ALLB; GROUP = 2;       /then jacknife the second group of replicates/
    IJACK = IJACK + 1;           /these are numbered IJACK = NA+1 to NA+NB/
    JB = B - ALLB;               /remove data from the ith sample/
    C01 = (A = 0)*(JB > 0);      /recalculate bits of the jaccard/
    C10 = (A > 0)*(JB = 0);
    C11 = (A > 0)*(JB > 0);
    OUTPUT;                                /store on data set/
  END;
  NB = IJACK - NA;               /NB = number of samples in 2nd group/
  CALL SYMPUT("NA",NA);          /store NA and NB in macro variables/
  CALL SYMPUT("NB",NB);

*** USER enters data here.  Format is: species name, list of abundances.
  First NA abundances are for the first group of samples, rest (NB) of abundances are
  for the second group.  Each species goes on one line of data, but you can modify
  the input command to read other forms of data. ***

CARDS;
alon3 0 0 0 0 0 0 0 1

  .....

;
PROC SORT; BY IJACK;                      /compute jaccard coefficient by
                                 sorting data by jackknife number/
PROC MEANS SUM NOPRINT;                   /then adding up (over species)
  BY IJACK;                      separately for each jackknife number/
  VAR C01 -- C11;                        /bits of the jaccard coefficient/
  ID GROUP;                              /include group number/
  OUTPUT OUT=SUMS SUM=B C A;             /store on a data set/
```

```
DATA JACCARD; SET SUMS;                    /use these sums to calculate Jaccard
                                                  for each jackknife sample/
    RETAIN SAMPLE;                         /remember J from original sample
                                           (needed to calculate pseudo-values/
    J = A/(A+B+C);                            /Jaccard coefficient formula/
    IF IJACK = 0 THEN DO;
        SAMPLE = J;                        /remember J if this was the original/
        CALL SYMPUT("SAMPLE",J);             /also store in a macro variable/
    END;
    ELSE DO;                               /otherwise, compute a pseudovalue/
        IF GROUP = 1 THEN                      /find N for this group/
            N = SYMGET("NA");
        ELSE N = SYMGET("NB");
        PJ = (N-0.5)*SAMPLE - (N-1)*J;        /and compute pseudovalue
    END;                                   using Smith et al. (1986) formula/
PROC PRINT; ID IJACK;                      /print out the values and pseudovalues/
    VAR GROUP A B C J PJ;
    TITLE 'JACKNIFE SAMPLES
        AND PSEUDOVALUES';
PROC MEANS MEAN STD VAR;                    /calculate the within group variances/
    WHERE GROUP NE .;
    BY GROUP; VAR PJ;
    OUTPUT OUT=SD STDERR=SE;               /store standard errors in a data set/
    TITLE 'VARIANCES OF
        PSUEDOVALUES FOR EACH GROUP';
DATA POOLED;                               /calculate standard error of the Jaccard/
    SET SD END=LAST;                       /LAST is set to 1 on the last obervations/
    RETAIN POOL 0;                         /POOL stores the sum of variance/N/
    POOL = POOL + SE*SE;                   /calculated by squaring the standard error/
    IF LAST THEN DO;
        POOL = SQRT(POOL);                 /calculate standard error of the Jaccard/
        OUTPUT;
    END;
PROC PRINT;                                /print results/
    TITLE "JACKNIFE SE FOR JACCARD
        COEFFICIENT (SAMPLE J = &SAMPLE)";
    VAR POOL;
```

14

Spatial Statistics: Analysis of Field Experiments

Jay M. Ver Hoef and Noel Cressie

14.1 Introduction

More and more, ecologists are turning to designed field experiments. Earlier in the history of ecology, it was enough to collect field observations to generate reasonable hypotheses (McIntosh, 1985). However, as these hypotheses multiplied, they needed to be rigorously tested. A glance at any current ecological journal will reveal that many ecologists are designing and analyzing field and laboratory experiments to test such hypotheses.

There are several difficulties associated with conducting field experiments. One is that they are usually expensive. Another is that true replication may be unattainable (e.g., Carpenter, 1990). Finally, there is often considerable "noise" in the data, both because the environment is heterogeneous at many scales (e.g., Dutilleul, 1993; Legendre, 1993) and because field measurements are often crude compared to those achieved under laboratory conditions, so there is greater measurement error in field studies (Eberhardt and Thomas, 1991). Even at smaller scales, it can be difficult to find "relatively homogeneous" areas. All of these factors contribute unwanted variability to the experiment. Hence, ecologists typically use as many experimental units as they can afford and hope for the best. Too often, natural variability simply swamps many of the treatment effects that we try to detect.

In light of the need for ecological experiments, and the expense and difficulty in conducting them, it is imperative that the designs and analyses of field experiments be conducted so that the ability to detect treatment effects is maximized. One of the common misuses of statistics is the use of a less powerful technique when more powerful techniques are available. Most ecological field experiments are spatial in nature, yet spatial information is not used in a classical analysis of variance (ANOVA) (Chapter 3). This chapter introduces what we will call the

GLS-variogram method, which uses the underlying spatial variation to estimate treatment contrasts (contrasts will be described in greater detail in Section 14.3.2) with greater precision than estimates using classical ANOVA.

The example that will be used in this chapter is a designed experiment to examine the effect of time of burn on plant species diversity. The data consist of the numbers of different vascular plant species, or species richness, in a 5×5 grid of 7×7 m contiguous plots (Fig. 14.1A). The data come from a glade in the Ozark plateau area of southeastern Missouri. Glades are grassy openings, usually caused by shallow and droughty soils, in a predominantly forested landscape (Kucera and Martin, 1957). This particular glade occurred on a relatively steep slope of 14% (based on the average of 25 measurements from the center of each plot), and on a dolomitic substrate that often weathers into exposed bedrock "benches," in a stairstep fashion. These 25 plots came from the center of the glade, and can be considered to be as homogeneous as possible from this particular glade. There were scattered small trees in the plots, but the main forest vegetation was at least 14 m from the edge of the plots in each direction.

Actually, the number of different plant species per plot was recorded from the plots *before* any fire treatments were added, so the underlying natural variability was obtained without any treatments actually applied (this is called a uniformity trial in the statistical literature). The overall mean was 24.08 different species per plot. Treatment effects were artificially added to the real data to simulate a real experiment for the purposes of demonstration, and to evaluate the classical ANOVA and GLS-variogram methods. A completely randomized design was employed (see Chapter 3). Thus, according to Table 14.1, we chose the number of species to change by $+6$ if the plot were randomly assigned a November burn, we chose the number of species to change by -3 if a plot were randomly assigned a May burn, etc. Then the treatment effect of a November burn is $\tau_4 = +6$, that of a May burn is $\tau_2 = -3$, etc. The reason for "simulating" the experiment, rather than actually conducting it, is to have known treatment effects. In a real experiment, the treatment effects are superimposed on the natural variability among the plots, and hence are unknown parameters to be estimated. Now, by analyzing the experiment and estimating the parameters (treatment effects) as if they were unknown, the "closeness" to the true values can be compared for a classical ANOVA and for the GLS-variogram method introduced in this chapter.

Although the effect of fire on species diversity was chosen to illustrate the spatial analysis of designed experiments, the method is not specific to any particular ecological problem. It may be applied to any field experiment where spatial coordinates are available (in one, two, or three dimensions) and where one would ordinarily use a classical ANOVA. Several three-dimensional examples include the assignment of treatments to fish in a lake or to insects in a tree, followed by a GLS-variogram analysis.

Column

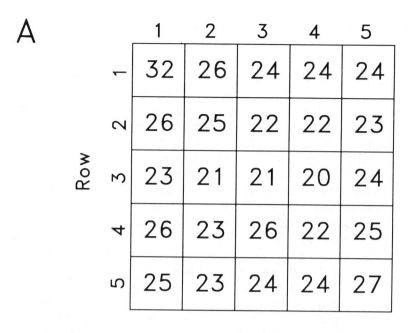

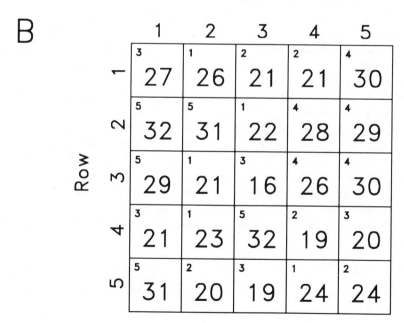

Figure 14.1. (**A**) Uniformity trial data set of species richness in a 5×5 grid of plots, each plot was 7×7 m. (**B**) Uniformity trial data set with treatment effects from Table 14.1 added randomly. The treatment number is given in the upper left corner of each plot.

14.2 Statistical Issues

A field experiment has spatial components; that is, the experimental units are located in one-, two-, or three-dimensional space. This spatial environment is usually heterogeneous. Typically, we observe that locations close together tend to have more similar values, or are positively correlated, than those that are farther apart; this tendency is termed spatial autocorrelation. (It is also possible to have temporal autocorrelation, often manifest as a time series, Chapter 7.) Notice that a cursory inspection of the data (Fig. 14.1A) seems to indicate that plots closer together have more similar values.

Well-designed experiments classically rely on three basic concepts: randomization, blocking, and replication (Chapter 3). In general, an experimental unit is the unit to which treatments are applied according to some design. In the example from the Ozarks (Fig. 14.1A), the experimental units are 7×7 m plots, to which fire treatments are applied. Replication is obtained by the application, several times, of the set of treatments to all of the experimental units. Replication allows estimation of treatment effects by averaging over the underlying variability in the experimental units (see discussion in Chapter 8). Blocking helps control natural heterogeneity by assigning experimental units to relatively homogeneous groups. Blocks may consist of experimental units that are spatially close, or related in some other way. For example, if a field has a shallow center drainage, it might be blocked into those areas that are higher and drier, and into those that are lower and wetter. Randomization is the process of assigning treatments, at random according to some design, to experimental units. Randomization helps to provide unbiased estimates of parameters.

Now we come to a subtle concept. Suppose that the experimental units themselves are random variables. For example, this would mean that if one were to go back in time, to say 10 years before the Ozark data were collected, and let time start over again, the number of species in each plot would be slightly different due to the random or stochastic processes of nature. If we could "generate" species numbers for the same set of plots again and again by going back in time repeatedly, then a statistical distribution would be obtained for each plot, as well as autocorrelation among all pairs of plots. In this case, the data from the experimental units are the result of what is termed a spatial stochastic process.

Table 14.1. Treatment Effects for Simulated Data

Treatment	Treatment symbol	Treatment effect
Control	τ_1	0
May burn	τ_2	-3
August burn	τ_3	-5
November burn	τ_4	$+6$
February burn	τ_5	$+6$

The alternative view is that if we went back in time again and again, we would always get exactly the same number of species per plot, i.e., nature is inherently deterministic. In the latter case, all of the "randomness" in an experiment comes from assigning treatments at random to experimental units. In other words, all of the probability statements (e.g., *P* values, tests of significance, confidence intervals) must come only from the randomness in the *design* of the experiment.

If nature is inherently stochastic, a randomized design has two sources of randomness, that in the design and that in the experimental units. These experimental units may be autocorrelated. Even Fisher (1935), the father of modern experimental design based on randomization, noted: "After choosing the area we usually have no guidance beyond the widely verifiable fact that patches in close proximity are commonly alike, as judged by the yield of crops, than those which are far apart." Randomization helps neutralize the effect of autocorrelation (Yates, 1938; Grondona and Cressie, 1991; Zimmerman and Harville, 1991). Intuitively, however, one knows that there is a cost to randomization. If the experimental units are considered to be random variables, with their own natural variability, then through randomization more variability is added. What is gained by the additional variability is the use of classical theory that relies on randomization.

An alternative approach to classical ANOVA is to *model* the spatial autocorrelation among the experimental units, making possible a more powerful analysis of the experiment. This is a relatively new approach to analyzing designed experiments in a spatial setting, and several recent papers show it is very powerful in a variety of data sets (Baird and Mead, 1991; Cullis and Gleeson, 1991; Grondona and Cressie, 1991; Zimmerman and Harville, 1991). In this chapter, we model the spatial autocorrelation with variograms that contain information on the spatial autocorrelation among the experimental units.

14.3 Example

Suppose we had conducted an experiment to examine the effect of fire, at different times of the year, on species richness in the Ozark glade (Fig. 14.1A). Also suppose that there were four treatments where plots were burned at four different times of the year: May, August, November, and February, and species richness was measured in the summer following the February burn. There was also a control, with no burn, so there were a total of five treatments. For illustration purposes, suppose that the true effects of burn time are those given in Table 14.1. These treatment effects allow our artificial experiment to be conducted as follows. To the 25 original plots (Fig. 14.1A), five replications of each treatment were assigned randomly (Fig. 14.1B). The only information that is available to the scientist is the plot location, the treatment applied to each plot, and the datum at each plot (with treatment effect added), i.e., the information in Fig. 14.1B.

Hence, we will analyze the data of Fig. 14.1B with a classical ANOVA and with the GLS-variogram method.

14.3.1 Statistical Model and Assumptions

A statistical model summarizes the experiment:

$$Y_{ijk} = \mu + \tau_k + \delta_{ij} \tag{14.1}$$

where Y_{ijk} is a random variable (species richness), i is the ith row ($i = 1,. . .,5$, numbered from top to bottom), j is the jth column ($j = 1,. . .,5$, numbered from left to right), and k is the kth treatment ($k = 1,. . .,5$); μ is the overall mean, τ_k is the kth treatment effect, and δ_{ij} is random error for the (i,j)th plot, with possible spatial autocorrelation among the δ_{ij}. It is the statistical distribution of δ_{ij} that provides probability statements about estimates of treatment effects or combinations of treatment effects (e.g., $\mu + \tau_1$ or $\tau_1 - \tau_2$). For a classical ANOVA, all δ_{ij} are assumed to be normally distributed, have expectation of zero, and have constant variance (σ^2). Finally, as mentioned in Section 14.2, randomization in the experimental design enables one to proceed as if δ_{ij} and δ_{st} are independent when $(i,j) \neq (s,t)$ (the errors at all locations are independent).

The randomization distribution yields random errors δ_{ij} that have small autocorrelation and hence allows use of the classical theory (Grondona and Cressie, (1991), as confirmed by several simulations (Besag and Kempton, 1986; Baird and Mead, 1991; Zimmerman and Harville, 1991). However, only one randomization actually occurs. If we ignore all possible randomizations, and just concentrate on the one that occurred, then natural variation in the plots implies that the δ_{ij} are autocorrelated (see Section 14.3.3). The problem then is how to model the autocorrelation and subsequently use this information to obtain the best parameter estimates and associated probability statements. This problem will receive further attention in Section 14.3.4.

14.3.2 Treatment Contrasts

Often, the ultimate goal of an experiment is to estimate a treatment *contrast* (sometimes called a planned comparison). If the initial hypothesis of no significant difference among treatment means is rejected, an important next step is to ask, "How different are the treatment means?" (Snedecor and Cochran, 1989, p. 224; Hicks, 1982, p. 46; Day and Quinn, 1989). Treatment contrasts express the *difference* in treatment effects, or the difference of a combination of treatment effects. For the Ozark example, we might be interested in how a "summer" burn affects the subsequent summer's species diversity. In terms of the treatments listed in Table 14.1, this is expressed as the average of May and August burns minus the control. Using the mathematical notation from Table 14.1, this contrast becomes

$$c_1 = \frac{1}{2}(\tau_2 + \tau_3) - \tau_1 \qquad (14.2)$$

From Table 14.1, the true value is

$$c_1 = \frac{1}{2}(-3-5) - 0 = -4$$

This means that the actual effect of a summer burn was to decrease by 4 the number of species per plot in the subsequent summer. In the case of c_1, the multiplying coefficients for all treatment effects τ_i; $i = 1,\ldots,5$, are $\{-1, 0.5, 0.5, 0, 0\}$. It is characteristic of treatment contrasts that the coefficients sum to zero.

As another example, consider how much a "winter" burn affects the subsequent summer's species diversity. In terms of the treatments (Table 14.1), this is expressed as the average of November and February burns minus the control. In mathematical notation:

$$c_2 = \frac{1}{2}(\tau_4 + \tau_5) - \tau_1 \qquad (14.3)$$

Again, from Table 14.1, the true value is

$$c_2 = \frac{1}{2}(6+6) - 0 = 6$$

This means that the true effect of a winter burn was to increase species richness by six per plot in the subsequent summer.

The difference between winter and summer burns may also be of interest:

$$c_3 = \frac{1}{2}(\tau_4 + \tau_5) - \frac{1}{2}(\tau_2 + \tau_3) \qquad (14.4)$$

The multiplying coefficients for c_3 are $\{0, -0.5, -0.5, 0.5, 0.5\}$. From Table 14.1, the true value is

$$\frac{1}{2}(6+6) - \frac{1}{2}(-3-5) = 10$$

or an increase of 10 species per plot by winter burns over summer burns. Other contrasts of interest are the difference between the two summer burns,

$$c_4 = (\tau_2 - \tau_3) = 2 \qquad (14.5)$$

and the difference between the two winter burns,

$$c_5 = (\tau_4 - \tau_5) = 0 \tag{14.6}$$

Next we shall see how a classical ANOVA compares to the GLS-variogram approach for estimating the five treatment contrasts, Eqs. 14.2–14.6. Recall that the known treatment effects (Table 14.1) were added to the original, untreated data (Fig. 14.1A) in a completely randomized design (Fig. 14.1B). Now the data in Fig. 14.1B will be used to estimate contrasts with classical ANOVA and GLS-variogram. Because the experiment was conducted artificially, the estimated values can be compared to the known values.

14.3.3 Classical ANOVA Compared to GLS-Variogram Method

Before going through the mechanics of the GLS-variogram method, we motivate its use by giving some results and comparing them to classical ANOVA. For the example data (Fig. 14.1B), a classical ANOVA provided contrast estimates of Eqs. 14.2–14.6 and their estimated standard errors. Table 14.2 shows that only for the second and third contrasts would the null hypothesis of zero be rejected as $P<0.05$ based on a classical ANOVA. For example, to test whether the first contrast is significantly different from zero, the t value is $(-2.4-0)/1.29 = 1.86$. Because

$$|-1.86|<t_{\alpha = 0.025, \, df = 20} = 2.07$$

the first contrast is not declared to be significantly different from zero for a two-tailed test at significance value level $\alpha = 0.05$. Of course, we would not want to reject the hypothesis that contrast c_5 equals zero (but we will, one out of twenty times on average, when $\alpha = 0.05$). Therefore, of the four contrasts that had true

Table 14.2. Contrast Estimates for Example Data (Fig. 14.1B)[a]

Contrast	True value	OLS est	OLS se	GLS est	GLS se	iter1 GLS est	iter1 GLS se
$c_1 = (\tau_2+\tau_3)/2-\tau_1$	-4.00	-2.40	1.29	$-2.80*$	0.96	$-2.82*$	0.94
$c_4 = (\tau_4+\tau_5)/2-\tau_1$	6.00	6.60*	1.29	6.69*	1.09	6.71*	1.07
$c_3 = (\tau_4+\tau_5)/2-(\tau_2+\tau_3)/2$	10.00	9.00*	1.05	9.49*	0.88	9.53*	0.86
$c_4 = (\tau_2-\tau_3)$	2.00	0.40	1.49	0.45	1.15	0.46	1.13
$c_5 = (\tau_4-\tau_5)$	0.00	-2.40	1.49	-2.07	1.60	-2.04	1.60

[a]The true values are given first. The next two columns contain classical ANOVA estimates (OLS est), along with their estimated standard errors (OLS se). The next two columns contain contrast estimates from the GLS-variogram method (GLS est) along with their estimated standard errors (GLS se). One more iteration is also given. Contrast estimates that test as significantly different from 0, at $P<0.05$, are marked with an asterisk.

values not equal to zero, the classical analysis had only enough power to detect two of them.

The contrast estimates using the GLS-variogram method, and their estimated standard errors, are also given in Table 14.2. Notice how the contrast estimates using the GLS−variogram method generally migrate from the classical ANOVA contrast estimates toward the true contrast values with each iteration, and the standard error decreases in size. Further iterations only changed the estimates past the second decimal place. The estimated standard errors in Table 14.2 reflect the fact that the contrast estimates using the GLS-variogram method are better than the classical contrast estimates. For the contrast estimates using the GLS-variogram method, the null hypothesis of zero contrast values can be rejected at $\alpha = 0.05$ for the first three contrasts. For example, to test whether the first contrast c_1 is significantly different from zero, the Z value is $(-2.8-0)/0.96 = -2.92$. Because

$$|-2.92|>Z_{\alpha/2 \, = \, 0.025} = 1.96$$

the first contrast is declared to be significantly different from zero at $\alpha = 0.05$. The inference procedure for the GLS-variogram method is only approximate, which is why we use the Z distribution as opposed to the t distribution used for the classical ANOVA. Our simulations indicate that if the residuals are roughly normally distributed, the Z distribution is appropriate for inference. Alternatively, a 95% confidence interval for c_1 is

$$-2.8\pm(1.96)(0.96) = (-4.68, -0.9)$$

Notice that the GLS-variogram method has more power than the classical ANOVA to detect contrast c_1. Also, notice that the absolute values of the true contrast values increase by increments of two (Table 14.2). Basically, the classical ANOVA had enough power to detect contrasts with absolute values greater than or equal to six. The GLS-variogram method had enough power to detect contrasts with absolute values greater than or equal to four. Neither method had enough power to detect contrast c_4 with a magnitude of two, and neither method committed the type I error of falsely rejecting the null (true) hypothesis of contrast c_5 as being equal to zero.

It is interesting and instructive to compare the standard error estimates of contrasts c_4 and c_5 for both classical ANOVA and GLS-variogram. For c_5, the GLS-variogram method gives a higher estimated standard error than classical ANOVA. Inspection of Fig. 14.1B shows why. The randomized assignments for treatments 4 are clustered in the upper right while treatments 5 are clustered toward the left. The GLS-variogram method accounts for the "poor" randomization for contrast c_5, that is, treatment effects may be confounded with local variation. On the other hand, for contrast c_4, treatments 2 and 3 are well-dispersed and intermixed throughout the plots—a "good" randomization (Hurlbert, 1984)—and consequently the GLS-variogram estimated standard error is lower. Because classical ANOVA relies on the average over *all* randomizations, it does not recognize the singular features of this particular randomization, and hence does not differ in the standard errors for the two contrast estimates. It is also apparent that contrast c_5 is farther from its true value than contrast c_4 is from its true value, which was reflected in the GLS-variogram standard error but not the classical ANOVA standard error.

14.3.4 Statistical Methods and Computer Programs

The problem of estimating the parameters and autocorrelation is that knowledge of one is required to estimate the other. However, we have neither; only the values for Y_{ijk} are known. From Eq. 14.1, it can be seen that if $\mu+\tau_k$ were known, then δ_{ij} is known by taking $\delta_{ij} = Y_{ijk}-\mu-\tau_k$. Then the autocorrelation can be estimated by modeling a variogram to the δ_{ij}. The variogram is a function that contains information on the spatial autocorrelation among the experimental units; more details are given later. On the other hand, if the distribution and autocorrelation for all δ_{ij} are known, then statistical methods that rely on the variogram can be used to estimate $\mu+\tau_k$ optimally (see Cressie, 1991, p. 328, and Appendix 14.1). Here, an iterative approach is used. First, a classical ANOVA is performed to estimate $\mu+\tau_k$. Then, the δ_{ij} are estimated with the residuals from the classical ANOVA, $\delta_{ij} = R_{ij}$, where $R_{ij} = Y_{ijk}-\hat{\mu}-\hat{\tau}_k$ and $\hat{\mu}+\hat{\tau}_k$ is the classical ANOVA estimate. Next, a variogram is modeled from the R_{ij}. The variogram can then be used to obtain better estimates of $\mu+\tau_k$ (using the formulas in Appendix 14.1). The whole procedure can go through another iteration by starting with $R'_{ij} = Y_{ijk}-\hat{\mu}'-\hat{\tau}'_k$ where $\hat{\mu}'+\hat{\tau}'_k$ is the new estimate of $\mu+\tau_k$. This GLS-variogram method is given in Table 14.3 as a series of steps (more detailed descriptions of each step follows), along with computer programs (Appendix 14.2).

In step one, a classical ANOVA is performed (Appendix 14.2, program 1); the results for the Ozark data are given in Table 14.2. Actually, this is a process much like regression in which ordinary least squares is used to estimate the parameters μ and τ_k in Eq. 14.1. Once the parameters have been estimated, denoted $\hat{\mu}$ and $\hat{\tau}_k$, the contrasts are estimated by replacing τ_k with $\hat{\tau}_k$ in Eqs. 14.2–14.6. Residuals (denoted R_{ij} for the plot in the ith row and jth column) are formed by taking

$$R_{ij} = Y_{ijk}-\hat{\mu}-\hat{\tau}_k$$

where Y_{ijk} are the observed values due to treatment k in the (i,j)th plot. Program 1 (Appendix 14.2) calculates the residuals. If the parameters have been estimated well, the residuals (Fig. 14.2B) should be close to the original data (Fig. 14.1A)

Table 14.3. *Steps in Performing Classical ANOVA and GLS-Variogram to Estimate Treatment Contrasts*[a]

Step	Input	Program	Description	Output
1	data file	OLS.PRG	Perform a classical ANOVA to estimate treatment contrasts and output residuals from fitted model	Contrast estimates to terminal and resid.out
2	resid.out	SEMIVGM.FOR	Calculate the empirical semivariogram on residuals from the classical ANOVA	semivgm.dat
3	semivgm.dat	FITVEXP.PRG	Fit an exponential semivariogram model to the empirical semivariogram values	Parameter estimates to terminal
5	data file and gamma.dat	GLS.PRG	Reestimate the treatment contrasts using GLS-variogram and output residuals from fitted model	Contrast estimates to terminal and resid.out
6			Iterate; go to step 2	

[a]The programs are given in Appendix 14.2; the extension *.PRG indicates a SAS program and the extension *.FOR indicates a FORTRAN program.

with the overall mean subtracted from it (Fig. 14.2A). Figure 14.2B does resemble, reasonably well, the spatial patterning in Fig. 14.2A. The residuals contain spatial patterning that may be modeled as autocorrelated, which will allow better parameter estimates.

Some intuitive reasoning is given as follows. Suppose that by chance, due to the completely randomized design, several of treatment τ_k were assigned to plots close together, e.g., observe treatment 5 in plots (2,1), (2,2), and (3,1) of Fig. 14.1B. Then, rather than estimating $\mu+\tau_5$ with the simple average of all plots assigned treatment 5, a weighted average is used that gives smaller individual weights to those plots close together because they are likely to have similar values. That is, they have little extra information beyond that of a single plot in the same local region. On the other hand, a plot assigned treatment 5 that is spatially isolated from the other plots assigned treatment 5, e.g., plot (4,3) in Fig. 14.1B, would get a higher weight because it "represents" a larger region. The optimal weights are obtained through formulas (Appendix 14.1) that use the semivariogram.

To use the GLS-variogram approach, the semivariogram of residuals among all pairs of plots must be estimated (step 2 in Table 14.3; program 2 in Appendix 14.2). The semivariogram can be used to estimate contrasts as shown by Cressie (1991, p. 328; Appendix 14.1). To estimate the variogram from the residuals, it is necessary for the residuals to exhibit approximate *intrinsic stationarity* (Matheron, 1963; Journel and Huijbregts, 1978; Cressie, 1991). Simply stated, this

Column

A

	1	2	3	4	5
1	7.92	1.92	−0.08	−0.08	−0.08
2	1.92	0.92	−2.08	−2.08	−1.08
3	−1.08	−3.08	−3.08	−4.08	−0.08
4	1.92	−1.08	1.92	−2.08	0.92
5	0.92	−1.08	−0.08	−0.08	2.92

Row

B

	1	2	3	4	5
1	6.40	2.80	0.00	0.00	1.40
2	1.00	0.00	−1.20	−0.60	0.40
3	−2.00	−2.20	−4.60	−2.60	1.40
4	0.40	−0.20	1.00	−2.00	−0.60
5	0.00	−1.00	−1.60	0.80	3.00

Row

Figure 14.2. (**A**) Uniformity trial data (Fig. 14.1A) with the overall mean subtracted from each plot. (**B**) Residuals from the classical ANOVA.

means that if the spatial data were collected repeatedly, the values at all locations would average toward some constant value, μ, and that the variogram between two sites does not depend on their exact locations, only the relative displacement between them. We shall assume further that the variogram actually depends only on the *distance* between any pair of locations, which is called the isotropy assumption. These assumptions are impossible to test, because it is impossible to go back in time again and again and generate the experiment each time to check whether each experimental unit has the same mean value, or whether the correlation is the same for all pairs of plots that are at some fixed distance from each other. However, any gross spatial trends in the residuals (e.g., high values at one end of the study area gradually shifting to low values at the other end) would cause suspicion that they are not stationary. In order to check these conditions, the methods of exploratory data analysis for spatial data (Cressie, 1991, Section 2.2; see also Chapter 2) are recommended. The residuals (Fig. 14.2B) for these data appear satisfactory, except possibly the residual at (1,1), which seems to be rather large (in absolute value) compared to most others.

For step 2 (Table 14.3), the empirical semivariogram is used:

$$\hat{\gamma}(h) \sum_{M(h)} (R_{ij} - R_{st})^2 / 2N(h), \tag{14.7}$$

where R_{ij} is the classical ANOVA residual (output from program 1, Appendix 14.2) in the ith row and jth column, $M(h) \equiv \{[(i,j),(s,t)]: \sqrt{(i-s)^2 + (j-t)^2} = h\}$ (the set of all pairs of data that are at a Euclidean distance of h apart), and $N(h)$ is the number of pairs in $M(h)$. An intuitive way to write Eq. 14.7 is

$$\hat{\gamma}(h) = (1/2) \text{ average } (R_{ij} - R_{st})^2$$

where R_{ij} and R_{st} are at a distance h (h is often called the lag) from each other.

An example will help illustrate how to use Eq. 14.7. Let $h = 1$. Then $M(1)$ is the set of all *pairs* of plots that are one unit apart, e.g., plots (1,1) and (1,2), plots (1,1) and (2,1), etc.; there are 40 such pairs in Fig. 14.2B. Thus, $N(1) = 40$. The empirical semivariogram $\hat{\gamma}(1)$ is 1/2 the average over all 40 pairs where each pair is differenced and squared. As another example, from Fig. 14.2B consider $R_{11} = 6.4$ and R_{55} eq 3.0, where $(R_{11} - R_{55})^2 = 11.56$. The distance between R_{11} and R_{55} is $(4^2 + 4^2)^{1/2} = 5.567$. There is only one more pair, R_{15} and R_{51}, that is also 5.567 units apart. Thus, $\hat{\gamma}(5.567) = 1/2$ average $(11.56, 1.96) = 3.38$. This value is plotted against lag in Fig. 14.3, along with all of the other values for $\hat{\gamma}(h)$. Figure 14.3 shows that plots close together have more similar residuals than those that are farther apart because $\hat{\gamma}(h)$ is smaller for smaller values of h.

In order to use the spatial method, a semivariogram matrix Γ is needed that contains the semivariogram values for all pairwise locations. However, the empirical values calculated for $\hat{\gamma}(h)$ (program 2, Appendix 14.2) must not be

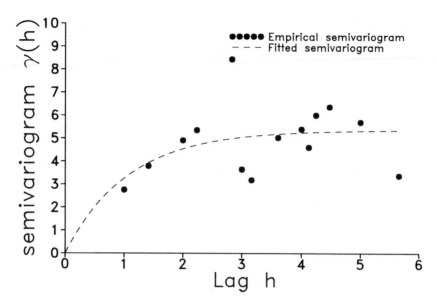

Figure 14.3. Empirical semivariogram and fitted semivariogram for residuals from the classical ANOVA.

used in the semivariogram matrix Γ, because Γ must satisfy certain positive-definiteness properties (Cressie, 1991, Section 2.5). For example, if empirical values were used, the variance estimate may be negative, which is wrong. Therefore, in step 3 (Table 14.3), a model was fit to $\hat{\gamma}(h)$ that satisfies the properties. Several models are possible (e.g., Cressie, 1991, Section 2.3), and here the exponential model was chosen after inspecting a plot of the empirical semivariogram values. Zimmerman and Harville (1991) indicate that the choice of model does not matter greatly as long as it fits the data reasonably well. A spherical model might also have been chosen for these data. But the model "linear and sill" cannot be used because it may yield negative variances for two-dimensional data; however, it may be used for one-dimensional (e.g., transect) data (Webster, 1985). Webster (1985) and Cressie (1991, Section 2.3) mention several "safe" models that can be used for spatial data in one, two, or three dimensions.

For step 3 (Table 14.3), the exponential semivariogram model was fit to the empirical semivariogram. The exponential semivariogram is

$$\gamma(h) = \begin{cases} \theta_0 + \theta_1(1 - \exp(-\theta_2 h)) & \text{for } h > 0 \\ 0 & \text{for } h = 0 \end{cases} \qquad (14.8)$$

where θ_0, θ_1, and $\theta_2 \geqslant 0$. Notice that h is always positive (because it is a distance) and $\gamma(h) = 0$ when $h = 0$. In step 3, nonlinear regression (weighted least squares) with weights proportional to $N(h)$ (the number averaged for each h in Eq. 14.7) was used, as recommended by Cressie (1985), to fit the exponential variogram model Eq. 14.8 to the empirical variogram Eq. 14.7 (program 3, Appendix 14.2); the following estimates were obtained: $\hat{\theta}_0 = 0.0$, $\hat{\theta}_1 = 5.3795$, and $\hat{\theta}_2 = 0.9332$. The empirical and fitted semivariogram are given in Fig. 14.3. The estimates $\hat{\theta}_0$, $\hat{\theta}_1$, $\hat{\theta}_2$ were inserted into Eq. 14.8 and used to generate a matrix of semivariogram values between all pairs of experimental units (step 4, Table 14.3; program 4, Appendix 14.2).

In step 5 (Table 14.3), the formulas of Cressie (1991, pp. 328; Appendix 14.1), that rely on the semivariogram matrix in step 4, were used. Contrasts and their standard errors were estimated (program 5, Appendix 14.2); the results are in Table 14.2. A test for a nonzero contrast can be obtained as follows. Construct a confidence interval around the contrast estimate by taking the standard error multiplied by some $Z_{\alpha/2}$, and if the confidence interval does not include zero, the contrast is declared significant at that α-level. It is possible to stop here, but residuals can again be formed from the current estimates of $\hat{\mu}' + \hat{\tau}'_k$ by taking $R'_{ij} = Y_{ijk} - \hat{\mu}' - \hat{\tau}'_k$ and then starting the procedure over again at step 2 (Table 14.3). A stopping rule can be chosen, such as when the contrast estimates stop changing at one decimal place less than the least significant digit.

14.4 Discussion

This chapter showed how a spatial analysis of the residuals from a classically designed field experiment can increase the precision of estimating treatment contrasts. It is worthwhile to discuss all of the assumptions for both methods here (Table 14.4). There are two main differences. First, as was mentioned in section

Table 14.4. Comparison of Assumptions When Performing a Classical ANOVA versus the Spatial Analysis Introduced in this Chapter

Classical ANOVA	GLS-variogram analysis
Experimental units fixed in value or random variables	Experimental units are random variables
Expectation of random errors $= 0$	Expectation of random errors $= 0$[a]
Constant variance for random errors	Constant variance for random errors[a]
Independent random errors	Autocorrelated random errors[a]—variogram depends only on the spatial displacement between errors
normally distributed random errors[b]	normally distributed random errors[b]

[a]Forms the intrinsic stationarity assumption.

[b]Not strictly necessary for estimation; only needed for inference, e.g., confidence intervals, tests of hypotheses, etc.

14.2, for a spatial analysis, it is assumed that the data are the result of a spatial stochastic process with possible autocorrelation. We believe that this is a realistic assumption for most ecological problems; in fact, ecologists also associate the word "process" to spatial pattern, beginning with Watt (1947). An advantage of classical ANOVA is that the data may be assumed either to be the result of a spatial stochastic process or to be fixed in value. The second main difference is that, for the GLS-variogram method, the random errors do not need to be assumed independent, only intrinsically stationary, which allows more precise contrast estimates.

The example presented in this chapter consisted of only 25 plots, which kept the illustration simple, and this is the minimum number of plots we consider necessary to do an adequate job in estimating the variogram from the residuals. We would recommend having more experimental units if possible, and as usual, the more the better. Also, in order to get residuals that reflect the underlying spatial variability, it is important that the initial (i.e., classical) parameter estimates are fairly accurate. Their accuracy depends on the number of replications of each treatment, and we recommend a minimum of about five replications; again, the more the better.

This chapter can be compared to the Mantel method of correcting for spatial effects described in Chapter 15. There are two main differences. First, the Mantel test is nonparametric in comparison to this one, and no assumptions about the distribution of variables or the form of their autocorrelation (e.g., the variogram) are necessary. Here, a variogram model must be chosen and the residuals need to be roughly normally distributed. No formal comparison of the two methods was conducted. However, one might expect the usual results when comparing nonparametric vs parametric methods, i.e., the parametric method will be more powerful if the data follow the assumed model, whereas the nonparametric method gives some protection against incorrect model assumptions. The second main difference from the Mantel test is that, here, we are *estimating* treatment contrasts. That is, we go beyond the question of "Are there differences among treatment effects?" to "What are the specific differences among treatment effects?"

In general, spatial methods that account for autocorrelation appear to be quite robust. Several authors have carried out simulations and tried spatial methods similar to the GLS-variogram approach on a variety of data sets, with good results. Zimmerman and Harville (1991) used restricted maximum likelihood instead of using residuals, and estimated covariance functions rather than variograms. They examined three different data sets with a total of 11 different blocking configurations, and found that their spatial method reduced estimation variance to 80–20% that of a classical analysis. This translates to much more powerful tests of hypotheses and shorter confidence intervals. Grondona and Cressie (1991) took one large data set, broke it into 6 subsets, and used a robust semivariogram estimator (Cressie and Hawkins, 1980). Their results indicated reduced estimation variance to 75% that of a classical analysis. It should be pointed out here that it is also possible to use the GLS-variogram method for blocked designs (Chapter 3). Both Zimmerman and Harville (1991) and Grondona and Cressie (1991) examined similar spatial methods for blocked designs. In this chapter we used a completely randomized design to keep the illustration simple.

Using a slightly different methodology of blocking by columns and using a time-series type of analysis within blocks, Cullis and Gleeson (1989), in a study of over 1000 variety trials, reduced estimation variance, on average, to 58% that of a classical analysis. Cullis and Gleeson (1991) extend their models to two dimensions. Baird and Mead (1991) simulated data from several models rather than using real data, so they controlled the variability and autocorrelation in the experimental units as well as the variability in the design. They found the spatial methods of Cullis and Gleeson (1989) to be valid over a wide range of simulation models. Additional references for the interested reader are Bartlett (1978), Kempton and Howes (1981), Wilkinson et al. (1983), Green et al. (1985), and Besag and Kempton (1986).

We conducted our own simulations for the Ozark data (Fig. 14.1A). By repeatedly applying treatments randomly (as was done to obtain Fig. 14.1B) 200 times to the uniformity data in Fig. 14.1A, and estimating the contrast Eqs. 14.2–14.6 each time with a classical ANOVA and the GLS-variogram method, we found that the GLS-variogram method reduced estimation variance to 58% that of a classical analysis for these data.

Besides the spatial analysis of an experiment, there is also the notion that for spatially correlated variables, there are better designs than the classical ones based on randomization. For example, Fig. 14.1B shows that all applications of treatment 4 were assigned to the upper right corner. This is undesirable, due to the spatial autocorrelation. A better design would spatially distribute the treatments evenly. For example, first-order nearest-neighbor balanced block designs (Kiefer and Wynn, 1981; Cheng, 1983; Grondona and Cressie, 1993) can be shown to be optimal under certain conditions (Kiefer, 1975; Grondona and Cressie, 1993), and contain good interspersion of treatments (Hurlbert, 1984). Despite the advantages, the theory is rather difficult, and there is no guarantee that an optimal design will exist for a given number of replications, treatments, and experimental units (see also the discussion of latin square designs in Chapter 3). The advantage of the classical designs is that they are very easy to construct and well-understood. The compromise presented in this chapter is to construct classical designs and analyze them spatially.

There are several modifications to the GLS-variogram method to make it more robust. One could robustly/resistantly remove treatment effects by subtracting treatment medians, rather than means, to obtain residuals. One could also estimate the variogram with a robust/resistant estimator (Cressie, 1991; Grondona and Cressie, 1991). If the data appear to be affected by outliers, such as unusually large or small residuals, such an approach is recommended.

In this chapter we presented a model-based method for analyzing classically designed experiments. That is, the random errors were considered to be the result of a spatially autocorrelated process, and hence were modeled with a variogram. The GLS-variogram method was used to estimate treatment differences as well as to test for differences in treatment effects. The GLS-variogram method did considerably better than a classical ANOVA. The increased power of the GLS-variogram method can allow ecologists to detect more real treatment effects for a limited amount of time and money.

Acknowledgments

Financial support for this work was provided by Federal Aid in Wildlife Restoration to the Alaska Department of Fish and Game, and by the Midwest Region of the National Park Service (Ver Hoef), and the National Science Foundation (Cressie). Drs. Jessica Gurevitch, Eric Rexstad, and Samuel Scheiner, and Ms. Susan Abbott reviewed the manuscript; their comments were greatly appreciated.

Appendix 14.1: Mathematical details of methods

From Eq. 14.1, all of the data can be written as a linear model $\mathbf{y} = \mathbf{X}\boldsymbol{\beta} + \boldsymbol{\delta}$, where $\mathbf{y}$ is a vector of random variables, $\mathbf{X}$ is the design matrix, $\boldsymbol{\beta}$ is a vector of parameters, and $\boldsymbol{\delta}$ is a vector of zero-mean random error. Let each row of the matrix Λ contain the coefficients for each contrast. Then classical ANOVA uses ordinary least squares to yield estimates of the vector of contrasts $\Lambda\boldsymbol{\beta}$:

$$\widehat{\Lambda\boldsymbol{\beta}}_{\text{OLS}} = \Lambda\hat{\boldsymbol{\beta}}_{\text{OLS}} = \Lambda(\mathbf{X}'\mathbf{X})^{-}\mathbf{X}'\mathbf{y},$$

where $(\mathbf{X}'\mathbf{X})^{-}$ is the generalized inverse of $\mathbf{X}'\mathbf{X}$. The exact from of the design matrix, $\mathbf{X}$, depends on the results of randomizing the treatment assignments. The variance of the estimate $\widehat{\Lambda\boldsymbol{\beta}}_{\text{OLS}}$ is estimated by

$$\text{vâr}(\widehat{\Lambda\boldsymbol{\beta}}_{\text{OLS}}) = \hat{\sigma}^2 \Lambda(\mathbf{X}'\mathbf{X})^{-}\Lambda',$$

where $\hat{\sigma}^2$ is $\mathbf{r}'\mathbf{r}/[n-\text{rank}(\mathbf{X})]$ and $\mathbf{r}$ is a vector of residuals $\mathbf{r} = \mathbf{y} - \mathbf{X}\hat{\boldsymbol{\beta}}_{\text{OLS}}$. The estimated standard errors for the contrast are the square roots of the diagonal elements of $\text{vâr}(\widehat{\Lambda\boldsymbol{\beta}}_{\text{OLS}})$.

The spatial method uses generalized least squares to yield estimates of the contrasts by the following equations:

$$\widehat{\Lambda\boldsymbol{\beta}}_{\text{GLS}} = \Lambda\hat{\boldsymbol{\beta}}_{\text{GLS}} = \Lambda(\mathbf{X}'\Gamma^{-1}\mathbf{X})^{-}\mathbf{X}'\Gamma^{-1}\mathbf{y}$$

where $\boldsymbol{\Gamma}^{-1}$ is the inverse of the matrix containing semivariogram values between all possible pairs of locations containing data $\mathbf{y}$ (in the same matrix arrangement as $\mathbf{yy'}$). The variance of the estimate $\boldsymbol{\Lambda}\widehat{\boldsymbol{\beta}}_{GLS}$ is estimated by

$$\text{vâr}(\boldsymbol{\Lambda}\widehat{\boldsymbol{\beta}}_{GLS}) = -\boldsymbol{\Lambda}(\mathbf{X'}\boldsymbol{\Gamma}^{-1}\mathbf{X})^{-}\boldsymbol{\Lambda'}$$

The estimated standard errors for the contrasts are the square roots of the diagonal elements of $\text{vâr}(\boldsymbol{\Lambda}\widehat{\boldsymbol{\beta}}_{GLS})$.

Appendix 14.2: Computer programs for the example data set

The exact format of these programs, such as some of the input/output statements, may have to be changed to conform with local system requirements. Also, programs may need to be modified to accommodate different experimental designs, contrasts, and variogram models.

The first entry in file EXAMPLE.DAT, from Fig. 14.2B, is the ith row, the second entry is the ith column, the third entry is treatment number, and the fourth entry is the data value, as follows:

```
1   1   3   27
1   2   1   26
1   3   2   21
        .
        .
        .
5   5   2   24
```

Program 1. SAS program to compute OLS estimate of treatment effects, and output residuals to a system file called resid.out.

```
DATA EXAMPLE;                                        /data input/
   INPUT I J TRT Y;        /ith row, jth column, treatment number, observed value/
PROC GLM;                                /classical ANOVA of treatment contrasts/
   CLASS TRT;
   MODEL Y = TRT;
   ESTIMATE 'CONTRAST EQ. (14.2)'
       TRT -1 .5 .5 0  0;
   ESTIMATE 'CONTRAST EQ. (14.3)'
       TRT -1  0  0 .5 .5;
   ESTIMATE 'CONTRAST EQ. (14.4)'
       TRT  0 -.5 -.5 .5  .5;
   ESTIMATE 'CONTRAST EQ. (14.5)'
       TRT  0  1 -1  0  0;
   ESTIMATE 'CONTRAST EQ. (14.6)'
       TRT  0  0  0  1 -1;
   OUTPUT OUT=RESID1 R=RES;           /new SAS data set including residuals/
DATA RESID2;                           /output residuals to system file 'resid.out'/
   SET RESID1;                          /new data set using resid1 from PROC GLM/
   FILE 'RESID.OUT';                    /system file name for residuals data set/
   PUT I 1-5 .0                                          /ith row/
       J 10-15 .0                                        /jth column/
       RES 20-35 .7;                                     /residual/
```

Program 2. FORTRAN program to calculate empirical semivariogram. The input file must have three columns; the first is the ith row, the second is the jth column, and the last one contains the data (here, residuals). Output is in a system file called semivgm.dat, where the first column is lag, the second column is the empirical semivariogram for the corresponding lag, and the last column is a weight, normalized so the last column sums to one.

```
      REAL Y(50),I(50),J(50),LAG(1225),VGM(1225),LAGN(1225)
      CHARACTER*60 FN
      WRITE(*,10)
10    FORMAT('    PRGM: CALCULATE EMPIRICAL SEMIVARIOGRAM')
      WRITE(*,11)
11    FORMAT(' INPUT FILE: '\)
      READ(*,12) FN
12    FORMAT(A)
      OPEN(05,FILE=FN)                           /open input file and read data/
      N=1                                        /data number arrays: I=row,
80    READ(05,*,END=90) I(N),J(N),Y(N)             J=column, Y=data at (I,J)/
      N=N+1
      GOTO 80
90    N=N-1
C                                                /calculate empirical semivariogram/
      NL=0                                       /number of unique lag values for
      DO 20 K=1,N-1                                all possible pairs of data/
        DO 21 L=K+1,N
          XL=SQRT((I(K)-I(L))**2+(J(K)-J(L))**2)     /euclidean distance (lag)
                                                        for each data pair/
          D2=(Y(K)-Y(L))**2              /difference squared for any pair of locations/
          IF (NL.EQ.0) THEN                                    /initialize arrays/
            LAG(1)=XL                               /array for all unique lags/
            VGM(1)=D2                               /array to sum D2 for each lag/
            LAGN(1)=1                               /N for each element in array V/
            NL=1                                    /one unique lag value so far/
            GOTO 21
          ENDIF
          DO 22 M=1,NL
            IF (LAG(M).EQ.XL) THEN                            /if previous lag/
              VGM(M)=VGM(M)+D2                  /add difference squared for lag/
              LAGN(M)=LAGN(M)+1                            /N for array V/
              GOTO 21                                       /jump out of loop/
            ENDIF
22        CONTINUE
C                                                            /if new lag value/
          NL=NL+1                          /add 1 to number of unique lag values/
          LAG(NL)=XL                                         /new lag value/
          VGM(NL)=D2                                         /D2 for new lag value/
          LAGN(NL)=1                                /N for new element in array V/
21      CONTINUE
20    CONTINUE

      DO 23 K=1,NL
        VGM(K)=VGM(K)/(LAGN(K)*2)          /empirical semivariogram is (1/2)*(average
                                             difference squared) for a given lag/
        LAGN(K)=LAGN(K)/FLOAT((N*(N-1))/2)     /weights (normalized) for each lag/
23    CONTINUE
C                                                            /write data to disk/
      OPEN(06,FILE='SEMIVGM.DAT')                           /open output file/
      DO 24 K=1,NL
C                           /output file: lag, empirical semivariogram, weight/
        WRITE(06,15)LAG(K),VGM(K),LAGN(K)
15      FORMAT(3(2X,F10.4))
24    CONTINUE
      END
```

Program 3. SAS program to fit an exponential variogram model to the output from the semivariogram FORTRAN program.

```
DATA FITVGM;                              /input empirical semivariogram
   INFILE 'SEMIVGM.DAT';                      values from Program 2/
   INPUT LAG GAMMA WEIGHT;

*** Nonlinear weighted least squares fits exponential
      semivariogram model to empirical semivariogram values ***

PROC NLIN;
   PARMS C0=0 TO 20 BY 5                   /initial parameter values/
      C1=0 TO 20 BY 5
      C2=0 TO 2 BY 0.5;
   BOUNDS C0>0, C1>0, C2>0;                /parameter bounds/
   MODEL GAMMA=C0+C1*(1-                   /semivariogram/
                    EXP(-C2*LAG));
   DER.C0=1;                               /partial derivative of parameters/
   DER.C1=1-EXP(-C2*LAG);
   DER.C2=C1*LAG*EXP(-C2*LAG);
   _WEIGHT_=WEIGHT;                        /weights/
```

Program 4. Using the parameter estimates from the SAS program that fit the semivariogram model, this FORTRAN program generates a matrix of semivariogram values between all locations. The original data file (here, example.dat, with four columns) must be provided. The output is a file called gamma.dat.

```
      REAL GAMMA(25,25),Y(25),I(25),J(25),LAG
      CHARACTER*60 FN
      WRITE(*,10)
10    FORMAT(' PRGM: MATRIX FOR EXPONENTIAL SEMIVARIOGRAM')
      WRITE(*,11)
11    FORMAT('          Original data file: '\)
      READ(*,12)FN
12    FORMAT(A)
      WRITE(*,13)
13    FORMAT(' C0, C1 and C2 for exponential model: '\)
      READ(*,*) C0,C1,C2
C                                                             /read data/
      OPEN(05,FILE=FN)                              /open original data file/
      N=1
80    READ(05,*,END=90) I(N),J(N),TRT,Y(N)
      N=N+1
      GOTO 80
90    N=N-1
      DO 21 K=1,N                         /gamma matrix for all pairwise data/
        DO 22 L=1,N
          LAG=SQRT((I(K)-I(L))**2+(J(K)-J(L))**2)    /euclidean distance (lag)/
          IF (LAG.EQ.0) THEN                                 /if on diagonal/
            GAMMA(K,L)=0                           /diagonal must contain zeros/
          ELSE                      /if off diagonal fitted semivariogram values/
            GAMMA(K,L)=C0+C1*(1-EXP(-C2*LAG))
          ENDIF
22      CONTINUE
21    CONTINUE
C                                                        /write matrix to disk/
      OPEN(06,FILE='GAMMA.DAT')                          /open output file/
      DO 23 K=1,N                            /output semivariogram matrix/
        WRITE(06,14)(GAMMA(K,L),L=1,N)
14      FORMAT(6(1X,F10.4)/6(1X,F10.4)/6(1X,F10.4)/6(1X,F10.4)/6(1X,F10.4))
23    CONTINUE
      END
```

Program 5. SAS program that computes a GLS estimate of treatment contrasts, using the matrix of semivariogram values. Residuals are printed out to the system file resid.out.

```
DATA EXAMPLE;
  INPUT I J TRT Y;
DATA GAMMA;                                  /matrix of semivariogram values/
  INFILE 'GAMMA.DAT';
  INPUT G1-G25;
PROC IML;                             /contrast estimates using gls-variogram method/
  RESET NOLOG NONAME SPACES=3;                       /set printing options/
  USE EXAMPLE;                        /example data as matrix into PROC IML/
  READ ALL INTO YD;
  USE GAMMA;                                    /gamma matrix into PROC IML/
  READ ALL INTO GAM;
  T=YD[,3];                                /vector of treatment assignments/
  X1=J(NROW(T),1,1);                                      /vector of N 1s/
  X2=DESIGN(T);                                            /design matrix/
  X=X1||X2;                                  /design matrix with overall mean/
  C=/0 -1 .5 .5 0 0,                                          /contrasts/
    0 -1  0  0 .5 .5,
    0 0-.5-.5 .5 .5,
    0 0  1 -1  0 0,
    0 0  0  0  1 -1/;
  Y=YD[,4];                                        /vector of data values/
  G=INV(GAM);                                       /invert the gamma matrix/
  A=GINV(X'*G*X);
  WTS=A*X'*G*Y;
  TAUGLS=C*WTS;                                         /contrast estimates/
  TGLSV=-C*A*C';                 /covariance matrix of contrast estimates/
  TAUSE=J(5,1,1);                              /initialize standard error vector/
  DO II=1 TO 5;
  TAUSE[II]=SQRT(TGLSV[II,II]);                         /standard errors/
  END;
  EQN=/"CONTRAST (14.2)",
       "CONTRAST (14.3)",
       "CONTRAST (14.4)",
       "CONTRAST (14.5)",
       "CONTRAST (14.6)"/;
  LABEL=/"CONTRAST ESTIMATE",
       "STANDARD ERROR"/;
  PM=TAUGLS||TAUSE;
  PRINT "GLS-VARIOGRAM
          CONTRAST ESTIMATES",,;
  PRINT PM[ROWNAME=EQN
          COLNAME=LABEL FORMAT=25.2];
  PRINT ,,"VARIANCE-COVARIANCE
          MATRIX OF CONTRAST ESTIMATES";
PROC PRINT NOOBS;                   /print the values for each confidence interval/
  TITLE1 "&CI PERCENT BOOTSTRAP
          CONFIDENCE INTERVALS";
  TITLE2 "SAMPLE VALUE: &SAMPLE";
  TITLE3 "&NBOOT BOOTSTRAP
          SAMPLES, CONSTANTS:
          Z0 = &Z0, A = &A";
  TITLE4 "PERCENTILES FOR THE
          BC INTERVAL ARE:
          &PLBC, &PUBC";
  TITLE5 "PERCENTILES FOR THE
          ACCELERATED INTERVAL
          ARE: &PLBCA, &PUBCA";
```

15

Mantel Tests: Spatial Structure in Field Experiments

Marie-Josée Fortin and Jessica Gurevitch

15.1 Ecological Issue

Landscapes are composed of mosaics of patches, with different degrees of spatial autocorrelation within and among them. The phenomenon of spatial autocorrelation, i.e., the spatial dependence of the values of a variable, has been widely reported (Chapter 14; Cliff and Ord, 1981; Upton and Fingleton, 1985; Legendre and Fortin, 1989). Positive spatial autocorrelation may result from microenvironment or from dispersal of offspring near the maternal parent, among other causes. Negative spatial autocorrelation may result, for example, from competition for resources. In the case of positive spatial autocorrelation, plants near one another are more similar than are distant plants. In this event, the results of an analysis of variance (ANOVA) will be affected by the spatial pattern since the data violate one of the basic assumptions of parametric inferential methods: the independence of the observations (Cliff and Ord, 1981; Legendre et al., 1990).

The use of nonindependent observations affects the estimation of degrees of freedom, since as each observation is not independent of the others, it does not contribute a full degree of freedom (Legendre et al., 1990). Furthermore, with positively spatially autocorrelated data, the differences within groups will appear small, which can result in a type I error in that the differences among groups are declared significant when in fact they are not (Cliff and Ord, 1981; Legendre et al., 1990; Sokal et al., 1993). Recently, the implications and problems of the use of spatially autocorrelated data for experiments and subsequent statistical analyses have been the focus of several studies (Sokal et al., 1993; Legendre et al., 1990).

With field experiments on plants, there is the additional problem that the degree of spatial heterogeneity already present in the field is rarely assessed in advance. This inherent spatial heterogeneity can result in unequal plant responses to the experimental treatments. Even when small scale preexperimental sampling or trials can be carried out, there is no guarantee that the nature of the spatial autocorrelation will be the same in response to the actual experimental treatments, or when the spatial scale is expanded for the full experiment. Furthermore, when experimental manipulations of resource availabilities or competitive effects are conducted on spatially autocorrelated plants and the results analyzed with conventional statistical methods such as ANOVA, it is difficult to disentangle whether the outcome is due to the responses to treatments or to the inherent spatial structure.

Problems caused by spatial and temporal heterogeneity are familiar to field experimentalists (Hurlbert, 1984; Mead, 1988). Randomized blocks are perhaps the most common experimental design used in ecology to minimize the effects of spatial and temporal heterogeneity (Chapter 3). However, even the use of randomized block designs does not guarantee that the block size employed matches the inherent spatial pattern of the plants or their spatial responses to the treatment (van Es et al., 1989; Chapter 3). In the worst cases, blocking reduces the power of the analysis without necessarily removing all or even some of the effects of spatial heterogeneity. Other types of experimental designs such as the GLS-variogram method of ANOVA (Chapter 14), nearest neighbor analysis (Wilkinson et al., 1983; Chapter 14), trend analysis (Tamura et al., 1988), analysis of covariance, and latin squares (Mead, 1988; Chapter 3) have been employed instead of the randomized blocks design.

The methods presented in this chapter, the Mantel and partial Mantel tests (Mantel, 1967; Sokal, 1979; Hubert, 1987; Legendre and Fortin, 1989; Manly, 1991), are based on distance matrices and permutation tests. These methods differ from those above primarily in that they are nonparametric and rely on fewer assumptions. Although in this chapter we discuss the use of these methods to account for the presence of underlying spatial autocorrelation of the data, they can also be used to account for other types of autocorrelation. In fact, autocorrelation exists not only in space but genetically, through species dispersal and clonal spread, and temporally, through daily, seasonal and yearly cycles (Chapter 7). In the present study, we will illustrate how these methods can be used to detect first the presence of inherent spatial autocorrelation, and then to distinguish the effects on plant growth of treatments (i.e., experimental design factors) from the effects of spatial autocorrelation in a field competition experiment.

15.2 Statistical Solution

Analysis of variance is usually the most powerful statistical method employed for testing whether there are significant treatment effects. However, in a field experiment, the appropriate statistical method to use depends first on whether the data satisfy the specific assumptions of ANOVA such as normality, homogeneity of the variances, and independence of observations. Thus preliminary tests should be carried out to verify that the data have normally distributed residuals and homogeneity of variances among groups (Sokal and Rohlf, 1981). Independence of the observations can be insured by gathering the data using random assignment of experimental units to treatments (Sokal and Rohlf, 1981; Chapters 3 and 14). However, randomization procedures do not ensure that neighboring units are spatially independent of one another (Fortin et al., 1989; van Es et al., 1989). A recent review by Day and Quinn (1989) pointed out the importance and statistical implications of deviations from the assumptions of parametric tests such as ANOVA. Although ANOVA has been found robust to some deviations from normality and from homogeneity of the variances, it is quite sensitive to the nonindependence of the observations, which is often the case with ecological data (Bradley, 1978; Kenny and Judd, 1986). When the data do not satisfy the assumption of normality, nonparametric procedures such as Mann–Whitney, Kruskal–Wallis, and Friedman tests, or resampling techniques as suggested by Dixon (Chapter 13) should be used; when the data are spatially autocorrelated, Mantel tests may be more appropriate.

When data from a field experiment on plants, for instance, are analyzed by an ANOVA and the treatment effects are not significant, this implies either that the treatments have no influence on the outcome, or that the treatment effects are canceled out by the spatial responses of the plants or by other unmeasured and uncontrolled factors (Sokal et al., 1993). However, when the differences are significant, there are three possible reasons for this outcome: (1) the plants show no significant spatial pattern, and the treatments really affect plant responses, (2) the degree of spatial autocorrelation of the plants is significant, and is inducing spurious significant treatment effects, or (3) both the degree of spatial autocorrelation of the plants and the treatment effects are significant. Mantel and partial Mantel tests enable the investigator to distinguish which of these three cases is occurring.

15.2.1 Mantel Test

To analyze disease patterns through space and time, Mantel (1967) developed a randomization test that takes the spatial and/or the temporal autocorrelation of the data into account by computing the relationship between two distance matrices. The null hypothesis is that the observed relationship between the two distance matrices could have been obtained by any random arrangement in space (time, or treatment assignment) of the observations through the study area. The Mantel test has been used by ecologists to evaluate the relationship between ecological data and their spatial structure (Douglas and Endler, 1982; Burgman, 1987; Legendre and Fortin, 1989), or to test the goodness-of-fit of data to a model or hypothesis (e.g., Sokal et al., 1987; Legendre and Troussellier, 1988; Sokal et al., 1990). Previous uses of Mantel and partial Mantel tests have concentrated on detecting existing patterns in sampled populations. We extend this approach to analyze the outcome of designed experiments, given that there is an underlying spatial pattern in the data.

A Mantel test works by examining the relationship between two matrices. The entries in the matrices are actually values (such as physical distances) measured between individual "points," which might be sampling or experimental units. In the example we will discuss below, the points are individual bunchgrass plants. By carrying out the Mantel test, we are testing the relationship between two different kinds of variables measured on the sampling units, much as in computing and testing a conventional correlation coefficient. What actually goes into these matrices? Rather than using the original measurements for the elements of each matrix, we calculate some measure of distance (or similarity) between each point and all of the others. The Mantel test computes a correlation between the two n by n distance matrices, where one matrix might represent spatial distances, for example, while the other represents differences between pairs of plants in some measure of plant status (e.g., mass).

The results of such a test will reveal whether small plants are located near other small plants, while large ones have large neighbors, as opposed to the null hypothesis of no relationship between spatial location and plant size. In calculating the Mantel statistic, the products of corresponding elements of the distance matrices are summed as follows,

$$Z = \sum_{i=1}^{n} \sum_{j=1}^{n} A_{ij}B_{ij}, \text{ for } i \neq j \tag{15.1}$$

where the variable distance matrix (**A**) might contain some measure that represents the differences in the outcome of the experiment among all n experimental units, and the spatial distance matrix (**B**) might contain the actual Euclidean (spatial) distances among the n experimental units. The matrices must be square (i.e., same number of rows as columns): when the square matrices are symmetric, the computation is carried out only on the lower or upper diagonal matrices. This is because the distance between points one and two is exactly the same as the distance between points two and one, so we need to include that distance only one time in the matrix. When one of the matrices is asymmetric, the computation has to be carried out on the whole matrices.

A major advantage of using distance matrices is that the values of the matrix elements can be computed using the distance measurements of one's choice. This allows for testing the effects of spatial structure or experimental treatments on various type of outcome measurements (qualitative or quantitative), as well for one or more variables at once. The difference in spatial locations can be thus compared with differences in size, or in genetic relatedness. The disadvantage of using distances is that by summing the cross-products of ecological and geographical distances, the Z statistic is unbounded and cannot be compared from one study to another. To overcome this, the Z statistic can be normalized (r) such that it behaves as a product-moment correlation coefficient (similar to Pearson's r, representing a linear relationship) varying between -1 and $+1$. The normalization of each distance matrix is carried out separately using the standard normal transformation, subtracting the mean of that matrix from each element and then dividing by the standard deviation of the elements in that matrix. This normalized Mantel statistic (r) can be used to compare results from different variables, or studies, by means of confidence limits as described by Manly (1986a, 1991). When the r statistic is calculated between a variable distance matrix and a geographic distance matrix, the value of r corresponds to the average magnitude of spatial autocorrelation of the variable for the entire study area.

Once the Mantel statistic has been calculated, one would usually wish to test its statistical significance. Unfortunately it is not possible to use the familiar statistical tests for this purpose. In fact, the Mantel statistic, Z or r, cannot be tested as an ordinary product-moment correlation because the distances in each matrix are not independent of one another. Therefore, the significance is assessed either by using a permutation test to construct a reference distribution, or by using an asymptotic t-approximation test (Mantel, 1967). In a permutation test, the statistic calculated on the actual data is compared with what happens when the elements of the matrices are shuffled around at random. If there is a strong spatial pattern in the data, shuffling around the data points will eliminate that pattern.

In practice, only one of the matrices needs to be reordered to accomplish the task. The reference distribution obtained by permutation is constructed by randomly reassigning the value of each observation to another location many times, each time computing a new Mantel correlation. By this randomization procedure a population of "experiments" is created, although only one experiment has actually been carried out (Manly, 1991). Under the null hypothesis of no relationship between the two distance matrices, the observed Mantel statistic is expected to have a value located near the mode of the reference distribution obtained by randomization of the data, that is, the correlation between the two matrices should have neither an extremely low nor an extremely high value. With no relationship between spatial values and sizes, one expects that a random reshuffling of the data will result in about as many higher values of the Z statistic as lower ones. On the other hand, if there is a strong relationship (positive or negative) between the two matrices, the observed Mantel statistic (on the actual data) is expected to be more extreme (either higher or lower) than most of the reference distribution values.

For small sample sizes ($n \leq 10$), all the possible permutations ($n!$) can be computed for the exact probability level. However, with a sample size that small, it is difficult to detect significant spatial patterns, and this is true not only for Mantel tests, but for other spatial analysis methods as well (Legendre and Fortin, 1989). Legendre and Fortin (1989) recommend sample sizes of at least 30 to detect significant spatial autocorrelation with Moran's I coefficients. Likewise, it is recommended that one use as large a sample size as possible (e.g., $n > 20$)

to detect significant spatial patterns with the Mantel test (Legendre, personal communication). Therefore, when patches and spatial autocorrelation are suspected to be important, the experimental design should incorporate a relatively small number of treatments so that there are as many replicates as possible (e.g., van Es et al., 1989). With such larger sample sizes, the observed Mantel statistic can be tested against a null distribution generated using a random sampling (with replacement) of all possible permutations of the data (Smouse et al., 1986). The minimum number of permutations recommended by Manly (1991) is 1000—the higher the number of permutations, the more accurate is the significance test.

When n is very large ($n \geq 40$), the Mantel statistic, Z or r, can by transformed into a t statistic as follows:

$$t = \frac{Z - \text{EXP}(Z)}{\text{SE}(Z)} \tag{15.2}$$

where

$$\text{EXP}(Z) = \frac{\Sigma \Sigma A_{ij} \Sigma \Sigma B_{ij}}{n(n-1)} \tag{15.3}$$

and

$$\text{SE}(Z) = \sqrt{\Sigma \Sigma A_{ij} A_{kl} \text{EXP}(B_{ij} B_{kl}) - \frac{(\Sigma \Sigma A_{ij})^2 (\Sigma \Sigma B_{ij})^2}{n^2(n-1)^2}} \tag{15.4}$$

The significance of the t statistic is achieved using an asymptotic t-approximation test (Mantel, 1967). The larger the number of observations (n), the more reliable is the significance level by the asymptotic approximation test.

15.2.2 Three-Matrix (Partial) Mantel Test

The Mantel test allows for a comparison between only two variables, or two sets of variables by mean of multivariate distance coefficients (Legendre and Legendre, 1983). This becomes quite a limitation in ecology, however, where several processes may interact with each other. Furthermore, as discussed above, with field data space can created spurious relations between two variables that are in fact driven by a spatial gradient or by a third variable that follows the spatial gradient. For example, a positive relationship between plant growth and geographic distances can exist simply because both are related to environmental conditions. But genetic relationships or experimentally imposed treatment factors might also be responsible for differences in plant response. For such cases, we need to disentangle the relative contributions of the various factors influencing the outcome measurement.

To address this issue, the Mantel test has been modified in different ways to allow the comparison among three or more distance matrices (Dow and Cheverud, 1985; Hubert, 1987; Smouse et al., 1986; Manly, 1986a, 1991). The third distance matrix, C, can either be another variable (species, environmental condition, genetic data) collected independently at the same locations, or a design matrix that refers to a structure imposed on the data by an experimental design or by a hypothesis to test (Legendre and Troussellier, 1988; Sokal et al., 1987).

The method that we use in this study was originated by Smouse et al. (1986). In this approach, a partial correlation, $r_{AB.C}$, that establishes the degree of relationship between two matrices is calculated keeping the effects of the third matrix constant. This partial correlation is computed by first constructing a matrix of residuals, A', of the regression between A and C, and a matrix of residuals B' of the regression between B and C. These are linear regressions, based on the standardized matrices. Then the partial Mantel test is computed as in Eq. 15.1 but using the two residuals matrices A' and B'. The resulting partial Mantel statistic, $r_{AB.C}$, corresponds to a partial product-moment varying from -1 to $+1$. Its significance is assessed as for the two-way Mantel test by either permutation or by the asymptotic t-approximation test. According to Oden and Sokal (1992), the method of Smouse et al. (1986) has the best statistical properties for analyzing the relationship among three distance matrices when spatial autocorrelation is present.

15.3 Example

Ideally, data should be obtained from an experiment where the spatial structure is known, so that an optimal experimental design could have been created and carried out to account for the spatial effects (Chapters 3 and 14). However, the real world is not ideal, and it is often difficult or impossible to account accurately for the underlying spatial structure in plant characteristics in nature. Field workers may be forced to compromise, doing their best to carry out the most appropriate experimental design according to the information available. The necessity for compromise being the most common case, the following example will illustrate how the Mantel and partial Mantel tests can be used to test for the presence of and to distinguish the relative contributions of the treatment effect and the effect of spatial structure.

15.3.1 Data

Data from an experiment on plant competition (Gurevitch, 1986) will be employed to illustrate these methods. Gurevitch used removal experiments carried out at three sites along a topographic gradient (near Sonoita, Santa Cruz County, Arizona), to test whether the growth of the C_3 grass *Stipa neomexicana* was limited by competition from perennial C_4 grasses. The response of *Stipa* was compared with that of a C_4 grass, *Aristida glauca*. The two species were subjected to several experimental treatments (partial or complete removal of neighbors) and followed over a period of 20 months. Here we will present the results from the mid-slope site, and for only two of the four experimental treatments used by Gurevitch (1986): all neighboring plants within a 0.5 m radius of the target plant (*Stipa* or *Aristida*) were removed (treatment), or neighboring plants were left alone (control). The size (basal area, cm^2) of each target plant was recorded at the beginning (January 1980, hereafter referred to as initial size) and at the end of the experiment (August 1981, hereafter referred to as final size).

A randomized block design was employed to assign at random two replicates per treatment within each block (Fig. 15.1). Note that the experimental design was less than perfect, as it would have been better to assign *Stipa* and *Aristida* plants as targets within the same blocks rather than in adjacent ones. We empha-

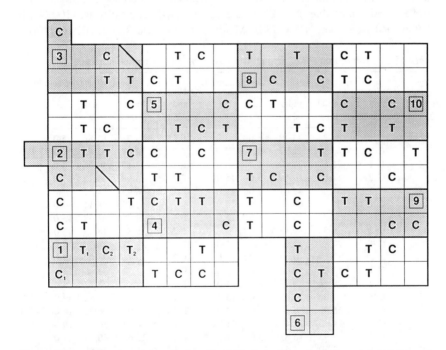

Figure 15.1. Experimental replicates. Each block (solid rectangle) contains eight replicates (1.0 × 1.0 m), where C indicates controls and T removal treatment replicates used in this study. In the gray blocks the target species is *Stipa neomexicana* and in the white ones it is *Aristida glauca*. The squares with a diagonal were not usable in the experiment. The blocks where *Stipa neomexicana* is the target species are numbered as in Table 15.1. In block 1, the replicate numbers are indicated by subscript.

size that the Mantel approach discussed in this chapter can be used with a variety of experimental designs, including completely randomized designs as well as randomized blocks, and with more than two levels per treatment. The experimental unit was an individual mature grass plant, located near the center of a 1.0 × 1.0 m plot. For each species, there were ten blocks, for a total of $n = 40$ replicates (20 controls and 20 removal treatments for each of the two species).

The following examples were computed using the *Stipa neomexicana* data as shown in Table 15.1. The Mantel (two-matrix) and partial Mantel (three-matrix) statistics, as well as their associated significance levels were computed by the MANTEL program of "The R package: multidimensional analysis, spatial analysis" of Legendre and Vaudor (1991). This package is distributed without cost by P. Legendre, and is available in CMS(IBM), VMS(VAX), and Macintosh versions (P. Legendre, Département de sciences biologiques, Université de Montréal, C.P. 6128, Succursale A, Montréal, Québec, Canada H3C 3J7). The Mantel (two-matrix) statistic can be computed as well as with the MXCOMP (for MatriX COMParison) program of the NTSYS-pc package (version 1.6 and higher) for IBM PC and compatible. NTSYS-pc is marketed by Exeter Software (100 North Country Rd., Setauket NY 11733). Manly (1991) provides FORTRAN subrou-

Table 15.1.   Initial and Final Size (Basal Area, cm²) of Stipa neomexicana[a]

Block	Treatment and replicate	Initial size	Final size	Geographic coordinated		Treatment and replicate	Initial size	Final size	Geographic coordinates	
				x	y				x	y
1	C 1	28.34	23.13	3.0	5.0	T 1	79.73	85.88	5.0	7.0
	C 2	30.63	22.97	7.0	7.0	T 2	12.02	32.45	9.0	7.0
2	C 1	75.87	39.80	3.0	13.0	T 1	29.15	58.61	5.0	15.0
	C 2	11.94	7.59	9.0	15.0	T 2	5.18	19.15	7.0	15.0
3	C 1	53.60	49.01	7.0	19.0	T 1	8.95	24.57	7.0	17.0
	C 2	39.25	68.61	3.0	25.0	T 2	32.99	78.11	9.0	17.0
4	C 1	4.32	3.41	11.0	11.0	T 1	23.28	61.38	13.0	11.0
	C 2	5.36	9.53	17.0	9.0	T 2	5.37	18.85	15.0	11.0
5	C 1	22.89	13.62	15.0	17.0	T 1	22.78	33.57	13.0	17.0
	C 2	17.59	0.07	17.0	19.0	T 2	26.14	50.08	17.0	17.0
6	C 1	13.21	30.04	23.0	3.0	T 1	4.90	33.60	23.0	3.0
	C 2	28.84	33.13	23.0	1.0	T 2	9.24	25.92	25.0	1.0
7	C 1	21.32	27.43	21.0	13.0	T 1	46.73	67.80	19.0	13.0
	C 2	15.90	6.48	25.0	13.0	T 2	6.03	26.30	25.0	15.0
8	C 1	4.40	14.42	21.0	21.0	T 1	6.44	50.80	19.0	23.0
	C 2	33.93	16.87	25.0	21.0	T 2	11.64	30.63	23.0	23.0
9	C 1	58.50	15.55	31.0	9.0	T 1	8.95	28.65	27.0	11.0
	C 2	26.30	23.50	33.0	9.0	T 2	27.57	53.85	29.0	11.0
10	C 1	7.66	5.93	27.0	19.0	T 1	14.07	29.99	27.0	17.0
	C 2	8.29	10.05	31.0	19.0	T 2	6.57	9.33	31.0	17.0

[a]T, removal treatment; C, controls. For geographic coordinates, (0,0) is the bottom left corner of the field pictured in Fig. 15.1.

tines to carry out the Mantel (two-matrix) statistic (MANT2), as well as a test for three matrices (MANCOR; Manly, 1986a, 1991) based on Manly's regression approach.

15.3.2 Building the Distance Matrices

Given the data from the above study, three general types of n by n distance matrices can be computed.

1. The *variable* distance matrix, **A**, refers to the differences in the measurements (here, size) of the plants between all possible pairs of replicates where i and j represent different experimental units (here, each plant is an experimental unit). Since we are interested in any treatment effects on size, we compute the distance as the absolute difference between all pair of replicates as follows:

$$\text{outcome}(i,j) = | \ \text{size}_i - \text{size}_j \ | \qquad (15.5)$$

For example, from Table 15.1, the difference between the final size of the first control plant, C_1, and the second control plant, C_2, in the first block (C_1, C_2) = $| \ 23.13 - 22.97 \ |$ = 0.16. The variable distance matrix can be computed not only for the final size but for the initial size as well. Indeed, the variable distance matrix of the initial size can be used to test whether there was already a significant spatial pattern in plant size at the beginning of the experiment.

According to the hypothesis under study, as well as the type of data available, different distance coefficients can be used to establish the variable (or outcome) distance matrix, **A**. To select the appropriate distance coefficient for the type of data (qualitative or quantitative), and for the biological context (community structure, genetic distance), one can consult the textbook of Legendre and Legendre (1983) or the paper of Gower and Legendre (1986). The actual computation of the distance matrix can be carried out with the program SIMIL of the "R package" (Legendre and Vaudor, 1991) which offers 15 distance (dissimilarity) coefficients. The input data for the SIMIL program is an ASCII file containing the raw data where the "rows" are the objects (here n rows for the n replicates) and the "column" holds the values of a given variable (here, the plant size). The output from SIMIL is the upper triangle of a n by n distance matrix written as a SIMIL binary format. This SIMIL binary file can be used directly with the MANTEL program. An ASCII file of the SIMIL binary file can be obtained by the LOOK program. If the SIMIL program does not have the specific distance coefficient desired, the distance matrix can be computed by another software program. In such cases, the ASCII distance matrix file can be rewritten into the SIMIL binary format using the IMPORT program.

2. The *geography* (or spatial) distance matrix, **B**, contains the physical location distances between each point (here, each plant) and all the others. It can be

computed using the Euclidean distance (i.e., the actual physical distance), between the spatial coordinates of all possible pairs of plants as follows:

$$\text{geographic } (i, j) = \sqrt{(x_i - x_j)^2 + (y_i - y_j)^2} \qquad (15.6)$$

So, the geographic distance between the two control replicates of the first block (Table 15.1) is

$$\text{geographic}(C_1, C_2) = \sqrt{(3.0 - 7.0)^2 + (5.0 - 7.0)^2} = 4.47$$

that is, the distance between plants C_1 and C_2 in the field is 4.47 m (Fig. 15.1).

When the relationship between plant response and geography is known not to be linear, other types of distance measures can be used. Choices might be the inverse of the Euclidean distance, which makes small distances more important than large ones (Estabrook and Gates, 1984), or simply relative position given by a nearest-neighbor network (Upton and Fingleton, 1985; Harvey et al., 1988).

3. The *design* matrix, **C**, expresses the differences in the treatments to which the plants (experimental units) were exposed. It allows one to test whether plants subjected to different treatments were any different than those with the same treatment. This matrix is coded analogously to a set of contrasts in ANOVA: when two replicates have received the same treatment, their difference is coded zero; when two replicates have received different treatments, their difference is coded one. A similar type of coding can also be used to test the differences in responses between the two species:

	C-treatment					C-species			
	C_1	C_2	T_1	T_2		Stipa	Stipa	Aristida	Aristida
C_1	0	0	1	1	Stipa	0	0	1	1
C_2		0	1	1	Stipa		0	1	1
T_1			0	0	Aristida			0	0
T_2				0	Aristida				0

As the coding of the design matrix is specific to the experiment hypothesis and the data, this matrix should be created as an ASCII file and then rewritten into a SIMIL binary format using the IMPORT program. Coding the treatment matrix requires careful thought so that the test really makes the comparison one wants. When there are more than two treatment levels, the coding can be as above where the code zero is given when replicates received the same level of a treatment, and the code one when replicates received different treatment levels, regardless of the level. This coding provides the same weight to all treatments, as does the null hypothesis of an analysis of variance. When directional differences among several treatment levels are to be tested, the coding has to be such that it gives appropriate weights to different pairs of treatment levels.

The interactions among factors, here the interaction between species and removal treatment, can also be tested by appropriately coding the design matrix. For example, within-species coding between the treatment type can be the same as above, while between-species coding is of opposite signs to create a contrast that tests for interactions between species and treatment:

		Stipa		*Aristida*	
		Control	Treatment	Control	Treatment
Stipa	Control	0	1	1	−1
	Treatment	1	0	−1	1
Aristida	Control	1	−1	0	1
	Treatment	−1	1	1	0

Here we used the same the code, 1, for the distance between the plants of different species but having same treatment, and for plants of the same species but with different treatments. Different weights could have been used to test other hypotheses. However, given that the design matrix, C, must be a distance matrix, it can be hard to code complex designs.

15.3.3 Computing the Mantel Statistic

The measure of the correlation between two matrices, such as between the variable and the geography matrices, can be computed using the Mantel test (Mantel, 1967; Sokal, 1979). By doing so, we will be able to evaluate whether there is significant spatial autocorrelation over the entire stand of plants. Thus we are interested in assessing the correlation between the distance (difference) in size among all pairs of plants, contained in the variable matrix (A), and th actual Euclidean distances among the respective pairs of plants contained in the geography matrix (B). If the spatial location of the plants does not affect plant size, the observed Mantel statistic (r) should not be statistically different than zero.

For example, to compute the Mantel statistic between the initial size and the geographic distances of the four replicates of *Stipa* in the first block (replicates C_1, C_2, T_1 and T_2; see Table 15.1), we used the following distance matrices:

	A-initial size					B-geography					$A \times B$		
	C_1	C_2	T_1	T_2		C_1	C_2	T_1	T_2		C_1	C_2	T_1	T_2
C_1	0.00	2.29	51.39	16.32	C_1	0.00	4.47	2.82	6.32	C_1	0.00	10.24	144.92	103.14
C_2		0.00	49.10	18.61	C_2		0.00	2.00	2.00	C_2		0.00	98.20	37.22
T_1			0.00	67.71	T_1			0.00	4.00	T_1			0.00	270.84
T_2				0.00	T_2				0.00	T_2				0.00

where the matrix $A \times B$ is the element-by-element cross-product of the values in the two distance matrices A and B. For example, the first element of A (above) multiplied by the first element of **B**, 0×0, equals the first element in $\mathbf{A} \times \mathbf{B}$, which is zero, while the second element in **A** multiplied by the second element in **B**, $2.29 \times 4.47 = 10.24$, which is the second element in the $\mathbf{A} \times \mathbf{B}$ matrix. The Mantel statistic Z, as mentioned in Section 15.2.1, is the summation of the cross-product elements of the matrix $\mathbf{A} \times \mathbf{B}$ excluding the diagonal ($i = j$) since the distance of a replicate with itself is zero:

$$Z = 10.24 + 144.92 + 103.94 + 98.20 + 37.22 + 270.84 = 665.07$$

Of course, ordinarily Z would be calculated across both species and all blocks, but this small part of the data set is useful for illustrating the method. Given that the distance between i and j equals that between j and i (in symmetric matrices), the Mantel statistic is evaluated by using only the upper (or lower) diagonal matrices without affecting the relative strength of the relation between the two matrices.

To compute the normalized Mantel statistic, r, each distance matrix is first standardized individually by subtracting the mean of all the elements in the matrix from each observation and then dividing by the standard deviation. The mean distance for this first block of the variable matrix **A** (initial size) is 34.24 cm^2 and the standard deviation is 25.38 cm^2, while for the distance matrix **B** (geographic distance) the mean is 3.60 m and the standard deviation is 1.68 m. So, for example, to standardize the second element in the first row of **A**, the calculation is $(2.29 - 34.24)/25.38 = -1.26$. The r is calculated as for the Z statistic by summing the cross-products of the standardized matrices (here abbreviated by "std") and by dividing by $(n - 1)$:

$$r = \frac{\Sigma \Sigma \, \text{std}A_{ij} \, \text{std}B_{ij}}{n - 1} \tag{15.7}$$

Thus, for the initial size, the normalized Mantel statistic is computed for the first block from the following matrices:

	A-initial size				B-geography				A×B			
	C_1	C_2	T_1	T_2		C_1	C_2	T_1	T_2		C_1	C_2	T_1	T_2
C_1	0.00	−1.26	0.68	−0.71	C_1	0.00	0.52	−0.46	1.62	C_1	0.00	−0.65	−0.31	−0.55
C_2		0.00	0.58	−0.62	C_2		0.00	−0.95	−0.95	C_2		0.00	−1.15	0.58
T_1			0.00	1.32	T_1			0.00	0.24	T_1			0.00	0.31
T_2				0.00	T_2				0.00	T_2				0.00

It is equal to

$$r = (-0.65 + -0.31 + -0.55 + -1.15 + 0.58 + 0.31)/(6 - 1) =$$
$$-1.77/5 = -0.354$$

if calculated for the first block only.

We can also test for treatment effects using the same approach. We would expect no significant correlation between the initial size and the treatments, because the treatments were assigned to plants at random and were not begun until after the initial sizes were measured. From the example,

	A-initial size					C-treatment					A×C			
	C_1	C_2	T_1	T_2		C_1	C_2	T_1	T_2		C_1	C_2	T_1	T_2
C_1	0.00	2.29	51.39	16.32	C_1	0	0	1	1	C_1	0.00	0.00	51.39	16.32
C_2		0.00	49.10	18.10	C_2		0	1	1	C_2		0.00	49.10	18.10
T_1			0.00	67.71	T_1			0	0	T_1			0.00	0.00
T_2				0.00	T_2				0	T_2				0.00

and $Z = 0.00 + 51.39 + 16.32 + 49.10 + 18.10 + 0.00 = 134.91$, while $r = -0.023$. As anticipated, this value is very close to zero.

15.3.4 Assessing the Significance of the Mantel Statistic

As for any statistical test, the significance of the observed value of the Mantel statistic, Z or r, is assessed by comparing it to the reference distribution obtained under the null hypothesis. In practice, the reference distribution can be obtained by permuting the arrangement of the elements of one of the distance matrices randomly at least 1000 times, each time computing the Mantel statistic. The significance of the observed Mantel value is obtained by comparison with the number of times that the permuted Mantel values are smaller, equal to or greater than the Mantel statistic for the actual data.

As mentioned above, the null hypothesis refers to the absence of a significant relationship between the two matrices. The alternative hypothesis can be evaluated using either a one- or two-tailed test. The one-tail probability of any positive observed Mantel statistic (i.e., right-hand tail probability) is given by first adding the number of permuted values of equal or greater magnitude than the calculated (actual) value and then dividing by the total number of permutations. Similarly, the one-tail probability of a negative observed Mantel value (i.e., left-hand tail probability) is given by adding the number of equal and smaller values and then dividing by the number of permutations. In the MANTEL program, the observed value is added to the number of equal values, so to have a reference distribution of 1000, one needs to ask for only 999 permutations. While MANTEL (Legendre and Vaudor, 1991) and MXCOMP (NTSYS-pc) programs provide only the number of permutations that are smaller, equal, and greater than the observed value, and hence one-tail probability levels, MANT2 provides one-tail as well as two-tail probabilities, where the two-tail probability is given by the number of permutations that in absolute values are greater than or equal to the observed statistic.

Using the MANTEL program and thus employing one-tailed tests, the following relationships were computed for the 10 blocks of *Stipa* across the entire midslope site:

	r	Smaller	Equal	Greater	P(r)
Initial × Geography	0.1381	977	1	22	0.023
Initial × Treatment	−0.0011	647	1	352	0.648
Final × Geography	0.0389	736	1	263	0.264
Final × Treatment	0.1396	996	1	3	0.004

The results reveal that spatial location (geography) had a significant ($r = 0.1381$, $P < 0.05$) effect on the initial size of *Stipa* plants (above and Table 15.2A). Note that each design factor is tested independently; each line above (and in Tables 15.2 and 15.3) represents a complete Mantel test of 999 permutations. As expected, competition treatments did not affect the initial size of *Stipa* plants, as treatments were assigned at random after initial size was measured (above and Table 15.2A, $r = -0.0011$, NS). By the end of the experiment, the treatment (removal of competitors) had a highly significant effect on *Stipa* plant size (above and Table 15.2B, $r = 0.1396$, $P \leq 0.004$).

In a more complete analysis, we included the entire data set for both *Stipa* and *Aristida*. There were large differences in initial size between the two species (Table 15.3A, $r = 0.2208$, $P < 0.001$). There were also highly significant differences between the two grass species in final plant size (Table 15.3B, $r = 0.2891$, $P < 0.001$), a highly significant species × treatment interaction on final

Table 15.2. *ANOVA and Mantel Tests on the Sizes (Basal Area, cm^2) of* Stipa neomexicana *Plants*

		ANOVA		Mantel tests	
	Factor	F	P	r	P
A.	Initial Size				
	Block	1.17	0.3629	0.0147	0.224
	Treatment	0.99	0.3323	−0.0011	0.648
	Treatment.Block			0.0024	0.312
	Treatment.Geography			0.0044	0.285
	Geography			0.1381	0.023
	Geography.Treatment			0.1382	0.027
B.	Final size				
	Block	1.31	0.2920	−0.0002	0.510
	Treatment	10.73	0.0038	0.1396	0.004
	Treatment.Block			0.1409	0.001
	Treatment.Geography			0.1414	0.001
	Geography			0.0389	0.264
	Geography.Treatment			0.0451	0.230

Table 15.3. ANOVA and Mantel Tests on the Sizes (Basal Area, cm²) of Stipa
neomexicana *and* Aristida glauca *Plants*

		ANOVA		Mantel tests	
	Factor	F	P	r	P
A.	Initial Size				
	Species	43.56	0.0001	0.2208	0.001
	Block	2.26	0.0284	0.0046	0.270
	Treatment	0.02	0.9003	−0.0092	0.081
	Species × Treatment	0.35	0.5541	−0.0065	0.281
	Geography			0.0071	0.389
	Treatment.Geography			0.0067	0.391
	Geography.Treatment			0.0069	0.425
B.	Final size				
	Species	70.94	0.0001	0.2891	0.001
	Block	0.84	0.5422	−0.0184	0.037
	Treatment	29.55	0.0001	0.0876	0.001
	Species × Treatment	10.61	0.0025	0.0496	0.001
	Geography			−0.0387	0.173
	Treatment.Geography			0.0501	0.001
	Geography.Treatment			−0.0371	0.173

size ($r = 0.0496$, $P < 0.001$), and a marginally significant effect of blocks at
the end of the experiment ($r = -0.0184$, $P < 0.05$).

15.3.5 Computing Partial Mantel Tests

The partial Mantel statistic, $r_{AB.C}$ (the correlation between matrix **A** and **B** given
C), can be computed to test whether there are significant treatment effects (design
matrix) on plant size (variable distance matrix) when the effects of the spatial
location are kept constant (geographic distance matrix). One might also wish to
investigate whether plant size (variable distance matrix) is spatially autocorrelated
(geographic distance matrix) while the treatment effects are kept constant (design
matrix). Thus, the partial Mantel test can be carried out to insure that the effects
of the treatments are not canceled out by spatial effects. We used the three-way
(partial) Mantel test to examine some of the questions that are probably of greatest
interest to the experimentalist; for example, what are the effects of the treatments
when spatial effects are taken into account? This may be thought of as being
analogous to an analysis of covariance, where spatial effects are "held constant"
while the effects of the experimental treatment are examined.

When the results were analyzed for *Stipa* alone, taking spatial location into
account strengthened the already strong effect of the treatment on final plant size
(Table 15.2B, Treatment.Geography, $r = 0.1414$, $P < 0.001$, as compared with
Treatment, $r = 0.1396$), while there was no significant effect of the treatments
on initial size when geography was held constant (Table 15.2A, Treatment.Geog-

raphy, $r = 0.0044$, NS). Partial Mantel tests could be used to test many relationships of interest: for example, one could test for the effect of the treatments on final plant size, holding initial plant size constant (Treatment.Initial, $r = 0.5518$, $P < 0.001$). For the entire data set including both *Stipa* and *Aristida* (Table 15.3), again the effects of the experimental treatment on final plant size remained statistically significant when spatial effects were held constant (Table 15.3B, $r = 0.0501$, $P < 0.001$). Spatial location (geography) did not affect plant size, even when treatment effects were taken into account (Table 15.3B, $r = -0.0371$, NS).

15.3.6 Mantel Tests vs ANOVA

The results of the above experiment using the Mantel tests were compared with the results from a standard ANOVA. The results were similar for the two tests (Tables 15.2 and 15.3), which is to be expected since strong spatial dependence was not detected in this data set. Therefore, the ANOVA assumption of no spatial autocorrelation (i.e., independence of observations) was not seriously violated in the present study. Nevertheless, the Mantel and partial Mantel tests can bring out complementary information that ANOVA cannot provide, such as the detection of a significant underlying spatial autocorrelation for *Stipa* at the beginning of the experiment (initial size, Table 15.2A), which did not hold at the end (final size, Table 15.2B). This loss of significant spatial pattern in final size offers insight into the spatial responses of *Stipa*, and indicates that the treatment effects were strong enough to cancel out the initial spatial pattern. Notice that the Mantel tests were able to detect the spatial pattern regardless of the block size used.

Furthermore, the Mantel approach seemed better than ANOVA at detecting block effects and the species × treatment interaction at the end of the experiment as indicated by small P levels, although these relationships were not strong (Table 15.3B). The same trend held for the partial Mantel tests (Tables 15.2 and 15.3). Recall that even though the spatial effects (geography) were not statistically significant, holding the effects of space constant using the partial Mantel test resulted in a higher r value (i.e., a stronger relationship) for treatment effects (Table 15.2B, $r = 0.1396$ for treatment alone, while $r = 0.1414$ for treatment with geography held constant). The stronger treatment effect when spatial effects were held constant was not apparent when *Stipa* and *Aristida* were analyzed together, however (Table 15.3B), perhaps because the spatial patterns for the two species differed.

15.4 Conclusions

In this study, we have emphasized the importance of detecting and taking into account the underlying spatial pattern in the analysis of field data to obtain a better understanding of the plants' responses. Our results emphasize the value of using larger and simpler experiments to reveal pattern that may otherwise be obscured (Chapter 17; Gurevitch et al., 1992). There is a trend in ecological experiments to include many experimental treatments and few replicates, but because the ability to detect spatial pattern increases with sample size, fewer treatments and a larger number of replicates may be preferable.

Other approaches to the design and analysis of ecological data when spatial autocorrelation is present have also been proposed (e.g., Chapter 14; Besag and Kempton, 1986; Legendre et al., 1990). Although we have used Mantel tests to analyze spatially autocorrelated variables, these tests are not restricted to that use and can be implemented to analyze other types of data or in other contexts (Sokal et al., 1990, 1993). The usefulness and flexibility of the Mantel and partial Mantel tests make them good exploratory tools to detect a posteriori the scale of the spatial pattern or to test the goodness-of-fit to an alternative hypothesis or model (Legendre and Troussellier, 1988; Legendre and Fortin, 1989; Sokal et al., 1987).

The extension of the Mantel and partial Mantel tests presented here offers a promising approach in designed experiments where spatial heterogeneity may pose problems for the analysis of the results. Using this approach, it is possible to distinguish the effects of spatial pattern from those of experimentally imposed treatment effects. Therefore, this offers an alternative to classical experimental design and parametric analysis of variance in ecology, especially when the extent and pattern of spatial heterogeneity are not known in advance.

Acknowledgments

Professor Robert Sokal provided invaluable suggestions on how to approach the problem. The work was supported in part by NSF Grant BSR 89-08112 to J.G., which is gratefully acknowledged.

16

Model Validation: Optimal Foraging Theory
Joel S. Brown

"A proof is that which suffices to convince."

16.1 Ecological Issues

Here, I will use optimal foraging theory as an exemplar of science as a process of building theories and testing models. In particular, I hope to illustrate how statistics fits into a larger research program of model formulation, the fit between models and experiments, and model reformulation. As such, this chapter uses foraging theory as a case study for the scientific processes envisaged in Chapter 1 (particularly, Section 1.2). In this process we will see how the interplay between experimentation and model validation uses statistics as the adjudicator among competing ideas, models, and hypotheses. And in doing so, we will see how experiments in foraging theory address many of the statistical techniques, considerations, and sins discussed in the previous chapters.

Foraging theory is well suited as a case study for the use of statistics in the scientific process. As a paradigm of hypotheticodeductive science, foraging theory arguably represents the tightest lock-step between model formulation and hypothesis testing of all the subdisciplines of ecology and evolution.

I will begin by briefly discussing the ecological and evolutionary context from which foraging theory emerges as part of the optimization research program (sensu Mitchell and Valone, 1990). This will be followed with a recipe for formulating models and generating testable predictions. Finally, I will discuss three experimental studies that test hypotheses of foraging-theory models. In these examples, a strong interface with statistics begins in the enterprise of experimental design and hypothesis testing. The end-product includes using statistical analyses to reject or accept hypotheses, the validation or refutation of a theory, and the reformulation of new models, theories, and hypotheses. If this sounds like a recipe for science in general, it is meant to. Hopefully this treatment of foraging theory provides a useful conclusion to the preceding chapters for how to profitably apply statistics within a scientific research program.

16.1.1 Foraging Theory as a Research Program

Births and deaths, and a need for energy to grow, function, and maintain homeostasis are two universal properties of life. Combine these properties with evolution by natural selection and a theory of adaptive feeding behaviors emerges in which organisms attempt to access food quickly and/or efficiently. Hence, foraging theory can be defined as the study of feeding behaviors as adaptations. I will view foraging theory as residing within the optimization research program as defined by Mitchell and Valone (1990). As a research program (Lakotos, 1978), foraging theory has a hierarchy of assumptions.

The hard core assumptions (sensu Lakotos, 1978) of foraging theory assume that existing feeding behaviors have been shaped by natural selection and that they are close to optimal in the sense that they maximize fitness given the circumstances. By themselves, the hard core assumptions neither form a model nor make testable predictions. Additional, auxiliary assumptions are required.

In foraging theory, these auxiliary assumptions determine three aspects of the ecological and evolutionary circumstances. (1) Some assumptions define the environmental context and include assumptions regarding the number and characteristics of food types, and the distribution and abundance of these foods. (2) Some assumptions postulate how harvesting food influences fitness via the rate of food harvest, metabolic effort of foraging, and exposure to predators. (3) Some assumptions consider the constraints on the set of feasible behaviors, and include assumptions regarding time available for foraging, the forager's ability to vary its search rate, food handling times, metabolic effort, and vigilance behaviors. These auxiliary assumptions can produce a model, but in the absence of the hard core assumptions the model generates no predictions.

16.1.2 Formulating Models and Generating Testable Predictions

In foraging theory, the auxiliary assumptions produce a model that has an *objective function* and a *behavior set*. When the model is conjoined with the hard core assumptions, predictions are generated by selecting the behavior from the behavior set that maximizes the objective function.

The auxiliary assumptions regarding the environmental context and the ways in which foraging influence fitness define the objective function and allow the researcher to incorporate hypotheses regarding the salient features of the organism's ecology. Ideally, this objective function is the expected fitness of the individual possessing the particular behavioral trait. More frequently, this function is assumed to be monotonically or linearly related to fitness. In either case, it is the objective function that determines the selective value of a given behavior. The objective function may be formulated as a mathematical or verbal model.

Auxiliary assumptions regarding the set of feasible behaviors define the behavior set and allow the researcher to make hypotheses regarding the organism's sensory, cognitive, and memory abilities, and make hypotheses regarding the organism's behavioral repertoire. In most models of foraging theory, the behavioral trait under selection includes the ability of the organism to assess its environment (often under imperfect information) and respond to this assessment. Constraints on the behavior set may be defined mathematically or verbally, the set may be finite or infinite, and the different behavioral options may lie on a quantitative continuum or be qualitatively distinct.

The above modeling recipe becomes a powerful tool for studying foraging behavior and its population or community level consequences. Observations and patterns contribute to the inductive reasoning that generates the auxiliary assumptions and the model. Once formulated, the predictions follow logically (deductively) from the model's assumptions. This recipe allows one to model a wide range of ecological and behavioral scenarios. While an exhaustive list would fill the rest of this chapter and be redundant to several excellent reviews (Stephens and Krebs, 1986; McNamara and Houston, 1987; Green, 1990), some feel for this scope of application is instructive.

Models are often characterized as patch use (Charnov, 1976) or diet choice (Pulliam, 1974) and all of their subsequent derivatives. Models may deal with time-independent strategies (static optimization) or time-dependent strategies that explicitly consider the temporal sequence and temporal consequences of particular behaviors (dynamic optimization: e.g., Ludwig and Rowe, 1990). Organisms may be sessile or mobile, and may have perfect or imperfect information (Iwasa et al., 1981). Inputs into fitness have included energy, safety (Gilliam and Fraser, 1987), and specific nutrients (Pulliam, 1975; Belovsky, 1981). Foraging behavior can be modeled as an evolutionary game where the forager's best strategy depends on the behaviors of other foragers (Mitchell, 1990; Mangel, 1991).

Modeling in foraging theory follows two lines (Green, 1990). The first evaluates the behavioral rules by which organisms assess the environment and make optimal decisions. The second considers the ecological consequence of foragers behaving optimally and makes predictions that link the behavior of the individual with population and community level phenomena. Examples of this include density-dependent habitat selection (e.g., the ghost of competition past, Rosenzweig, 1981), short-term apparent competition in predator-prey systems (Holt and Kotler, 1987), and the functional response of optimal foragers (Chapter 8; Mitchell and Brown, 1990).

Given the scope of modeling and application, it is not surprising that optimal foraging theory has provided a tight relationship between theory and empiricism and between model formulation and hypothesis testing.

16.1.3 Optimal Foraging Theory and the Refutation of Predictions

Either because of or in spite of its success and popularity, optimal foraging theory has drawn diverse criticisms (Gray, 1987; Rapport, 1991); these fall into two categories. The first category questions the lack of suitable null or alternative hypotheses based on models of nonoptimal behavior. How can a research program be justified if its hard core assumptions are never tested? A second category views optimal foraging theory as ad hoc and infinitely pliable. Any data can be made to fit a model of optimal foraging through suitable adjustments of the objective and constraint functions. Worse still, any refutation of an optimal foraging model is met with ad hoc adjustments of the offending auxiliary hypotheses. Reformulated models may become progressively more complex through additional terms that may have no useful biological meaning.

With respect to the first criticism, a number of authors make the important point that foraging theory does not test the hard core assumption of optimality (Stephens and Krebs, 1986; Mitchell and Valone, 1990; Parker and Maynard Smith, 1990). Why is this? Philosophically this comes from making the distinction between what is selected and what is selected for (sensu Sober 1984). The hard core assumes that the behavior results from natural selection and hence has been selected. The auxiliary assumptions involve an explanation for why the behavior has been selected for. In refuting an optimality model, the researcher questions these auxiliary assumptions that provide a functional explanation for the behavior.

The second criticism concerns the rapid substitution of alternative assumptions to explain the refutation of a prior model. But this is as it should be. Any model is the sum total of its assumptions. In this case, these assumption include the hierarchy of hard core and auxiliary assumptions (we are more likely to drop auxiliary rather than hard core assumptions), assumptions regarding logic and deductive reasoning (we assume there are no errors in our modeling computations), assumptions regarding the experimental methodology (we assume that the balance provides an accurate measure of mass, or that the oxygen analyzer is operating properly), and statistical assumptions.

Corroboration of the model's predictions builds confidence in the theory but may, in fact, emerge as the right answer for the wrong reasons. A refuted prediction means that one or more of the assumptions are false. Reevaluating the assumptions in response to a refuted prediction is often the most informative and rewarding aspect of the scientific process. Consider the statistical assumptions. A failure to corroborate a prediction may come from a failure to meet the assumptions of the statistical model. For instance, were cell-variances homogeneous in the ANOVA (Chapter 3)?, heteroscedasticity in your regression (Chapters 4 and 9)?, nonrandomness or nonindependence in your sampling regime (Chapters 7, 14, and 15)? In addition, the experimental methodology may not have measured the intended quantity. After convincing ourselves that we have satisfied these statistical and methodological assumptions, we gain biological insights by questioning our auxiliary assumptions. Foraging under predation risk (Sih, 1980) questions the assumption that animals forage so as to maximize net energy gain, and risk-sensitive foraging (Caraco, 1979) questions the assumption that fitness increases linearly with net energy gain.

Reevaluating possibly flawed assumptions in the face of a refuted prediction is not necessarily ad hoc nor is it necessarily circular (so long as you do not reuse the data as a confirmation of your new assumptions). It may be the most vital activity of a research program. Such a process generates a quick turnaround of models from formulation to test to reformulation. In fact, the researcher may often have several alternatives operating at once. Such parallel alternatives represent a strong inference (Platt, 1964) approach to hypothesis testing as discussed in Chapter 1. Mitchell and Valone (1990) say it best:

> A refuted prediction leads to modifying the strategic [auxiliary] hypotheses rather than the hard-core ones. In a progressive research programme, the new, modified hypotheses avoid being characterized as ad hoc by virtue of the fact that they generate new testable predictions.

In summary, refutation of a model's predictions can result because of either

1. Flawed statistical assumptions;
2. Flawed methodological assumptions;
3. Misidentification of the organism's objective function or environmental context;
4. Misidentification of the set of feasible behaviors available to the organism; or
5. The organism is not optimal (e.g., history and phylogenetic accidents, or satisfycing; see Ward, 1992).

The experimenter is likely to opt for (3) and (4) as the explanation for the failure of the model's predictions, and the experimenter endeavors to avoid (2) as the explanation for the model's failure. A central purpose of this book regards how to recognize and deal with statistical assumptions (1).

16.2 Statistical Issues in Model Validation

16.2.1 Patch Use

I will use models of patch use as examples; but the interplay between models, statistics, and experimentation applies equally well to any hypotheticodeductive inquiry. In Charnov's (1976) formulation, the forager's objective is to maximize its rate of energy harvest. To do this, it can alternatively engage in two mutually exclusive and equally costly activities: harvesting food from patches and traveling between patches. The harvest rate from a patch is a deterministic process and may be a function of patch quality. Furthermore, as a result of food depletion, the forager's harvest rate within a patch declines with time spent using the patch. The forager encounters patches of different quality in proportion to their environment-wide frequencies. On encountering a patch, the forager knows the quality of the patch and its harvest rate function for the patch. Finally, while a forager depletes the food of a given patch, the forager does not alter the overall quality of the environment (sometimes referred to as an infinitely repeating environment).

The objective function—which is simply the expected harvest from patches divided by the sum of travel time and the expected amount of time spent in patches—follows directly from these assumptions.

$$\text{net energy gain} = \frac{\Sigma\, p_i H_i(t_i)}{T + \Sigma\, p_i t_i} \tag{16.1}$$

where p_i is the frequency of patch type $i = 1, \ldots, n$, $H_i(t_i)$ is the cumulative harvest from patches of type i as a function of time spent in the patch, t_i, and T is travel time between patches.

The behavior set is the vector of times, $\mathbf{t} = (t_1, \ldots, t_n)$, devoted to food patches of different types (with the proviso that time cannot take on a negative value, $t_i \geq 0$). The predictions of this model, known as the marginal value theorem, are the logical consequences of the objective functions and the constraint set.

16.2.2 Quantitative vs Qualitative Predictions

If all of the parameter values of the model are specified, then the patch use model makes several quantitative predictions. One can predict the amount of time to devote to patches of a given quality, the amount of food harvested from a given patch, and the forager's environment-wide harvest rate. These predictions follow from the main result that a patch should be abandoned and a new one sought when the harvest rate within the patch drops to the average harvest rate of the forager in the environment at large.

Quantitative predictions often connote a bold and risky theory (for such a position see Gray, 1987; Green, 1990). Saying that the temperature will be higher tomorrow hardly represents a risky proposition; but saying that the high for tomorrow will be exactly 26.4°C is provocative. Testing a quantitative prediction, however, poses statistical problems. A quantitative prediction provides only a statistical null against which one tests alternative hypotheses. It is important to keep in mind that the type I error rate or P-value given for statistical tests does *not* represent the probability that the null hypothesis (quantitative prediction) is correct, rather it is the probability of obtaining such an extreme result *assuming* that the null hypothesis and associated statistical assumptions are correct.

Furthermore, we must be willing to permit a certain amount of slop around a quantitative prediction, and this permissible slop (tolerance) is somewhat subjective and situation dependent. Most of us would be impressed by someone's prediction of temperature that always fell within two degrees of the daytime high. This becomes an important issue in optimal foraging, because generally there exists a range of behaviors in the neighborhood of the optimal behavior that perform as well as the optimum within a selected tolerance range. With the inclusion of tolerance the quantitative prediction is simultaneously improved, obscured, and transformed into a range of values.

Does this mean that quantitative predictions are at once to be praised as gutsy but disdained as statistically unsound? Not entirely. From the perspective of strong inference (see below) a model such as the marginal value theorem may provide a valid null against which to test alternative hypotheses (witness the success of the Hardy-Weinberg Law in population genetics). In the absence of alternative hypotheses, we need to become less concerned with type I errors (the statistical level of significance) and more interested in type II errors and the power of the test. The surest way to save one's pet quantitative prediction against an alternative is to collect a small sample of highly variable data. Statistical evaluation of a quantitative prediction requires raising the power of the test to the point where the probability of not rejecting a false null hypothesis is brought below some critical threshold. At this point, applying the significance value of a statistical test becomes a more convincing means of corroborating one's quantitative prediction.

The patch use model described above also makes a number of qualitative predictions. A forager should spend more time in high quality patches than low-quality patches. Increasing travel time will increase time spent in a patch. Increasing the quality of the environment will decrease the amount of time spent in a patch of a given quality. These predictions can be generated without knowledge of the specific parameter values. In addition, they are far from unique to the marginal value theorem and are applicable to more general classes of patch use models.

Qualitative predictions are often denigrated as weak. Given two experimental treatments and the collection of sufficient data (i.e., sufficient statistical power) some difference, for whatever reason, is usually detectable at the $P < 0.05$ level of significance. This difference may be so small as to be biologically meaningless, and there is still a fifty-fifty chance anyway that the qualitative property of the difference will be in the predicted direction.

While qualitative predictions may be viewed as weak they are the quintessential alternative hypothesis of statistical inference. The mere statement of a qualitative prediction presupposes a null hypothesis of either no difference or no change. While small differences may in fact be biologically meaningless, the researcher is not necessarily interested in showing very large effects or explaining a high proportion of the observed variance. Rather, the statistical issue is one of specifying and achieving some level of statistical significance as a way of rejecting the null hypothesis and corroborating the alternative hypothesis (qualitative prediction). Another nice property of most qualitative predictions is that they predict the direction of change or the rank ordering of treatment effects. As a result, one-tailed tests become statistically appropriate.

How can qualitative predictions throw off the stigma of being unrisky and weak? There are at least two ways. The first involves using the model to make a battery of predictions. For instance, the marginal value theorem makes at least three qualitative predictions in response to varying different aspects of the model (patch quality, travel time, and environmental quality). Correctly calling the direction of change for three qualitative predictions (flips of a coin) may be a more impressive display of predictive prowess than calling just one quantitative prediction. The second way involves generating a range of treatment levels rather than just pairwise treatments. By generating four different magnitudes of patch quality, the researcher has a rank ordering of predictions from the lowest to highest patch residence times. This is not only a more risky prediction but it also has nice statistical ramifications. Not only does the researcher want to test for a difference in treatment means, but the researcher has a priori reasons for making particular pairwise comparisons. This allows the researcher to make particular comparisons without unduly compromising the experimentwise error rate.

16.2.3 Strong Inference

Experiments in optimal foraging theory lend themselves to strong inference. Based on observations or hypotheses regarding the foraging scenario, the researcher may have several valid models. These varied models could be tested sequentially, each experiment pitting the prediction of a given model against an appropriate null hypothesis. Platt (1964) suggests that it is more efficient to test simultaneously among the alternative hypotheses of the different models. In this way, one model's prediction may act as the statistical null for another and vice versa. In fact, if the models' predictions are mutually exclusive and the predictions cover all possible outcomes, then the statistical test will conclude with the refutation of two of the models and support for the third. Critics of strong inference, on the other hand, distrust an approach that purports to have an answer for every eventuality (Quinn and Dunham, 1983). Furthermore, strong inference may lose potency when the alternative hypotheses share a narrow realm of auxiliary assumptions.

I will use patch use theory to illustrate strong inference. The marginal value theorem, as do more general models of patch use, predicts that patches should be abandoned at the same quitting harvest rate. This presupposes that foragers can assess and respond to the quality of a patch. This has led to three general classes of patch use strategies based on the forager's ability to assess patch quality

(Iwasa et al., 1981; Valone and Brown, 1989). An ignorant forager may use a fixed search time strategy and devote the same amount of search time to each encountered patch (prediction: the same proportion of food will be harvested from each patch regardless of quality). A forager meeting some physiological constraint may use a fixed amount strategy and harvest the same amount of food from each patch (prediction: a higher proportion of food will be consumed from poor than rich patches). Or, as in the marginal value theorem, the forager may try to leave each patch at the same quitting harvest rate (prediction: a higher proportion of food will be consumed from rich than poor patches). If we envisage an experiment in which foragers are offered patches that are identical save for their initial abundance of food, then a test for how the proportion of food harvested changes with initial abundance will refute two of the above hypotheses and support the third.

16.3 Experimental and Statistical Solutions

This section reviews three experiments in which the researchers tested aspects of patch use theory. In doing so, I hope to provide a feel for the models under consideration, the fit between model and experimental design, the appropriateness of the statistical model in the experimental design, and the appropriate reformulation of model and statistics. By necessity, the following descriptions are brief and incomplete. By selecting some of my own work I can be critical and offend no one but myself (and my collaborators?).

16.3.1 Field Experiment on Testing among Patch Use Strategies

Valone and Brown (1989) illustrate the use of strong inference to test qualitative predictions in a field experiment using a very simple statistical model (sign test; Siegel, 1956) that makes no assumptions regarding normality or homoscedasticity in the data. As discussed earlier, patch leaving rules can be lumped into three mutually exclusive categories: fixed time, fixed amount, and fixed quitting harvest rate. If we assume that a forager's harvest rate from a patch is a function only of the remaining abundance of food, then in response to an increase in the initial abundance of food a forager should harvest the same, a lower, and a higher proportion of food, respectively, with respect to the three patch use strategies.

To test this prediction we established two 4 × 4 grids in a desert habitat near Tucson, Arizona. At each of the 16 points on a grid or station, we arranged three feeding patches side by side. Each feeding patch was an aluminum tray (45 × 45 × 2 cm) filled with a measured amount of millet seed mixed into three liters of sifted dirt. Trays received, with equal probability, either 2.7, 3.0, or 3.3 g of millet. To collect data from round-tailed ground squirrels (*Spermophilus tereticaudus*), trays were filled in the morning, left out all day, and the remaining seeds recovered in the evening. To collect data from kangaroo rats (*Dipodomys merriami*), trays were filled in the evening, left out all night, and the remaining seeds recovered in the morning. We collected data for 5 consecutive days and nights.

The data were the mass of millet seeds remaining (giving-up density, GUD, sensu Brown, 1988) in the trays following a day or night of foraging by squirrels or kangaroo rats, respectively. Data came as triplets (the three trays) and we considered only those triplets where at least two different initial abundances were present. (Due to the randomization there was a one in nine chance that each tray of a triplet had the same initial amount.) From these triplets we selected pairs of trays that differed in initial abundance (we selected one pairing randomly from the possible permutations of pairwise comparisons). We then counted the number of times (x and y, respectively) in which the tray with the higher initial abundance had the higher or lower proportion of seeds harvested ($x = 69$ and $y = 35$, for kangaroo rats; $x = 63$ and $y = 49$ pairs for squirrels).

As a statistical model, we used a sign test where the null hypothesis was no difference in the probability of obtaining a higher or lower proportion of seeds harvested ($P[x>y] = P[x<y] = 0.5$). The probabilities associated with a sign test can be calculated from the binomial distribution. Because of large numbers of samples, we used a normal approximation of the binomial distribution (Siegel, 1956). The value for Z of the normal distribution is given by

$$Z = \frac{2x - (x+y)}{\sqrt{(x+y)}} \tag{16.2}$$

Under these statistical assumptions, obtaining 69 higher proportions in 104 trials yields $Z = 3.33$ ($P < 0.001$) for kangaroo rats, and obtaining 63 higher proportions in 112 trials yields $Z = 1.32$ ($P < 0.19$) for squirrels. These are two-tailed tests in that the alternative hypotheses allow for either an over- or underrepresentation of the outcome.

For kangaroo rats we reject the null hypothesis and conclude that they do not use a fixed time or a fixed amount patch use strategy. For the squirrels we do not reject the null hypothesis of a fixed time strategy. The results for the squirrels provide no support for a fixed amount strategy. However, lack of statistical power and small sample size may have precluded our rejection of a fixed time strategy in favor of a fixed quitting harvest rate strategy [a sample size of $n = 278$ would be required to conclude that the observed value (0.56) differed significantly from the null hypothesis (0.50) at the $P < 0.05$ level]. The strong inference approach of the research used the experiment to refute the kinds of auxiliary assumptions described under (4) of Section 16.1.3.

A Critique: The statistical and experimental designs have two striking weaknesses. First, the sign test is a much less powerful test compared to either a paired *t*-test or a Wilcoxon signed-ranks test. Second, over one third of the data were not used in the analyses. An alternative experimental design might have offered a different initial abundance in each of the patches of a triplet on each day of night. This alternative would be a three-way fully-crossed design with station (random effects), day (random effects), and initial abundance (fixed effects) as independent variables. The data (proportion of food harvested) could be analyzed as a mixed-effects model, three-way ANOVA without replication (Chapter 3), or, if one were concerned about temporal nonindependencies, as a repeated-measures analysis (Chapters 6) with station and initial abundance as independent variables.

Why was this not done? For the purposes of testing these (and other) models of patch assessment, it was important that the initial abundances of trays be independent of each other. By placing different initial abundances in each tray of a triplet, the forager can, in theory, gain information about the other trays while harvesting its present tray. The sign test (or a paired t-test) does incorporate some of the structure of a multifactorial design. By making paired comparisons, the sign test factors out the effects of station and day. For the sign test to do this successfully requires additional assumptions regarding the absence of confounding two-way interactions between the variables of day and station, and the variable of initial abundance.

The sign test also makes the assumption that all pairwise comparisons are independent; but this may not be so, particularly in light of repeated measures of presumably the same individuals across nights and days. Comparing across nights or days within a station basically determines whether an animal (or group of animals—there is no guarantee that only one forager visited each station) conforms to a particular patch use strategy. Comparisons across stations determines whether the population as a whole conforms to a particular strategy. Insofar as an individual faithfully represents the characteristics of the group there should be no station by initial abundance interaction. But if individuals within a species vary with respect to the three categories of patch use strategy, the results are compromised by not considering interaction effects. Such an interaction between individual and patch use strategy would emerge as a significant station by initial abundance interaction in the three-way ANOVA. With the sign test, this interaction could be tested by evaluating the data on a station by station basis using days as replicates.

Though not practical in our system, an alternative experimental design could be a fully-crossed design (different initial abundance in each tray of a triplet) with fewer replicates per site and movement of the entire experimental array to new locations for additional data. Such a design allows for a three-way ANOVA without replication and avoids giving animals sufficient experience to learn of the non-independence in initial abundances among trays of a triplet.

16.3.2 A Laboratory Test of the Marginal Value Theorem

Cassini et al. (1990) illustrate a laboratory experiment to test both the quantitative and qualitative predictions of the marginal value theorem. Armadillos (*Chae-*

tophractus vellerosus) were given 15 minutes in which to obtain food from two feeders (designed to mimic depleting food patches) at either end of a U-shaped alley. Each of two armadillos was tested six times under each of two experimental settings. Experimental treatments included providing a good and a poor patch, and treatments differed by varying the quality of the poor patch. Independent measurements of handling times and travel times were used in conjunction with the feeder delivery rates to make quantitative predictions regarding how many food items should be harvested from each patch.

The data consisted of two dependent variables: food items harvested per visit and patch residence time. The independent variables were patch quality (good and poor) and quality of poor patch (high or low). Separate one-way ANOVAs were used as statistical models to test the qualitative predictions. For the analyses, each visit to a feeder represented a data point, and visits for the separate bouts were lumped together. This statistical model is given by Eq. 3.1.

ANOVAs with experimental treatment as the independent variable and residence time (or prey per visit) in good patches as the dependent variable tested the prediction that increasing the quality of poor patches should decrease time spent in good patches. An ANOVA with experimental treatment as the independent variable and residence time (or prey per visit) in poor patches as the dependent variable tested the prediction that increasing the quality of poor patches should increase time spent in poor patches. All of these ANOVAs rejected the null hypotheses at the $P < 0.001$ level of significance. As Cassini et al. (1990) note: "The trends observed between treatments were as predicted by the MVT [marginal value theorem]. Within treatments more prey were taken from good than from poor patches. Between treatments exploitation of good patches decreased and of poor patches increased when the quality of poor patches was increased."

To test the quantitative predictions, the authors calculated the mean and modal numbers of prey per visit for each patch type and each experimental treatment. These were then compared (without statistics) to the quantitative predictions. This yielded eight comparisons (two subjects × two treatments × two patch qualities). In all but one of the cases, the modal value from the data equaled the predicted value. In all cases the mean number of prey per visit was as high or higher than the predicted value.

A critique: The model under consideration is specific and has a demanding list of assumptions regarding the forager's environment [see Dunning (1990) for a review on testing the marginal value theorem].

Testing the model's qualitative predictions is straightforward, although the one-way ANOVAs raise several concerns. Since two dependent variables were simultaneously under consideration, it may have been preferable to use MANOVAs instead (Chapter 5). A more complete (and complicated) analysis could reduce the need for many separate ANOVAs that inevitably inflate the type I error rate (for multiple independent t-tests the type I error rate is $1 - [1-\alpha]^c$ where c is the number of tests; Rice, 1990). In such a design, the independent variables would be subject (two armadillos), patch quality, experimental treatment, and bout (each armadillo was tested for six 15-minute sessions). Such a design would be partially hierarchical and include both the nesting of some variables and the full crossing of other variables (see Brownlee, 1965). To see this, note that bouts (each patch visit is a replicate within a bout) are nested within subject and experimental treatment, and that patch quality is a variable fully-crossed with subject, experimental treatment, and bout (see next section for a more detailed discussion of partially-hierarchical designs).

By fitting the qualitative predictions with one-way ANOVAs, the authors assumed that bouts did not vary significantly. Had there been a strong effect of bout then much of the treatment effect on time spent in good or poor patches could be the result of bout rather than the quality of poor patches. The concordance of the results across both armadillos and patch types suggests that the results are not simply an effect of bout.

Another concern is homogeneity of cell variances in the ANOVA. Not surprisingly, their data show a tight correlation between means and variances. A t-test with unequal variances could have been substituted for the one-way ANOVAs.

Data transformations provide another means of satisfying the assumptions of normality and homoscedasticity. A logarithmic transformation (treats data geometrically rather than arithmetically) frequently rids the data of correlations between means and variances. Other useful rules include using an arcsine transformation (taking the arcsine of the square root of data that lie on the interval between zero and one) on data that are proportions, and using a square root transformation on data that are counts and may fit a Poisson distribution. Transformations allow us to break from our arithmetic world view and to better satisfy the assumptions of parametric statistics. Transformations are *not* a smorgasbord from which to shop for the best results by reanalyzing the data many times (the convenience of statistical software and computers makes this an irresistible temptation).

Another important issue in this study concerns the number of subjects: are two animals enough? The answer depends on how much of the results were idiosyncrasies of the individual, idiosyncrasies of armadillos, or representative of a larger class of species. By using two individuals the researchers are not in a position to partition effects along the hierarchy of individuals nested within species and species nested within higher taxa. But the two individuals do serve a purpose. The concordance of results by the two individuals suggests that if one of the animals expressed aberrant behavior then coincidentally both of the subjects exhibit the same aberration. Variation among individuals is an important issue. On the one hand we might expect very little given that individuals should be genetically and behaviorally similar, on the other hand with respect to foraging the animal's age, experience, and sex may strongly influence its behavior.

The authors conclude that the armadillos tended to stay somewhat longer in a patch then predicted. If so, then one can reformulate the model by considering which assumptions of the model were incorrect or lacking. For instance, hard core assumption (5) of Section 16.1.3 may be incorrect. However, if the assumption of optimality is retained then auxiliary assumptions found under (3) of 16.1.3 may be seen as incorrect. For instance, if travel is energetically more costly, or perceived as more risky by the armadillos, then patch residence time should be greater than predicted by the marginal value theorem. Furthermore, the armadillos may perceive a stochastic component to their patch harvest rate. Based on the cost of deviating from the best behavior it may be better to err by overstaying than understaying at a patch. The armadillos may not perceive the environment as infinitely repeating and may anticipate that patches will eventually disappear. While the list of possible alternatives is legion, deviations from the quantitative predictions suggest new avenues for research and model reformulation that incorporate different or more specific assumptions about the foraging scenario.

In conclusion, a critic would view the qualitative results as ho-hum and the quantitative results as a refutation of the marginal value theorem and the hard core assumption of optimality (see Gray 1987). An advocate would use this paper's results to argue that the overall tenets of patch use theory have been supported via the qualitative predictions and would argue that the quantitative results are very encouraging and suggest new research initiatives.

16.3.3 An Aviary Experiment Testing for the Effect of Owls on Patch Use by Kangaroo Rats

Brown et al. (1988) illustrate an experiment in a seminatural enclosure (somewhere between a field and laboratory experiment) designed to use foraging theory to study the ecology of desert rodents. The model under consideration generalizes the marginal value theorem to include predation risk and alternative activities (Brown, 1988) and makes the prediction that GUDs in food patches should increase in response to perceived predation risk. I will use the results for the Arizona pocket mouse (*Perognathus amplus*) as an example of the experiments.

We used an aviary to simulate a desert environment, which we populated with 20 pocket mice. To quantify foraging behavior, we measured GUDs each night in 16 food patches identical to those discussed in Section 16.3.1. We had four experimental treatments resulting from all combinations of the presence and absence of three barn owls (predators) and the presence and absence of artificial illumination set to mimic the light level of a full moon (predation risk on nocturnal rodents is thought to be higher on moonlit nights). We ran the experiment for 16 nights and we randomized the sequence of the four treatments (on any given night the aviary could only be subjected to one combination of experimental treatments).

The design can fit two types of ANOVA models. The first ANOVA model considers all variables in a design that is partially hierarchical (like the previous example). The variable night is nested within the variables of owls (presence and absence) and lights (presence and absence), and the variable patch (16 different patches distributed within the aviary) is a variable fully crossed with nights, owls, and lights. The dependent variable was the arcsine transformation of the proportion of seeds removed from a patch during a given night. The statistical model for this experiment is similar to and somewhat more complicated than the split-plot design of Eq. 3.5, which is also partially hierarchical.

The general linear model for the data of this experiment is

$$X_{ijkl} = \mu + \alpha_i + \beta_j + \alpha\beta_{ij} + \varepsilon_{k(ij)} + \gamma_l + \qquad (16.3)$$
$$\alpha\gamma_{il} + \beta\gamma_{jl} + \alpha\beta\gamma_{ijl} + \varepsilon'_{k(ijl)}$$

where α_i is the effect of the ith level of owls (a between night factor), β_j is the effect of the jth level of light (a between night factor), $\alpha\beta_{ij}$ is the effect of the interaction between the ith owl level and the jth light level (between nights), $\varepsilon_{k(ij)}$ is the between night error term that designates the effect of night k within the ij level of owls and lights, γ_l is the effect of the lth food patch (the within night factor), $\alpha\gamma_{il}$ is the interaction between the ith owl level and the lth patch, $\beta\gamma_{jl}$ is the interaction between the jth light level and the lth patch, $\alpha\beta\gamma_{ijl}$ is the three-way interaction among the ith owl level, jth light level, and the lth patch, and $\varepsilon'_{k(ijl)}$ is the within-night error term (see Appendix 16.1 for the SAS commands).

The second ANOVA model is simpler and focuses on our interest in the effects of owls and lights. For this model, owls and lights represent variables in a fully-crossed and balanced two-way ANOVA where the dependent variable is the proportion of seeds removed from food patches on a given night (arcsine transformed). In this way each night reduces to a single data point and nights represent replicates (a total of 16 nights). The statistical model is given by Eq. 3.3; see Appendix 3.3 for the SAS commands.

Regardless of which ANOVA is used to analyze the data owls and lights had statistically significant effects (at $\alpha = 0.05$ level for rejecting the null hypothesis). The owl by light interaction was not significant, and in the first ANOVA model the effects of days and patches were significant (the owl by patch and light by patch interactions were not significant). Of interest to the study was the strong increase of GUDs in response to owls (from 2.82 g of millet to 4.07 g) and increase in GUDs in response to lights (from 3.15 to 3.74 g). Both the direct and indirect cues of predation risk resulted in the predicted increase in GUD. From this we concluded more generally that pocket mice can make nightly assessments of predation risk and respond accordingly. These results represent tests of auxiliary assumptions found in (3) and (4) of Section 16.1.3.

A critique: We may be requiring a rather large leap of faith to use the results from 20 pocket mice in an aviary to determine that this species assesses and responds to predation risk under natural conditions. We assume that if they can respond in a seminatural enclosure they should respond similarly in the wild. While the effects of owls and lights were well documented in the aviary we do not know whether this was simply an idiosyncrasy of the aviary (we would need several more aviaries to test for an aviary by treatment interaction) or whether it was an interaction between some biased sampling regime of the 20 individuals (we did not measure each individual's GUD but rather the GUDs of whomever was the last forager in each patch). With respect to biases from the population effects, we repeated the experiments using different populations of the same and different species (the results did not change—although the effect of nights and lights were not always significant).

Much of our sampling regime was dictated by the difficulty in obtaining aviaries and owls, by a limited number of nights available for data collection, and by the ease of setting out many patches per night. From the results (in hindsight) the number of patches was overkill. The lack of a treatment by patch interaction meant that a single or many fewer patches would have sufficed. The large number of patches did not increase our ability to detect the effects of owls and lights but simply increased our ability to detect an effect of nights independent of experimental treatment. The large effect of night diminishes the statistical model's ability to detect the effects of owls and lights. To compensate for a strong effect of night, the experimenters' only solution is to either increase the number of nights or build and operate more aviaries. We were lucky that the effect of owls and lights was so strong as to still be detectable above the strong effect of night. A nice experiment would have been to use four aviaries. In four nights of experiments the researcher could achieve a latin square design in time and space (see Section 3.3.4 and Appendix 3.4).

16.4 Conclusion

We have now come full circle back to the initial points and aims of Chapter 1. As illustrated by Fig. 1.2, we can view science as a general flow from ideas and observations distilled into a model, from model to predictions, from predictions to experimentation and statistical model, from experimentation to statistical analysis, and from analysis to conclusions and model reformulation. Just as parliamentary procedures structure and formalize political debates, statistics structures and formalizes the debate between competing ideas and hypotheses. We use statistics as a balance for weighing evidence as either for or against an idea. As we have seen, the particular statistical analyses introduce an additional set of assumptions into the model.

In this hierarchy of assumptions, the predictions of a model include hard core assumptions, (5) of Section 16.1.3, auxiliary assumptions, (3) and (4), methodological assumptions, (2), and statistical assumptions, (1). When a model's prediction fails a test any or all of these assumptions may be at fault. It is for this reason that there is no universally acceptable or critical test of either a research program or even a minor auxiliary assumption. When a prediction is falsified (and we have confidence in the methodological and statistical assumptions), we conclude that one or several of the auxiliary assumptions are incorrect. Once corrected, these new assumptions form a new model with additional testable predictions.

So, within foraging theory how does one falsify the hard core assumptions, and why not use alternative hard core assumptions to develop alternative models? Alternative hard core assumptions might include viewing the feeding behaviors as phylogenetic baggage rather than as adaptations (e.g., historical ecology research program; Brooks and McLennan, 1991) or developing rigorous assumptions for the theory of satisfycing (Ward, 1992). In general, testing two sets of hard core assumptions through a single test of their different predictions is problematic. What are reasonable auxiliary assumptions in one program may make no sense in another. Hence, a strong inference test of two such models is likely to include not only different hard core assumptions but different auxiliary assumptions as well.

According to Lakatos (1978), two sets of hard core assumptions can and should compete. But this competition occurs on a higher plane in which each attempts to explain and predict phenomena of interest. Falsified models are met with ingenious or practical modifications of auxiliary assumptions. Insofar as these modifications produce new and interesting insights and suggest new avenues for experimentation, the research program is vigorous. Insofar as these modifications are desperate attempts to explain away anomalies on a case by case basis and in a manner devoid of interesting insights, the research program is degenerate.

At present, foraging theory probably manifests a vigorous research program. To maintain vigor or to expose serious anomalies requires astute observations, novel modeling endeavors, clever experiments, and the application of sound statistics to judge experimental results.

Acknowledgments

I am grateful to J. Gurevitch, B. Manly, R. Morgan, S. Scheiner, and C. Whelan for their careful reading and many useful comments on this manuscript. I also thank W. Mitchell, he has been a continual source of inspiration and good ideas regarding optimal foraging theory and the role of statistics in a research program.

Appendix 16.1

SAS program code for a partially-hierarchical design (Eq. 16.3).

```
PROC GLM;
    CLASS O L N F;                              /O = owl level, L = light level,
                                                 N = night, F = food level/

    MODEL X = O L O*L N(O*L) F
           O*F L*F O*L*F/SS3;
```

*** If the design is unbalanced use the following command
    to obtain the correct *F* tests ***

```
    RANDOM N(O*L)/TEST;        /Declare night to be a random effect. The owl and
                                light and owl-light interaction effects will be tested
                                over the night effect. Food and the owl-food, light-
                                food, and owl-light-food interactions will be tested
                                over the error term. SAS will compute the
                                Satterthwaite correction for an unbalanced design/
```

*** If the design is balanced alternatively use the following
    commands to obtain the correct *F* tests ***

```
    TEST H= O  E= N(O*L);      /test the owl treatment over nights/
    TEST H= L  E= N(O*L);      /test the light treatment over nights/
    TEST H= O*L  E= N(O*L);    /test the owl-light interaction over nights/
```

17

Meta-analysis: Combining the Results of Independent Experiments

Jessica Gurevitch and Larry V. Hedges

17.1 Ecological Issues

The preceeding chapters discuss the best ways to design and analyze the results of individual experiments. In this chapter, we will examine methods to statistically combine the results of many separate studies to reach general conclusions. The statistical synthesis of the results of separate, independent experiments is known as *meta-analysis*. These somewhat controversial techniques have only recently been introduced to the field of ecology (Gurevitch et al., 1992), although they have had considerable influence on research synthesis in medicine (e.g., Chalmers et al., 1989; Sacks et al., 1987) and the social sciences (e.g., Glass et al., 1981; Hyde and Linn, 1986). The recent volume edited by Cooper and Hedges (1993) provides a reference on both statistical and other aspects of the subject.

At the heart of meta-analysis is the concept that the progress of science depends on the ability to reach general conclusions from a body of research. In the first chapter, Scheiner raised the basic question: what, after all, are the purposes of carrying out experiments in ecology? An experiment is designed to test a particular hypothesis or set of hypotheses. The results formally provide a test of these hypotheses only for specific individual organisms in one place at one time. The information provided by the experiment is limited to those particular circumstances. Experiments can be exactly replicated in different laboratories in some sciences, confirming the general applicability of the results. This is not possible in ecology, but that does not make it any less pressing to evaluate the generality of the findings of an experiment.

Basic and applied research in ecology differ in that in applied work, information about particular local systems that leads to better management, etc. may be useful in itself. In basic ecological research, the results of testing hypotheses with individual experiments are of no interest if they cannot be generalized. What can the outcome of any one experiment tell us about nature? To what extent can we extrapolate from the results? Do they seem to support or refute the conclusions of previous studies? Under what conditions do other studies come to the same or to contradictory conclusions? There are widely divergent views among ecologists on how to answer such questions.

The need to generalize is implicit in the way experimental results are interpreted in research papers, in textbooks, and in communicating the findings of ecological research to the general public. Textbooks commonly generalize from the results of single experiments to general truths about ecological processes and interactions, or to entire species or even trophic levels. Some ecologists are hesitant to extrapolate results beyond individual experiments, or across studies. Raudenbush (1991), however, points out that "summarizing may be regarded as the essential subject matter of statistical science," and that whether the summary is based on the results of a single study or many studies, and is qualitative or numerical, summarizing evidence is a fundamental task in all research. Raudenbush raises interesting issues about the tension between completeness and parsimony in generalizing results from data, suggesting that if parsimony were not important, the best summary would be the raw data itself. Most importantly, we risk missing the patterns and truths that may be contained in the results of a group of studies when we concentrate only on the individual details in which they differ.

Meta-analysis reaches general conclusions about a domain of research differently than the manner with which most ecologists are familiar. In a conventional review, one compares the results of different studies verbally in a research review, yet it is often desirable to compare and synthesize results quantitatively. When a number of independent studies have been carried out on a particular question, it might be important to assess, for example, the overall impact of doubling atmospheric CO_2 levels on plant growth, or to evaluate whether the survivorship of juveniles is affected on average by competitors. Usually one would like to know the magnitude and direction of the effect (does the experimental treatment have a positive or negative effect? is this effect significantly greater than zero? is the effect small or large?), and how variable that effect is among studies (do the results of the studies seem to agree, or are there statistically significant differences among them?). In many cases, differences among classes of studies in the magnitude of the effect are of the most interest. Do carnivores compete more than herbivores? Do C_3 plants respond differently to CO_2 enrichment than C_4 plants, as predicted by theory? Is competition among plants greater in productive habitats than in unproductive habitats? In the field of ecology, these kinds of questions have been addressed using a number of different approaches.

17.2 Statistical Issues

At first, it might seem obvious that one could just count up the number of statistically significant results in the various studies to get some idea of how important an effect is (i.e., how large it is), how frequently it occurs, and to compare the magnitude of the effect among different kinds of studies. Quantitative ecological reviews have commonly employed this method in recent years. Unfortunately, this "vote-counting" approach is subject to serious flaws, because the significance level of a study is a function not only of the magnitude of the effect, but also of its sample size. Studies with small sample sizes are less likely to have statistically significant results than large studies, even when both have effects of exactly the same magnitude. In fact, because small studies are less likely to produce significant results (i.e., they have low power), it has been shown that vote counting is strongly biased toward finding no effect (e.g., Hedges and Olkin, 1985).

The problems associated with vote counts are particularly acute when effects and sample sizes are modest—almost always the case in ecological experiments. Counting up the number of significant results is not a reliable indication of whether an effect is real (that is, different than zero), how important the effect is, how frequently or under what circumstances it exists, or whether the studies are in agreement in their assessment of the magnitude of the effect (e.g., Hunter et al., 1982; Hedges and Olkin, 1985). Narrative reviews may also suffer from a subtle form of vote counting, in that they are also likely to base their conclusions about the existence and frequency of ecological phenomena on the statistical significance of the outcomes, or on authors' summaries of their own results based on probability levels, without considering sample size and statistical power.

Typically, a meta-analysis begins by representing the outcome of each experiment by a quantitative index of the effect size. This effect size is chosen to reflect differences between experimental and control groups, or the degree of relationship between the independent and dependent variables in a way that is independent of sample size and of the scale of measurement used in the experiment. Meta-analytic techniques most commonly serve to estimate the average magnitude of the effect across all studies, to test whether that effect is significantly different than zero, and to examine differences in the effect among studies. Because effect size is not dependent on sample size (in contrast to the dependence of significance level on sample size), meta-analysis procedures are not subject to the problems of vote counting.

Meta-analysis offers a potentially invaluable tool for ecological research. While ecological experiments are essentially never replicated in any strict sense, it has become increasingly critical to be able to understand what an entire body of experimental data addressing conceptually similar questions tells us. How important and how consistent is a particular effect across a wide variety of systems? Does the effect differ substantially in aquatic and terrestrial habitats, or in disturbed and undisturbed environments, or in large vs small patches? Most ecologists can probably think of questions such as this for systems with which they are familiar where a substantial body of data already exists or could be collected. Furthermore, it is likely to become increasingly urgent for ecologists to be able to generalize from the results of studies conducted in different systems as we are increasingly called on to address the consequences of global change. Data collected in diverse locations with similar research objectives will have to be summarized to understand emerging patterns and to detect outliers. Quantitative generalizations can help us track both stability and change; outliers may be populations, species, or ecosystems in imminent danger, or those that are unusually resilient.

Some of the specificity and the fine detail of individual studies are necessarily overlooked in a meta-analysis for the sake of being able to generalize across studies. Depending on the questions being asked, this may or may not be appropriate. For example, in one of the studies included in the example of a meta-analysis below (Gurevitch, 1986; Chapter 15), the effects of competition on two grass species were examined at three positions along an environmental gradient. While the contrast between these responses was an important part of the primary study, it was ignored in the meta-analysis. Because the meta-analysis seeks to estimate the effect of competition on the growth of terrestrial plants and all of the measurements offer information on this, the particular details about the influence of the environmental gradient are sacrificed. It is always true that when summarizing the behavior of a population, the individual characteristics of members of that population are overlooked, whether the population consists of a group of individual studies, or the population is a group of individual animals whose responses are summarized using conventional statistics.

17.3 Statistical Solution

17.3.1 Statistical Models for Meta-analysis: Fixed Effect and Mixed Models

Ecological data are in some ways structurally similar to data in educational psychology, a field in which there has been considerable development of meta-analytic models, and here we follow the approaches outlined in that work. An *effect size* is calculated for each experiment as the difference between the means of two groups of individuals (typically, an experimental group and a control group), divided by their pooled standard deviation to standardize the effect among studies. The effect size then is the difference in standard deviation units between the experimental and control groups. Cohen (1969) provides a conventional interpretation of the magnitude of effect sizes: 0.2 is a "small" effect, 0.5 is "medium" in magnitude, 0.8 is "large," and presumably any effect greater than 1.0 standard deviations difference between experimental and control groups would be "very large." One can also calculate the standard error of the effect size as a measure of its sampling uncertainty. The data needed from each study are the means of the two groups (e.g., experimental and controls), the standard deviations about these means, and the number of individuals in each group. Thus, it is not necessary to have access to the original raw data to perform a meta-analysis, and the basic information and simple statistics that should be reported in published papers are sufficient.

Unfortunately, papers published in even the most selective ecological journals may be lacking this basic information on the outcome of the experiments on which they report; for example, sample size is often omitted, and the nature of the sampling unit left to readers' imaginations. It is good practice to include means, standard deviations (or other measures of variation from which standard deviations can be derived), and sample sizes (describing the units, i.e., $n = 30$ snails, or $n = 5$ cages, each based on the mean mass of six snails) when publishing the results of experiments, and to look for this information when reviewing papers for publication. It cannot be overemphasized that probability levels alone are insufficient to evaluate the results of an experiment.

Most meta-analyses have been based on *fixed effects* models. That is, it is assumed that a class of studies with similar characteristics shares a common true effect size. (Another assumption made in the models discussed here is that the data in the experimental and control groups are normally distributed.) Differences among studies in the actual effect size measured are assumed to be due to sampling error. This assumption is probably rarely justified in ecology, and in many instances may not be very reasonable in other fields either. In a *mixed model*, it is assumed that the studies within a class share a common mean effect, but that there is also random variation among studies in a class, in addition to sampling variation. Another way of picturing a mixed model is to think of the effect size of a particular study as being composed of various components: part of the effect is "fixed," or characteristic of all studies in a class (all herbivores share certain characteristic responses, for example), part of the effect is due to the particular characteristics of that one study in which it differs at random from other studies, and part of the effect is due to "error," or sampling variation. Mixed models in meta-analysis are analogous to mixed models in ANOVA (for a discussion of fixed and random effects in primary studies, see Chapter 3). While mixed models have not yet been used in published meta-analyses, they are often preferable to fixed effects models because the assumptions of the mixed model are more likely to be satisfied. Strict random models (in which all variation among studies is random variation) have been previously used in meta-analysis (Hedges, 1983), but mixed models combining fixed differences among classes with random variation among studies within classes have only recently been developed (Hedges and Gurevitch, unpublished). Both fixed effect and mixed models in meta-analysis will be demonstrated in this chapter.

17.3.2. Conducting the Meta-analysis: Gathering the Data

The process of gathering references and making decisions regarding the studies to include in a meta-analysis involves complex issues that we cannot hope to do justice to in a short chapter. In many ways, it is no different than in any other review of research literature, although the quantitative nature of the review brings certain problems into sharper focus. Many substantive publications offer specific suggestions on gathering and handling the data for a meta-analysis (e.g., Cooper, 1989; Cooper and Hedges, 1993; Light and Pillemer, 1984). If the meta-analysis is not to include every paper published on a topic, the criteria for inclusion should clearly be reasonable and scientifically defensible. Publication bias (e.g., if papers that demonstrate no effect are less likely to be published than those that come up with statistically significant effects, a meta-analysis—or any other review—of the published literature will overestimate that effect) is another issue to consider carefully in carrying out a meta-analysis, because it can clearly result in inaccurate conclusions. Problems associated with publication bias have been discussed at length among meta-analysts, and various different approaches have been proposed to estimate its magnitude and counter its influence (e.g., Hedges and Olkin, 1985; Cooper and Hedges, 1993).

The extent to which one can extrapolate the conclusions drawn from a research review, whether quantitative or narrative, will depend in part on the quality and nature of the data that are collected. As every ecologist who has gathered data is aware, carefully and properly setting up the experiments or observations and collecting the measurements are of the greatest importance in getting good results. Good analysis cannot lead to sound conclusions from bad data. This is no different when carrying out a single study, a conventional research review, or in meta-analytic research synthesis. If the collection of studies is biased or incomplete, the conclusions drawn from a meta-analysis (or any other analysis) of that collection will be suspect, or at least limited. Some controversy exists on whether all studies that have been located on a topic should be included in the analysis, or whether "low quality" studies (e.g., medical studies in which the treatments were not assigned to subjects randomly) should be excluded. One option is to include all studies, noting which are of low quality by some a priori criteria, and to test whether the results of these studies differ from those of higher quality, rejecting them if they do.

Critics of meta-analysis are sometimes suspicious of how data are gathered from studies, often expressing concern that this will be carried out blindly or automatically in some way. In fact, meta-analysis involves a series of subjective decisions, and at least in ecology cannot easily be automated. It is unlikely that two meta-analyses on the same topic will be exactly the same, and in some cases they may reach very different conclusions (Abrami et al., 1988). Personal experience (J.G.) has shown that ecological data are reported so eccentrically in published papers that it would be impossible to cumulate (combine) casually, even were that desirable. It is an eye-opening experience to attempt to extract information from a paper that you have read carefully and thoroughly understood, only to be confronted with ambiguities, obscurities, and gaps in the data that only an attempt to quantify the results reveals.

Ecological data are often published in the form of graphs rather than tables.

The best way to access such data is by digitizing the graph using a digitizing tablet or other device connected to a personal computer. For example, we used a Science Accessories Corporation Graf/Pen and Graf/Bar acoustic digitizer with DS-DIGIT software (Slice, 1990) to convert position on the graph to actual values. Determining the values for sample sizes and standard deviations from ecological papers is often needlessly difficult. Some suggestions on how to obtain this information when it is less than obvious have been offered elsewhere (Gurevitch et al., 1992). In particular, it may be possible to calculate standard deviations from other measures of error, such as standard errors or the mean square error term in an analysis of variance (Gurevitch et al., 1992). Once collected, descriptive information and data can be organized and the analysis carried out in various ways. Perhaps the most straightforward is to record information from each study on detailed data sheets, from there entering data onto a spreadsheet on a personal computer. Even more than in a primary data analysis, careful checking and rechecking of the data sheets and computer entries are critical; it is almost impossible not to make mistakes in a meta-analysis involving more than a few studies. The analysis demonstrated in this chapter was carried out completely on Microsoft EXCEL spreadsheets. The formulas needed can be entered directly onto the spreadsheet. No additional software is needed.

Careful thought must be given as to how each experiment should be recorded. The "control" identified by the author may not be what the meta-analyst identifies as a control for the purposes of research synthesis. In the example below, the "control group" was those organisms with natural densities of competitors, while the "experimental group" had experimentally manipulated densities of competitors, regardless of what the authors called the groups (this more uniform organization of the data allows one to address additional questions; Gurevitch et al., 1992). As competitor densities can be reduced or increased, the expected sign of the response to the manipulation will vary accordingly, if competition has been demonstrated. To make it possible to combine the estimates of the effect of competition, the sign of the response was reversed if competitor density was increased (reduction is much more common). While in many cases it is the effect on individual organisms that is summarized in the effect size (as measured by characteristics of the organisms such as growth or seed production), in other cases the "individual" is a plot, or other unit (such as when survivorship or density is the measure of outcome). The meta-analyst must decide on the units for which the effects are being evaluated, and identify the appropriate measures from each study. All of these issues must be considered and evaluated carefully in a meta-analysis.

Another potential problem is that more than one response may be measured on the same organisms. These measurements are not independent, and should not be included in the same meta-analysis. Not only may several different kinds of measurements (e.g., height, biomass, and increase in biomass) be made on each organism, but the same measurements may be repeated over time (Chapter 7). In the latter case, it is not correct to use more than one measurement per individual as if they were independent observations on different individuals, and so we may choose to take the final measurement made at the end of the experiment (e.g., Gurevitch et al., 1992). In complex experiments in which various manipulations are compared to a single control (e.g., competition of a target species with several different species of competitors), there may be no way to avoid including the same control group a number of times as the basis for calculating more than one effect size (the effect of each of the competing species) even though this is not ideal. If the proportion of comparisons in which there is nonindependence (e.g., use of a control group mean for more than one comparison) is small, this will have little effect on the conclusions drawn from the meta-analysis. If there is a great deal of nonindependence, this will inflate the significance levels of statistical tests, and underestimate confidence intervals. Currently nothing analogous to a Bonferonni correction exists for meta-analysis, but it would be possible to devise such a correction if there was substantial nonindependence in the data set. Alternatively, one could use multivariate methods (Hedges and Olkin, 1985, Chapter 10), although the calculations are considerably more difficult than the ones discussed here.

The results of an experiment may be tested and reported using many different statistical approaches. One would usually not include results reported using different kinds of statistics (e.g., means, slopes, correlations) in a single meta-analysis. The most common forms of data reported in ecological studies are means and some measure of variance based on data that are assumed to be normally distributed, and it is these data that are used here to calculate effect sizes. Most meta-analysis approaches, including the models presented here, assume normally distributed data, although some work has been done on nonparametric methods for calculating effect sizes (Hedges and Olkin, 1985).

Methods for combining correlation coefficients and odds ratios are also well established in meta-analysis (see any of the texts cited in this chapter). Methods do not yet exist for cumulating results reported as more complex types of response parameters, such as failure times (Chapter 12). When a sufficient literature accumulates in which such statistics are the norm, it may become worthwhile to develop meta-analytic methods to compare and summarize the results of such studies quantitatively.

While the scale of the measurements is standardized in calculating effect sizes (so that the responses of large organisms become comparable with those of small ones), one may choose not to combine experiments using very different sorts of measurements (such as biomass and survivorship) in a single meta-analysis. It is assumed that the measurements are linearly equatable among the studies, and this may not be a biologically reasonable assumption for very different sorts of responses. The decision about what to combine is a substantive one that depends on the generality of the questions the summary is designed to answer. If a very general summary of many different kinds of evidence is the goal, then results measured in many different ways may be combined. Most verbal reviews and vote counts in ecology have taken this approach, combining all measures of outcome in a single summary of the effects. However, choosing to combine only those studies measuring similar quantities (e.g., biomass, growth, or survivorship) allows one to ask more specific questions. There may also be biological reasons to expect that different kinds of responses may reveal different information about the phenomena being investigated.

17.3.3 The Example

The data were collected as part of a larger study on the effects of competition as studied in field experiments. The larger study included all measures of outcome (survivorship, reproductive output, density, etc.) in response to the manipulation of competitors for organisms in a wide range of trophic levels and systems. An analysis of organisms' responses measured as biomass has been published (Gurevitch et al., 1992), and that paper includes a detailed description of how papers were selected for inclusion in the study and how the data were collected. For the example presented here, we chose a small part of the larger data set for convenience in demonstrating the methods involved. The data that we examine are for the responses of primary producers to competition, where the responses were measured as recruitment (an increase in the number of individuals) or growth (an increase in the size of individuals). The data set is original and previously unpublished, and has not been edited other than arbitrarily selecting organisms of only one trophic level. Much more can be done with the data in a full meta-analysis than is attempted here, where we address only a small number of issues so that we can focus on how to actually carry out the analyses.

The studies were categorized a priori into three classes (terrestrial, lentic, and marine) that we wished to compare. The basic information taken from the studies can be found in columns E–J in Table 17.1.

First, an unbiased effect size (d_{ij}) must be calculated for each study,

$$d_{ij} = \frac{\overline{X}_{ij}^{E} - \overline{X}_{ij}^{C}}{s_{ij}} J \tag{17.1}$$

where d is calculated for the jth study in the ith class, and
$\overline{X}_{ij}^{C}$ = mean of the control group
$\overline{X}_{ij}^{E}$ = mean of the experimental group
s_{ij} = the pooled standard deviation of the control and experimental groups

$$s_{ij} = \sqrt{\frac{(N_{ij}^{E} - 1)(s_{ij}^{E})^{2} + (N_{ij}^{C} - 1)(s_{ij}^{C})^{2}}{N_{ij}^{E} + N_{ij}^{C} - 2}} \tag{17.2}$$

where

N^C_{ij} = the total number of individuals in the control group

N^E_{ij} = the total number of individuals in the experimental group

s^C_{ij} = the standard deviation of the individuals in the control group

s^E_{ij} = the standard deviation of the individuals in the experimental group. (Note that we use the terms study, experiment, and comparison interchangeably to indicate a comparison between a single experimental group and its control; several such comparisons may be used from a single published paper.) J is a term that corrects for bias due to small sample size:

$$J = 1 - \frac{3}{4(N^E_{ij} + N^C_{ij} - 2) - 1} \tag{17.3}$$

(Hedges and Olkin, 1985). J approaches 1 as sample size increases. Finally, the variance in the effect for the jth study in the ith class is approximated by

$$v_{ij} = \frac{N^E_{ij} + N^C_{ij}}{N^E_{ij} N^C_{ij}} + \frac{d^2_{ij}}{2(N^E_{ij} + N^C_{ij})} \tag{17.4}$$

(Hedges and Olkin, 1985). This term may be used for comparing effect sizes among studies directly, for evaluating the magnitude of the effect size (e.g., is the effect significantly greater than zero?), and for combining effect sizes across studies (see below). The values calculated from the example can be found in Table 17.1.

Combining effect sizes in the fixed effects model. The (fixed effect) cumulated effect size across studies within the ith class, d_{i+}, is a weighted average of the effect size estimates for the studies in that class. The weights w_{ij} used for combining effect sizes are the reciprocals of the sampling variances, $w_{ij} = 1 / v_{ij}$. The weighted estimate of the true effect size δ_i is

$$d_{i+} = \frac{\sum\limits_{j=1}^{k_i} w_{ij} d_{ij}}{\sum\limits_{j=1}^{k_i} w_{ij}} \tag{17.5}$$

where k_i equals the number of comparisons in class i (Hedges and Olkin, 1985). The cumulated effect size is a weighted average in which larger studies are counted more heavily than smaller studies, on the assumption that larger sample sizes will yield more precise results (Hedges and Olkin, 1985). The variance in d_{i+}, $s^2(d_{i+})$ is

Table 17.1. Data and Calculations for the Fixed-Effects Model[a]

	A	B	C	D	E	F	G	H	I	J	K	L	M	N	O	P	Q
	Publication	Species	Category	Source	N^C	N^E	$\bar{X}^C$	$\bar{X}^E$	s^C	s^E	J	d	v_{ij}	w_{ij}	wd	wd^2	k
	Fowler 1986	*Bouteloua rigidiseta*	Terrestrial	Table 2	7	7	78.14	79.71	40.650	40.650	0.936	0.036	0.286	3.499	0.127	0.005	1
	Fowler 1986	*Aristida longiseta*	Terrestrial	Table 2	7	7	18.86	26	9.170	9.170	0.936	0.729	0.305	3.282	2.392	1.744	1
	Platt and Weis 1985	*Mirabilis hirsuta*	Terrestrial	Table 1	6	6	-1.8	-2.1	0.490	0.490	0.923	0.565	0.347	2.885	1.631	0.922	1
	Platt and Weis 1985	*Verbena stricta*	Terrestrial	Table 1	5	5	-2.2	-2.8	0.224	0.447	0.903	1.533	0.517	1.932	2.962	4.540	1
	Platt and Weis 1985	*Solidago rigada*	Terrestrial	Table 1	7	7	-2.1	-3	0.265	0.529	0.936	2.014	0.431	2.322	4.677	9.421	1
	Platt and Weis 1985	*Asclepias syriaca*	Terrestrial	Table 1	6	6	-2.3	-4.2	0.490	1.225	0.923	1.880	0.481	2.081	3.912	7.356	1
	Gross and Werner 1982	*Verbascum thapsus*	Terrestrial	Table 8	3	3	85.3	285.7	115.008	153.806	0.800	1.181	0.783	1.277	1.508	1.780	1
	Gross and Werner 1982	*Oenothera biennis*	Terrestrial	Table 8	3	3	0	3	0.000	2.425	0.800	1.400	0.830	1.205	1.687	2.361	1
	Gross and Werner 1982	*Verbascum thapsus*	Terrestrial	Table 8	3	3	0	2.00	0.000	2.078	0.800	1.352	2.152	0.465	0.628	0.849	1
	Gross and Werner 1982	*Oenothera biennis*	Terrestrial	Table 8	3	3	0	1.67	0.000	1.732	0.800	1.545	2.199	0.455	0.703	1.086	1
	Pons and van der Toorn 1988	*Plantago major*	Terrestrial	Table 4	5	5	17	17	7.603	5.367	0.903	0.000	0.400	2.500	0.000	0.000	1
	Pons and van der Toorn 1988	*Plantago lanceolata*	Terrestrial	Table 4	5	5	47	37	10.286	9.391	0.903	-0.917	0.442	2.262	-2.075	1.903	1
	Burton and Mueller-Dombois 1984	*Metrosideros polymorpha*	Terrestrial	Table 2	4	4	87	272	37.712	183.532	0.870	1.214	0.592	1.689	2.051	2.490	1
	Gurevich 1986	*Stipa neomexicana*	Terrestrial	Fig. 4 (log)	18	20	-0.113	0.294	0.255	0.215	0.979	1.694	0.143	6.977	11.821	20.029	1
	Gurevich 1986	*Stipa neomexicana*	Terrestrial	Fig. 4 (log)	20	20	-0.163	0.412	0.588	0.218	0.980	1.273	0.120	8.316	10.586	13.476	1
	Gurevich 1986	*Stipa neomexicana*	Terrestrial	Fig. 4 (log)	18	20	0.140	0.632	0.380	0.359	0.979	1.303	0.128	7.818	10.190	13.283	1
	Gurevich 1986	*Aristida glauca*	Terrestrial	Fig. 4 (log)	20	20	-0.184	0.259	0.326	0.238	0.980	1.519	0.129	7.762	11.789	17.904	1
	Gurevich 1986	*Aristida glauca*	Terrestrial	Fig. 4 (log)	20	20	-0.075	0.354	0.487	0.182	0.980	1.144	0.116	8.595	9.829	11.240	1
	Gurevich 1986	*Aristida glauca*	Terrestrial	Fig. 4 (log)	20	20	0.147	0.541	0.340	0.299	0.980	1.206	0.118	8.461	10.206	12.310	1
	McCreary et al. 1983	*Eleocharis acicularis*	Lentic	Fig. 1A&B	4	4	281.11	-201.03	158.038	27.520	0.870	3.696	1.354	0.739	2.730	10.091	1
	McCreary et al. 1983	*Juncus pelocarpus*	Lentic	Fig. 1A&B	4	4	187.31	-155.32	80.163	41.252	0.870	4.674	1.865	0.536	2.506	11.711	1
	Stimson 1985	*Acropora* spp.	Marine	Table 7	7	7	11.8	16.0	3.080	3.370	0.936	1.218	0.339	2.953	3.596	4.380	1
	Stimson 1985	*Pocillopora verrucosa*	Marine	Table 7	3	10	0.4	9.5	1.470	7.230	0.930	1.288	0.497	2.011	2.592	3.339	1
	Reed and Foster 1984	*Pterygophora californica*	Marine	Table 1	20	20	0	14.1	7.603	7.603	0.980	1.818	0.141	7.077	12.864	23.384	1

Reference	Species	Habitat	Source													
Reed and Foster 1984	*Macrocystis pyrifera*	Marine	Table 1	20	20	0	7.1	3.130	3.130	0.980	2.223	0.162	6.182	13.742	30.547	1
Reed and Foster 1984	*Desmarestia ligulata*	Marine	Table 1	20	20	0	1.4	1.789	1.789	0.980	0.767	0.107	9.315	7.145	5.481	1
Reed and Foster 1984	*Desmarestia ligulata*	Marine	Table 3	10	10	82.2	94	29.093	9.171	0.958	0.524	0.207	4.834	2.533	1.327	1
Reed and Foster 1984	*Desmarestia kurilensis*	Marine	Table 3	10	10	8.3	10.5	14.546	11.068	0.958	0.163	0.201	4.983	0.812	0.132	1
Reed and Foster 1984	*Nereocystis luetkeana*	Marine	Table 3	10	10		20	42.691	42.691	0.958	0.449	0.205	4.877	2.188	0.982	1
Johnson and Mann 1988	*Laminaria longicruris*	Marine	Table 2	2	2	3.63	18.5	3.352	4.257	0.571	2.218	1.615	0.619	1.373	3.046	1
Johnson and Mann 1988	*Laminaria longicruris*	Marine	Table 2	2	2	0	0.25	0.000	0.354	0.571	0.571	1.041	0.961	0.549	0.314	1
Johnson and Mann 1988	*Laminaria longicruris*	Marine	Table 2	2	2	3.63	2.25	3.352	0.707	0.571	-0.326	1.013	0.987	-0.321	0.105	1
Turner 1985	*Rhodemela larix*	Marine	Table 4	4	4	0	34.8	0.000	58.200	0.870	0.735	0.534	1.873	1.378	1.013	1
Turner 1985	*Cryptosiphonia woodii*	Marine	Table 4	4	4	0	25.3	0.000	35.800	0.870	0.869	0.547	1.827	1.588	1.380	1
Turner 1985	*Phaeostrophion irregulare*	Marine	Table 4	4	4	5.4	23.6	10.880	47.000	0.870	0.463	0.513	1.948	0.902	0.417	1
Turner 1985	*Odonthalia floccosa*	Marine	Table 4	4	4	1.8	10.5	5.200	24.200	0.870	0.432	0.512	1.954	0.845	0.365	1
Turner 1985	*Microcladia borealis*	Marine	Table 4	4	4	0	10.3	0.000	17.400	0.870	0.728	0.533	1.876	1.365	0.994	1
Turner 1985	*Fucus distichus*	Marine	Table 4	4	4	0	8.7	0.000	17.000	0.870	0.629	0.525	1.906	1.199	0.755	1
Turner 1985	*Iridaea heterocarpa*	Marine	Table 4	4	4	0	5.7	0.000	14.000	0.870	0.501	0.516	1.939	0.971	0.486	1
Turner 1985	*Bossiella plumosa*	Marine	Table 4	4	4	10.8	5.4	15.800	8.800	0.870	-0.367	0.508	1.967	-0.722	0.265	1
Dungan 1986	*Ralfsia pacifica*	Marine	Table 2	4	4	21.25	37.25	9.540	22.020	0.870	0.820	0.542	1.845	1.513	1.240	1
Dungan 1986	*Ralfsia pacifica*	Marine	Table 2	4	4	40.25	20.25	8.780	9.000	0.870	-1.956	0.739	1.353	-2.646	5.177	1
Dungan 1986	*Ralfsia pacifica*	Marine	Fig. 3B	5	5	15.8445	11.9533	10.787	6.240	0.903	-0.399	0.408	2.451	-0.978	0.390	1

Sums:

		d_+		Q
Terrestrial	Lines 3–27	73.783	84.623	122.697	19
Lentic	Lines 29–30	1.275	5.236	21.802	2
Marine	Lines 32–57	65.739	52.488	85.520	22
Grand total	Lines 3–57	140.797	142.347	230.018	43

	d_+	$s^2(d_+)$	Q
Terrestrial	1.147	0.014	25.641
Lentic	4.107	0.784	0.297
Marine	0.798	0.015	43.612

[a]Symbols are as in text.

$$s^2(d_{i+}) = \cfrac{1}{\sum\limits_{j=1}^{k_i} w_{ij}} \tag{17.6}$$

The lower and upper limits for the 95% confidence interval for d_{i+}, d^L and d^U, respectively, are:

$$d^L = d_{i+} - [Z_{\alpha/2}\, s(d_{i+})] \tag{17.7a}$$
$$d^U = d_{i+} + [Z_{\alpha/2}\, s(d_{i+})] \tag{17.7b}$$

where Z is the two-tailed critical value of the standard normal distribution. (The tests presented throughout this chapter for comparing effect sizes are all two-tailed, but calculation of one-tailed tests of particular hypotheses would be straightforward.) Taking the sums needed in Eq. 17.5 from the bottom of columns N and O in Table 17.1, the values for the cumulated effect sizes for the three classes in the example are

$$d_{i+} \text{ (terrestrial producers)} = 84.623/73.783 = 1.15$$
$$d_{i+} \text{ (lentic producers)} \quad = 5.236/1.275 \quad = 4.11$$
$$d_{i+} \text{ (marine producers)} \quad = 52.488/65.739 = 0.80$$

and from Eqs. 17.6, 17.7a, and 17.7b, using the sums at the bottom of column N in Table 17.1, their variances and 95% confidence limits,

Terrestrial $= 1/73.783 = 0.014;\ 1.147 \pm 1.96\ \sqrt{0.014} = 0.919$ to 1.375
Lentic $\quad = 1/1.275\ = 0.784;\ 4.107 \pm 1.96\ \sqrt{0.784} = 2.371$ to 5.843
Marine $\quad = 1/65.739 = 0.015;\ 0.798 \pm 1.96\ \sqrt{0.015} = 0.557$ to 1.040

We conclude from these results that when examined across all studies, terrestrial, lentic, and marine plants all exhibited statistically significant effects of competition on average. The values are significantly greater than zero (at $P < 0.05$) because the 95% confidence limits for the effect sizes do not overlap zero. The effects of competition (the d_{i+}) are considered large for marine plants and very large for terrestrial and lentic plants (Cohen, 1969).

The grand mean effect size across all classes, d_{++}, is

$$d_{++} = \cfrac{\sum\limits_{i=1}^{m} \sum\limits_{j=1}^{k_i} w_{ij} d_{ij}}{\sum\limits_{i=1}^{m} \sum\limits_{j=1}^{k_i} w_{ij}} \tag{17.8}$$

where m is the total number of classes (here, $m = 3$: terrestrial, lentic, and marine). The variance of the grand mean,

$$s^2(d_{++}) = \frac{1}{\displaystyle\sum_{i=1}^{m} \sum_{j=1}^{k_i} w_{ij}}$$ (17.9)

can also be calculated. In the example,

$$d_{++} = 142.347/140.797 = 1.01$$
$$s^2(d_{++}) = 1/140.797 \qquad = 0.007$$

Thus, the mean effect of competition across all studies was large.

The null hypothesis that all effect sizes are equal, versus the alternative hypothesis that at least one of the true effect sizes in a series of comparisons differs from the rest, can be tested with the homogeneity statistic Q which has approximately a χ^2 distribution with degrees of freedom equal to one less than the total number of studies. The greater the value of Q, the greater the heterogeneity in effect sizes among comparisons. The total heterogeneity, Q_T, can be partitioned into within-class heterogeneity, Q_W, and between-class heterogeneity, Q_B, much as one can partition variance in an ANOVA,

$$Q_B + Q_W = Q_T$$

Within-class homogeneity, Q_W,

$$Q_W = \sum_{i=1}^{m} \sum_{j=1}^{k_i} w_{ij}(d_{ij} - d_{i+})^2$$ (17.10)

is a measure of the variation among studies within classes (across all of the classes), while the between-class heterogeneity, Q_B, is a measure of the variation between classes in mean effect size,

$$Q_B = \sum_{i=1}^{m} \sum_{j=1}^{k_i} w_{ij}(d_{i+} - d_{++})^2$$ (17.11)

which is distributed as χ^2 with degrees of freedom equal to the number of classes minus 1 (Hedges and Olkin, 1985). A computational formula for the Q_T statistic is

$$Q_T = \sum_{i=1}^{m} \sum_{j=1}^{k_i} w_{ij} d_{ij}^2 - \frac{\left(\sum_{i=1}^{m} \sum_{j=1}^{k_i} w_{ij} d_{ij}\right)^2}{\sum_{i=1}^{m} \sum_{j=1}^{k_i} w_{ij}} \qquad (17.12)$$

The within-class homogeneity statistic across all studies, Q_W, is calculated as the sum of the within-class statistics Q_{W1}, Q_{W2}, . . . , Q_{Wm} for the m classes. Each within-class statistic Q_{Wi} can be calculated using the computational formula:

$$Q_{Wi} = \sum_{j=1}^{k_i} w_{ij} d_{ij}^2 - \frac{\left(\sum_{j=1}^{k_i} w_{ij} d_{ij}\right)^2}{\sum_{j=1}^{k_i} w_{ij}} \qquad (17.13)$$

with $(k - 1)$ df, and Q_B is obtained by subtracting Q_W from Q_T.

The values for the fixed effect Q_{Wi} statistics within each class in our example are found using Eq. 17.13 and the sums at the bottom of Table 17.1:

Terrestrial, $Q_{W1} = 122.697 - (84.623)^2/73.783 = 25.641$, 18 df, $P < 0.25$
Lentic, Q_{W2} $= 21.802 - (5.236)^2/1.275 = 0.297$, 1 df, $P < 0.75$
Marine, Q_{W3} $= 85.520 - (52.488)^2/65.739 = 43.612$, 21 df, $P < 0.005$

The statistical significance of these statistics is evaluated using a standard χ^2 table. We interpret these results to mean that the studies in the first two classes are homogeneous—within these classes, the effect sizes differ by no more than would be expected due to random sampling variation. (While it is possible that true differences between the two lentic studies could not be detected due to lack of power, we have both observed many instances in which a small number of studies resulted in significant values for the Q statistic.) Studies in the third class (marine producers) apparently exhibited more variation than can be attributed to sampling error. The variation among studies within classes, even for marine plants, is actually quite small for ecological studies (J. Gurevitch, personal observation) particularly in light of the rather stringent assumption here that the studies within a class share a common true effect size.

We continue with the example to calculate

$$Q_T = 230.018 - (142.347)^2/140.797 = 86.105, \text{ 42 df}$$
$$Q_W = 25.642 + 0.297 + 43.612 = 69.550, \text{ 40 df}$$
$$Q_B = 86.105 - 69.550 = 16.555, \text{ 2 df}$$

across the three classes of primary producers. There is a highly significant difference between the three classes ($Q_B = 16.56$, 2 df, $P < 0.001$). Therefore we conclude that in these studies there is a statistically significant difference among the responses of terrestrial, lentic, and marine producers to competition, when measured in terms of growth and recruitment. The confidence limits for terrestrial and marine producers overlapped, but the confidence interval of lentic producers did not overlap either of the other classes, suggesting that lentic producers experienced greater competitive effects than terrestrial or marine producers, which did not differ from one another. In addition to an informal evaluation based on confidence limits, formal procedures for constructing contrasts for mean effect sizes have been developed (Hedges and Olkin, 1985, Chapter 7, Section E).

The mixed model analysis. The fixed effect variance of d_{ij}, or v_{ij}, is actually a conditional variance, because it depends on assuming that there is one true effect size, δ, shared by all of the studies in the same class. The unconditional variance of d_{ij}, v_{ij}^* (which can also be thought of as the mixed model variance), assumes that there is random variation among studies in the effect of interest, which therefore do not share a common true effect size. (The asterisk here indicates the mixed model version of a term.) To get v_{ij}^* we need to add to the usual v_{ij} a term for the pooled within-class variance component, $\hat{\sigma}_{pooled}^2$.

To carry out the mixed model analysis (Table 17.2), several additional terms must be calculated.

First, we calculate a constant c_i for each class,

$$c_i = \sum_{j=1}^{k_i} w_{ij} d_{ij}^2 - \frac{\sum_{j=1}^{k_i} w_{ij}^2}{\sum_{j=1}^{k_i} w_{ij}} \qquad (17.14)$$

where i is the class and k_i is the number of experiments in class i. Then we compute the estimate of σ_{pooled}^2 via

$$\sigma_{pooled}^2 = \frac{Q_W - \sum_{i=1}^{m} (k_i - 1)}{\sum_{i=1}^{m} c_i} \qquad (17.15)$$

where m is the number of classes and Q_W is the within-class heterogeneity from the fixed effects analysis. From our example, taking the data from the sums at the bottom of Table 17.2, for the terrestrial class,

$$c_i = 73.783 - 447.185/73.783 = 67.722$$

Table 17.2 Calculations for the Mixed Model[a]

Publication	d	v_{ij}	w_{ij}	w_{ij}^2	v^*	w^*	w^*d
Fowler 1986	0.036	0.286	3.499	12.246	0.514	1.946	0.070
Fowler 1986	0.729	0.305	3.282	10.772	0.533	1.876	1.368
Platt and Weiss 1985	0.565	0.347	2.885	8.322	0.575	1.740	0.983
Platt and Weiss 1985	1.533	0.517	1.932	3.734	0.746	1.341	2.056
Platt and Weiss 1985	2.014	0.431	2.322	5.393	0.659	1.518	3.057
Platt and Weiss 1985	1.880	0.481	2.081	4.329	0.709	1.411	2.653
Gross and Werner 1982	1.181	0.783	1.277	1.632	1.011	0.989	1.168
Gross and Werner 1982	1.400	0.830	1.205	1.452	1.058	0.945	1.323
Gross and Werner 1982	1.352	2.152	0.465	0.216	2.380	0.420	0.568
Gross and Werner 1982	1.545	2.199	0.455	0.207	2.427	0.412	0.637
Pons and van der Toorn 1988	0.000	0.400	2.500	6.250	0.628	1.592	0.000
Pons and van der Toorn 1988	−0.917	0.442	2.262	5.117	0.670	1.492	−1.368
Burton and Mueller-Dombois 1984	1.214	0.592	1.689	2.852	0.820	1.219	1.480
Gurevitch 1986	1.694	0.143	6.977	48.679	0.372	2.691	4.560
Gurevitch 1986	1.273	0.120	8.316	69.148	0.348	2.870	3.653
Gurevitch 1986	1.303	0.128	7.818	61.120	0.356	2.808	3.660
Gurevitch 1986	1.519	0.129	7.762	60.249	0.357	2.801	4.254
Gurevitch 1986	1.144	0.116	8.595	73.875	0.345	2.902	3.319
Gurevitch 1986	1.206	0.118	8.461	71.593	0.346	2.887	3.482
McCreary et al. 1983	3.696	1.354	0.739	0.546	1.582	0.632	2.336
McCreary et al. 1983	4.674	1.865	0.536	0.287	2.093	0.478	2.233
Stimson 1985	1.218	0.339	2.953	8.717	0.567	1.764	2.148
Stimson 1985	1.288	0.497	2.011	4.045	0.725	1.379	1.776
Reed and Foster 1984	1.818	0.141	7.077	50.083	0.370	2.706	4.919
Reed and Foster 1984	2.223	0.162	6.182	38.213	0.390	2.564	5.700
Reed and Foster 1984	0.767	0.107	9.315	86.767	0.336	2.980	2.286
Reed and Foster 1984	0.524	0.207	4.834	23.369	0.435	2.298	1.204
Reed and Foster 1984	0.163	0.201	4.983	24.835	0.429	2.332	0.380
Reed and Foster 1984	0.449	0.205	4.877	23.788	0.433	2.308	1.036
Johnson and Mann 1988	2.218	1.615	0.619	0.383	1.843	0.543	1.203
Johnson and Mann 1988	0.571	1.041	0.961	0.923	1.269	0.788	0.450
Johnson and Mann 1988	−0.326	1.013	0.987	0.974	1.241	0.805	−0.262
Turner 1985	0.735	0.534	1.873	3.510	0.762	1.312	0.965
Turner 1985	0.869	0.547	1.827	3.340	0.775	1.290	1.121
Turner 1985	0.463	0.513	1.948	3.794	0.742	1.348	0.624
Turner 1985	0.432	0.512	1.954	3.820	0.740	1.352	0.584
Turner 1985	0.728	0.533	1.876	3.518	0.761	1.313	0.956
Turner 1985	0.629	0.525	1.906	3.632	0.753	1.328	0.836
Turner 1985	0.501	0.516	1.939	3.761	0.744	1.344	0.673
Turner 1985	−0.367	0.508	1.967	3.869	0.737	1.357	−0.498
Dungan 1986	0.820	0.542	1.845	3.404	0.770	1.298	1.064
Dungan 1986	−1.956	0.739	1.353	1.830	0.967	1.034	−2.022
Dungan 1986	−0.399	0.408	2.451	6.009	0.636	1.572	−0.627

	w_{ij}	w_{ij}^2	c_i	w_{i+}^*	$w_{i+}^* d_{i+}^*$	$w_{i+}^* d_{i+}^{*\,2}$
Terrestrial	73.783	447.185	67.722	33.859	36.921	40.260
Lentic	1.275	0.833	0.621	1.110	4.569	18.809
Marine	65.739	302.582	61.136	35.016	24.517	17.166
Grand sum	140.797	750.600				

[a]The effect size, d, as well as v_{ij} and w_{ij} are taken from Table 17.1. Symbols are as defined in the text.

and $c_i = 0.621$ for lentic and 61.136 for marine producers. Continuing with the example,

$$\sigma^2_{pooled} = (69.550 - 40)/129.480 = 0.228$$

Now we can find v^*_{ij} for each study,

$$v^*_{ij} = v_{ij} + \sigma^2_{pooled}$$

From the example, in line 1 of Table 17.2, $v^*_{ij} = 0.514$ (taking v_{ij} in the third column $+ 0.228$). The weights used to carry out the meta-analysis are the reciprocals of the random effects variance estimates (just as the weights in the fixed effects model are the reciprocal of the fixed effects variance estimates),

$$w^*_{ij} = 1/v^*_{ij}$$

and are found in the seventh column of Table 17.2. These are multiplied by the effect sizes (d_{ij} from the second column), as in the last column of Table 17.2.

The cumulated effect sizes for each class in the mixed model, d^*_{i+}, and their variances, $s^2(d^*_{i+})$, are calculated as in the fixed effects model (Eqs. 17.5 and 17.6). From the example (Table 17.2), the values for the mixed model cumulated effect sizes (d^*_{i+}) for the three classes are

Terrestrial producers $= 36.921/33.859 = 1.09$
Lentic producers $= 4.569/1.110 = 4.12$
Marine producers $= 24.517/35.016 = 0.70$

and their variances and 95% confidence limits,

Terrestrial $= 1/33.859 = 0.030; 0.754$ to 1.427
Lentic $= 1/1.110 = 0.901; 2.256$ to 5.977
Marine $= 1/35.016 = 0.029; 0.370$ to 1.031

It is reassuring that the values are very similar to those for the fixed effects model, and, as would be expected, the confidence intervals are somewhat larger because an additional variance component is included. The conclusions regarding the effects of competition are essentially the same for the mixed model as for the fixed effect model, except that in this case we have not made the conventional assumption that all studies in a class share a common true effect size.

The grand cumulated effect size, d^*_{++}, and its variance, $s^2(d^*_{++})$, are calculated as

$$d^*_{++} = \frac{\sum\limits_{i=1}^{m} w^*_{i+} d^*_{i+}}{\sum\limits_{i=1}^{m} w^*_{i+}} \tag{17.16}$$

and

$$s^2(d^*_{++}) = \frac{1}{\sum\limits_{i=1}^{m} w^*_{i+}} \tag{17.17}$$

(Note that this is slightly different than in the fixed effects model.) From the example,

$d^*_{++} = (36.921 + 4.569 + 24.517)/(33.859 + 1.110 + 35.016) = 0.943$
$s^2(d^*_{++}) = 1/(33.859 + 1.110 + 35.016) = 0.014$

The effect of competition across all studies was large. Finally, the homogeneity among classes can be tested using

$$Q^*_B = \sum\limits_{i=1}^{m} w^*_{i+} d^{*2}_{i+} - \frac{\left(\sum\limits_{i=1}^{m} w^*_{i+} d^*_{i+}\right)^2}{\sum\limits_{i=1}^{m} w^*_{i+}} \tag{17.18}$$

which is also calculated somewhat differently than in the fixed effect model. From the example,

$Q^*_B = (40.260 + 18.809 + 17.166) - (36.921 + 4.569 + 24.517)^2$
$\qquad /(33.859 + 1.110 + 35.016)$
$\quad = 13.980$

with 2 degrees of freedom, which is significant at $P < 0.001$. The conclusions again are the same as for the fixed effect model, that is, the effects of competition are not the same for terrestrial, lentic and marine primary producers. In the mixed model, we do not calculate a Q_w, because we no longer are making the assumption that all studies in a class share a common true effect. The test of the homogeneity of the effect sizes within a class, using Q_w from the fixed effects model, can, however, be interpreted in the mixed model as a test that the within-class variance component σ^2 is larger than zero. This test would rarely be useful because it will typically lead to rejection of the hypothesis that $\sigma^2 = 0$, and so is not likely to be informative.

17.4 Interpretation and Conclusions

There was a large effect of competition on growth and recruitment of primary producers, with lentic plants experiencing very large effects and terrestrial and marine organisms lesser, but still large effects. The results here contrast to some degree with the results of a meta-analysis of the effects of competition on biomass (Gurevitch et al., 1992). In that study, primary producers shared a common, small to medium effect of competition (d_{++} = 0.34) that did not differ among terrestrial, freshwater or marine organisms. It seems reasonable that a change in size and number, as in the present study, might be expected to show a greater response to an experimental manipulation than the total size of the organisms, as in the previous study. The number of experimental comparisons in the three systems was similar in the previous and present studies for primary producers. It would be interesting to investigate lentic plants further, because the large effects of competition in freshwater are based on only two experimental comparisons by a single author in the present meta-analysis. However, we can be fairly confident about the conclusions regarding the generally substantial effects of competition on primary producers found here.

These results are difficult to compare with the ones of narrative and vote-count reviews of competition in field experiments, for the reasons discussed at the beginning of the chapter. Nevertheless, the results agree in broad terms with those of Goldberg and Barton (1992), Connell (1983), and Schoener (1983) in concluding that competition among primary producers was common. Previous reviewers have been unable to accurately assess the magnitude of competitive effects across studies, or to compare the intensity of competition among categories of studies.

It is reassuring to note the general agreement between the results of the more conventional fixed effects model and the mixed model presented here. As one would expect, the major difference between the two is the larger confidence intervals in the mixed model, which reflect the additional source of error included in that model. The mixed model assumption of random variation between studies within a class is in many cases a much more reasonable one than the fixed model assumption of all studies within a class sharing a single true effect size. For that reason, the more widespread use of the mixed model is to be strongly encouraged.

17.5 An Overview of the Potential Role of Meta-Analysis in Ecology

Underlying the motivation for this volume has been the explosion of interest in experimental ecology, particularly in the field, within the past 15 years or so. The preceeding chapters all attempt to assist experimentalists engaged in designing and analyzing their experiments. The sophistication of the statistical tools available to ecologists for analyzing the results of experiments has increased exponentially in recent years, and this book is an effort to make some of those tools more widely available. To date, the tools that exist to make the best use of the wealth of ecological data provided by the outcome of these efforts have been crude, inaccurate, and inadequate to the task. Meta-analysis establishes a new standard in the tools available to synthesize data gathered in different studies. However, meta-analysis itself, particularly as applied to ecological research, is in its infancy. There is no question that great potential exists to develop and improve the models used to conduct meta-analyses in ecology and in other fields.

We predict that meta-analysis will have a substantial impact on the field of ecology over the next few years. There are several reasons for believing this. First, ecologists are just beginning to become aware of these techniques, and the methods are not difficult to use. In fact, the statistics presented here are perhaps the easiest ones to compute in this entire volume. More substantively, the number of experimental studies in various subdisciplines of ecology has probably reached a critical mass: there is abundant material to summarize. Most critically, the ever-increasing impact of humans on the natural world has made the need to make sense of this growing body of data increasingly urgent. Meta-analysis presents compelling opportunities for revealing the broad patterns contained in the accumulated body of experimental ecological research.

Acknowledgments

This work was supported by NSF Grant BSR 9113065 to J.G. Todd Postol, Janet Morrison, Alison Wallace, Tom Meagher, and Sam Scheiner offered valuable suggestions on the manuscript. The generous help of D. Slice, who wrote a digitizing program that enabled us to obtain data from graphs, was very much appreciated. Alison Wallace and Joe Walsh provided invaluable assistance with the literature survey and data sheets. The methods for the collection and interpretation of the data were developed in part by Laura Morrow, whose efforts and insight in sweating the details substantially improved the quality of the work.

References

Abrami, P. C., Cohen, P. A., and d'Appollonia, S. (1988) Implementation problems in meta-analysis. *Review of Education Research*, 58, 151–179.

Abrams, P. A. (1990) The effects of adaptive behavior on the type–2 functional response. *Ecology*, 71, 877–885.

Agresti, A. (1990) *Categorical Data Analysis*. Wiley, New York.

Aiken, L. S., and West, S. G. (1991) *Multiple Regression: Testing and Interpreting Interactions*. Sage Publications, Newbury Park.

Antonovics, J. (1976) The input from population genetics: The new ecological genetics. *Systematic Botany*, 1, 233–245.

Antonovics, J., and Fowler, N. L. (1985) Analysis of frequency and density effects on growth in mixtures of *Salvia splendens* and *Linum grandiflorum* using hexagonal fan designs. *Journal of Ecology*, 73, 219–234.

Arnold, S. J. (1972) Species densities of predators and their prey. *American Naturalist*, 106, 220–236.

Austin, M. P. (1990) Community theory and competition in vegetation. In *Perspectives on Plant Competition* (eds J. B. Grace and D. Tilman). Academic Press, San Diego, pp. 215–239.

Ayala, F. J., Gilpin, M. E., and Ehrenfeld, J. G. (1973) Competition between species: Theoretical models and experimental tests. *Theoretical Population Biology*, 4, 331–356.

Ayres, M. P., and Thomas, D. L. (1990) Alternative formulations of the mixed-model ANOVA applied to quantitative genetics. *Evolution*, 44, 221–226.

Baird, D., and Mead, R. (1991) The empirical efficiency and validity of two neighbour models. *Biometrics*, 47, 1473–1487.

Baker, H. G. (1983) An outline of the history of anthecology, or pollination biology. In *Pollination Biology* (ed. L. Real). Academic Press, Orlando, FL, pp. 7–28.

Bard, Y. (1974) *Nonlinear Parameter Estimation*. Academic Press, New York.

Barker, H. R., and Barker, B. M. (1984) *Multivariate Analysis of Variance (MANOVA)*. University of Alabama Press, Tuscaloosa.

Barlow, R. E., Bartholomew, D. J., Bremer, J. M., and Brunk, H. D. (1972) *Statistical Inference Under Order Restrictions*. Wiley, New York.

Barnett, V. (1976) The ordering of multivariate data. *Journal of the Royal Statistical Society, Series A*, 139, 318–354.

Bartlett, M. S. (1978) Nearest neighbor models in the analysis of field experiments with large blocks. *Journal of the Royal Statistical Society, Series B*, 40, 147–174.

Bautista, L. M., Alonso, J. C., and Alonso, J. A. (1992) A 20-year study of wintering common crane fluctuations using time series analysis. *Journal of Wildlife Management*, 56, 563–572.

Bayes, T. (1763) An essay towards solving a problem in the doctrine of chances. *Philosophical Transactions of the Royal Society*, 53, 370–418.

Becker, R. A., Chambers, J. M., and Wilks, A. R. (1988) *The New S Language*. Wadsworth and Brooks, California.

Belovsky, G. E. (1981) Food plant selection by a generalist herbivore: The moose. *Ecology*, 62, 1020–1030.

Belovsky, G. E. (1987) Extinction models and mammalian persistence. In *Viable Populations for Conservation* (ed. M. E. Soule). Cambridge University Press, New York, pp. 35–58.

Benditt, J. (1992) Women in science, 1st annual survey. *Science*, 255, 1363–1388.

Bentler, P. M. (1985) *Theory and Implementation of EQS: A Structural Equations Program*. BMDP Statistical Software, Los Angeles.

Berger, J. O. (1985) *Statistical Decision Theory and Bayesian Analysis*. Springer-Verlag, New York.

Besag, J. E., and Kempton, R. A. (1986) Statistical analysis of field experiments using neighboring plots. *Biometrics*, 42, 231–251.

Bierzychudek, P. (1982) The demography of jack-in-the-pulpit, a forest perennial that changes sex. *Ecological Monographs*, 52, 335–351.

Bollen, K. A. (1989) *Structural Equations with Latent Variables*. Wiley, New York.

Bookstein, F. L. (1986) The elements of latent variable models: a cautionary lecture. In *Advances in Developmental Psychology* (eds M. E. Lamb, A. L. Brown and B. Rogoff). Lawrence Erlbaum, Hillsdale, NJ, pp. 203–230.

Bormann, F. H., and Likens, G. E. (1979) *Pattern and Process in a Forested Ecosystem*. Springer-Verlag, New York.

Box, G. E. P. (1954) Some theorems on quadratic forms applied in the study of analysis of variance problems: I. Effect of inequality of variance in the one-way classification. *Annals of Mathematical Statistics*, 25, 290–302.

Box, G. E. P., and Cox, C. R. (1964) An analysis of transformations. *Journal of the Royal Statistical Society, Series B*, 26, 211–243.

Box, G. E. P., Hunter, W. G., and Hunter, J. S. (1978) *Statistics for Experimenters: An Introduction to Design, Data Analysis, and Model Building.* Wiley, New York.

Box, G. E. P., and Jenkins, G. M. (1976) *Time Series Analysis: Forecasting and Control,* rev. ed. Holden-Day, San Francisco.

Box, G. E. P., and Pierce, D. A. (1970) Distributions of residual autocorrelations in autoregressive-integrated moving average time series models. *Journal of the American Statistical Association,* 65, 1509–1526.

Box, G. E. P., and Tiao, G. C. (1975) Intervention analysis with applications to economic and environmental parameters. *Journal of the American Statistical Association,* 70, 70–79.

Bradley, J. V. (1978) Robustness? *British Journal of Mathematical and Statistical Psychology,* 31, 144–152.

Bray, J. H., and Maxwell, S. E. (1985) *Multivariate Analysis of Variance.* Sage Publications, Newbury Park.

Breckler, S. J. (1990) Applications of covariance structure modeling in psychology: Cause for concern? *Psychological Bulletin,* 107, 260–273.

Brezonik, P. L., Baker, L. A., Eaton, J., Frost, T. M., Garrison, P., Kratz, T. K., Magnuson, J. J., Perry, J., Rose, W., Sheperd, B., Swenson, W., Watras, C., and Webster, K. (1986) Experimental acidification of Little Rock Lake, Wisconsin. *Water Air Soil Pollution,* 31, 115–121.

Brody, A. K. (1992) Oviposition choices by a pre-dispersal seed predator (*Hylemya* sp.): I. Correspondence with hummingbird pollinators, and the role of plant size, density and floral morhpology. *Oecologia,* 91, 56–62.

Brooks, D. R., and McLennan, D. A. (1991) *Phylogeny, Ecology, and Behavior: A Research Program in Comparative Biology.* University of Chicago Press, Chicago, IL.

Brown, J. H., Davidson, D. W., Munger, J. C., and Inouye, R. S. (1986) Experimental community ecology: The desert granivore system. In *Community Ecology* (eds J. Diamond and T. J. Case). Harper & Row, New York, pp. 41–61.

Brown, J. S. (1988) Patch use as an indicator of habitat preference, predation risk, and competition. *Behavioral Ecology and Sociobiology,* 22, 37–47.

Brown, J. S., Kotler, B. P., Smith, R. J., and Wirtz, W. O. (1988) The effects of owl predation on the foraging behavior of heteromyid rodents. *Oecologia,* 76, 408–415.

Brownlee, K. A. (1965) *Statistical Theory and Methodology in Science and Engineering.* 2nd ed. Wiley, New York.

Buckland, S. T. (1985) Calculation of Monte Carlo confidence intervals. *Applied Statistics,* 34, 296–301.

Buckland, S. T., and Garthwaite, P. H. (1991) Quantifying the precision of mark-recapture estimates using the bootstrap and related methods. *Biometrics,* 47, 255–268.

Bull, J. J., Thompson, C., Ng, D., and Moore, R. (1987) A model for natural selection of genetic migration. *American Naturalist,* 129, 143–157.

Buonaccorsi, J. P., and Liebhold, A. M. (1988) Statistical methods for estimating ratios and products in ecological studies. *Environmental Entomology,* 17, 572–580.

Burgman, M. A. (1987) An analysis of the distribution of plants on granite outcrops in southern Western Australia using Mantel tests. *Vegetatio*, 71, 79–86.

Burton, P. J., and Mueller-Dombois, D. (1984) Response of *Metrosideros polymorpha* seedlings to experimental canoy openings. *Ecology*, 65, 779–791.

Campbell, B. D., and Grime, J. P. (1992) An experimental test of plant strategy theory. *Ecology*, 73, 15–29.

Campbell, D. R., Waser, N. M., Price, M. V., Lynch, E. A., and Mitchell, R. J. (1991) Components of phenotypic selection: pollen export and corolla width in *Ipomopsis aggregata*. *Evolution*, 45, 1458–1467.

Caraco, T. (1979) Time budgeting and group size: A theory. *Ecology*, 60, 611–617.

Carpenter, S. R. (1989) Replication and treatment strength in whole-lake experiments. *Ecology*, 70, 453–463.

Carpenter, S. R. (1990) Large scale perturbations: Opportunities for innovation. *Ecology*, 71, 2038–2043.

Carpenter, S. R., Frost, T. M., Heisey, D., and Kratz, T. (1989) Randomized intervention analysis and the interpretation of whole ecosystem experiments. *Ecology*, 70, 1142–1152.

Cassini, M. H., Kacelnik, A., and Segura, E. T. (1990) The tale of the screaming hairy armadillo, the guinea pig and the marginal value theorem. *Animal Behaviour*, 39, 1030–1050.

Caswell, H. (1986) Matrix models and the analysis of complex plant life cycles. *Lectures on Mathematics in Life Sciences*, 18, 171–234.

Caswell, H. (1989) *Matrix Population Models*. Sinauer, Sunderland, MA.

Caswell, H., Naiman, R. J., and Morin, R. (1984) Evaluating the consequenes of reproduction in complex salmonid life cycles. *Aquaculture*, 43, 123–134.

Cerato, R. M. (1990) Interpretable statistical tests for growth comparisons using parameters in the von Bertalanffy equation. *Canadian Journal of Fisheries and Aquatic Sciences*, 47, 1416–1426.

Chalmers, I., Hetherington, J., Elhourne, D., Keirse, M. J. N. C., and Enkin, M. (1989) *Effective Care in Pregnancy and Childbirth*. Oxford University Press, Oxford.

Chambers, J. M., and Hastie, T. J. (eds) (1992) *Statistical Models in S*. Wadsworth and Brooks, Pacific Grove, CA.

Charlesworth, B. (1980) *Evolution in Age-Structured Populations*. Cambridge University Press, Cambridge.

Charnov, E. L. (1976) Optimal foraging, the marginal value theorem. *Theoretical Population Biology*, 9, 129–136.

Chatterjee, S. and Price, B. (1991) *Regression Analysis by Example*. 2nd ed. Wiley, New York.

Cheng, C. S. (1983) Construction of optimal balanced incomplete block designs for correlated observations. *Annals of Statistics*, 11, 240–246.

Cleveland, W. S. (1979) Robust locally weighted regression and smoothing scatterplots. *Journal of the American Statistical Association*, 74, 829–836.

Cleveland, W. S. (1985) *The Elements of Graphing Data*. Wasdworth Advanced Books and Software, Monterey, CA.

Cliff, A. D., and Ord, J. K. (1981) *Spatial Processes: Models and Applications*. Pion, London.

Cloninger, C. R., Rao, D. C., Rice, J., Reich, T., and Morton, N. E. (1983) A defense of path analysis in genetic epidemiology. *American Journal of Human Genetics*, 35, 733–756.

Cochran, W. G., and Cox, G. M. (1957) *Experimental Designs*. 2nd ed. Wiley, New York.

Cochran-Stafira, D. L. (1993) Food web interactions and microbial community structure in *Sarracenia purpurea* pitchers. PhD thesis, Northern Illinois University, DeKalb.

Cock, M. J. W. (1977) Searching behaviour of polyphagous predators. PhD thesis, Imperial College, London.

Cohen, J. (1969) *Statistical Power Analysis for the Behavioral Sciences*. Academic Press, New York.

Cohen, J., and Cohen, P. (1983) *Applied Multiple Regression/Correlation Analysis for the Behavioral Sciences*. Lawrence Erlbaum, Hillsdale, CA.

Cohen, J. E. (1979a) Comparative statics and stochastic dynamics of age-structured populations. *Theoretical Population Biology*, 16, 159–171.

Cohen, J. E. (1979b) Ergodic theorems in demography. *Bulletin of the American Mathematical Society*, 1, 275–295.

Coley, P. D. (1983) Herbivory and defensive characteristics of tree species in a lowland tropical forest. *Ecological Monographs*, 53, 209–233.

Collier, R. O. Jr., Baker, F. B., Mandeville, G. K., and Hayes, T. F. (1967) Estimates of test size for several test procedures based on conventional variance ratios in the repeated measures design. *Psychometrika*, 32, 339–353.

Connell, J. H. (1983) On the prevalence and relative importance of interspecific competition: evidence from field experiments. *American Naturalist*, 122, 661–696.

Connell, J. H. (1990) Apparent versus "real" competition in plants. In *Perspectives on Plant Competition* (eds J. Grace and D. Tilman). Academic Press, San Diego, pp. 9–26.

Connolly, J. (1986) On difficulties with replacement-series methodology in mixture experiments. *Journal of Applied Ecology*, 23, 125–137.

Connor, E. F., and Simberloff, D. (1979) Assembly of species communities: Chance or competition? *Ecology*, 60, 1132–1140.

Cooper, H. M. (1989) *Integrating Research: A Guide for Literature Reviews*, 2nd ed. Sage Publications, Newbury Park, CA.

Cooper, H. M., and Hedges, L. V. (eds) (1993) *Handbook of Research Synthesis*. Russell Sage Foundation, New York.

Crespi, B. J. (1990) Measuring the effect of natural selection on phenotypic interaction systems. *American Naturalist*, 135, 32–47.

Crespi, B. J., and Bookstein, F. L. (1989) A path-analytic model for the measurement of selection on morphology. *Evolution*, 43, 18–28.

Cressie, N. (1985) Fitting variogram models by weighted least squares. *Journal of the International Association for Mathematical Geology*, 17, 563–586.

Cressie, N. (1991) *Statistics for Spatial Data*. Wiley, New York.

Cressie, N., and Hawkins, D. M. (1980) Robust estimation of the variogram. *Journal of the International Association for Mathematical Geology*, 12, 115–125.

Crouse, D. T., Crowder, L. B., and Caswell, H. (1987) A stage-based populaltion model for loggerhead sea turtles and implications for conservation. *Ecology*, 68, 1412–1423.

Crowder, M. J., and Hand, D. J. (1990) *Analysis of Repeated Measures*. Chapman and Hall, London.

Cryer, J. D. (1986) *Time Series Analysis*. Duxbury, Boston.

Cullis, B. R., and Gleeson, A. C. (1989) The efficiency of neighbour analysis for replicated variety trials in Australia. *Journal of Agricultural Science, Cambridge*, 113, 233–239.

Cullis, B. R., and Gleeson, A. C. (1991) Spatial analysis of field experiments—an extension to two dimensions. *Biometrics*, 47, 1449–1460.

Czárán, T., and Bartha, S. (1992) Spatiotemporal dynamic models of plant populations and communities. *Trends in Ecology and Evolution*, 7, 38–42.

Day, R. W., and Quinn, G. P. (1989) Comparisons of treatments after an analysis of variance in ecology. *Ecological Monographs*, 59, 433–463.

de Kroon, H., Plaisier, A., van Groenendael, J., and Caswell, H. (1986) Elasticity as a measure of the relative contribution of demographic parameters to population growth rate. *Ecology*, 67, 1427–1431.

Diamond, J. (1986) Overview: Laboratory experiments, field experiments, and natural experiments. In *Community Ecology* (eds J. Diamond and T. J. Case). Harper & Row, New York, pp. 3–22.

Diamond, J. M., and Gilpin, M. E. (1982) Examination of the "null" model of Connor and Simberloff for species co-occurrences on islands. *Oecologia*, 52, 64–74.

Digby, P. G. N., and Kempton, R. A. (1987) *Multivariate Analysis of Ecological Communities*. Chapman and Hall, London.

Diggle, P. J. (1990) *Time Series: A Biostatistical Introduction*. Clarendon Press, Oxford.

Dixon, P. M., Weiner, J., Mitchell-Olds, T., and Woodley, R. (1987) Bootstrapping the Gini coefficient of inequality. *Ecology*, 68, 1548–1551.

Dorfman, J. H., Kling, C. L., and Sexton, R. J. (1990) Confidence intervals for elasticities and flexibilities: Reevaluating the ratios of normals case. *American Journal of Agricultural Economics*, 72, 1006–1017.

Douglas, M. E., and Endler, J. A. (1982) Quantitative matrix comparisons in ecological and evolutionary investigations. *Journal of Theoretical Biology*, 99, 777–795.

Dow, M. M., and Cheverud, J. M. (1985) Comparison of distance matrices in studies of population structure and genetic microdifferentiation: Quadratic assignment. *American Journal of Physical Anthropology*, 68, 367–373.

Draper, N. R., and Smith, H. (1981) *Applied Regression Analysis*, 2nd ed. Wiley, New York.

Dungan, M. L. (1986) Three-way interactions: Barnacles, limpets, and algae in a Sonoran desert rocky intertidal zone. *American Naturalist*, 127, 292–316.

Dunning, J. B. (1990) Meeting the assumptions of foraging models: An example using tests of avian patch choice. *Studies in Avian Biology*, 13, 462–470.

Dutilleul, P. (1993) Spatial heterogeneity and the design of ecological field experiments. *Ecology*, 74 (in press).

Dykhuizen, D. (1990) Experimental studies of natural selection in bacteria. *Annual Review of Ecology and Systematics*, 21, 378–398.

Eberhardt, L. L. (1976) Quantitative ecology and impact assessment. *Journal of Environmental Management*, 4, 27–70.

Eberhardt, L. L., and Thomas, J. M. (1991) Designing environmental field studies. *Ecological Monographs*, 61, 53–73.

Edwards, A. W. F. (1972) *Likelihood*. Cambridge University Press, Cambridge.

Edwards, W., Lindman, H., and Savage, L. J. (1983) Bayesian statistical inference for psychological research. *Psychological Review*, 70, 193–242.

Efron, B. (1982) The jackknife, the bootstrap, and other resampling plans. *Society of Industrial and Applied Mathematics*, CBMS-NSF Monograph 38.

Efron, B. (1987) Better bootstrap confidence intervals (with discussion). *Journal of the American Statistical Association*, 82, 171–200.

Efron, B. (1992) Six questions raised by the bootstrap. In *Exploring the Limits of Bootstrap* (eds R. LePage and L. Billard). Wiley, New York, pp. 99–126.

Efron, B., and LePage, R. (1992) Introduction to bootstrap. In *Exploring the Limits of Bootstrap* (eds R. LePage and L. Billard). Wiley, New York, pp. 3–10.

Efron, B., and Tibshirani, R. (1986) Bootstrap methods for standard errors, confidence intervals, and other measures of statistical accuracy. *Statistical Science*, 1, 54–77.

Efron, B., and Tibshirani, R. (1991) Statistical data analysis in the computer age. *Science*, 253, 390–395.

Ellison, A. M. (1992) Statistics for PCs. *Bulletin of the Ecological Society of America*, 73, 74–87.

Ellison, A. M., and Bedford, B. L. (1991) *Response of a Wetland Vascular Plant Community to Disturbance: A Simulation Study*. Ecosystems Research Center, publication ERC243, Cornell University, Ithaca.

Ellison, A. M., Denslow, J. S., Loiselle, B. A., and Brens M. D. (1993) Seed and seedling ecology of neotropical Melastomataceae. *Ecology* 74 (in press).

Endler, J. A. (1986) *Natural Selection in the Wild*. Princeton University Press, Princeton, NJ.

Epling, C., Lewis, H., and Ball, F. M. (1960) The breeding group and seed storage: a study in population dynamics. *Evolution*, 14, 238–255.

Estabrook, G. F., and Gates, B. (1984) Character analysis in the *Banisteriopsis campestris* complex (Malpighiaceae), using spatial auto-correlation. *Taxon*, 33, 13–25.

Evans, J. (1983) The ecology of *Ailanthus altissima*: an invading species in both early and late successional communities of the Eastern Deciduous Forest. BS thesis, Cornell University, Ithaca.

Feinsinger, P., and Tiebout, H. M. I. (1991) Competition among plants sharing hummingbird pollinators: Laboratory experiments on a mechanism. *Ecology*, 72, 1946–1952.

Firbank, L. G., and Watkinson, A. R. (1987) On the analysis of competition at the level of the individual plant. *Oecologia*, 71, 308–317.

Fisher, R. A. (1935) *The Design of Experiments*. Oliver and Boyd, Edinburgh.

Fortin, M.-J., Drapeau, P., and Legendre, P. (1989) Spatial autocorrelation and sampling design in plant ecology. *Vegetatio*, 83, 209–222.

Fowler, N. L. (1986) Density dependent population regulation in a Texas grassland. *Ecology*, 67, 545–554.

Fowler, N. L. (1990a) Disorderliness in plant communities: Comparisons, causes, and consequences. In *Perspectives on Plant Competition* (eds J. Grace and D. Tilman). Academic Press, San Diego, pp. 291–307.

Fowler, N. (1990b) The 10 most common statistical errors. *Bulletin of the Ecological Society of America*, 71, 161–164.

Fox, G. A. (1989) Consequences of flowering time in a desert annual: Adaptation and history. *Ecology*, 70, 1294–1306.

Fox, G. A. (1990a) Components of flowering time variation in a desert annual. *Evolution*, 44, 1404–1423.

Fox, G. A. (1990b) Perennation and the persistence of annual life histories. *American Naturalist*, 135, 829–840.

Fretwell, S. D., and Lucas, H. L. (1970) On territorial behaviour and other factors influencing habitat distribution in birds. *Acta Biotheoretica*, 19, 16–36.

Freund, R. J., and Littell, R. C. (1986) *SAS System for Regression, 1986 Edition*. SAS Institute Inc., Cary.

Frost, T. M., DeAngelis, D. L., Allen, T. F. H., Bartell, S. M., Hall, D. J., and Hurlbert, S. H. (1988) Scale in the design and interpretation of aquatic community research. In *Complex Interactions in Lake Communities* (ed. Carpenter, S. R.). Springer-Verlag, New York, pp. 229–258.

Fry, J. D. (1992) The mixed-model analysis of variance applied to quantitative genetics: Biological meaning of the parameters. *Evolution*, 46, 540–550.

Futuyma, D. J., and Philippi, T. E. (1987) Genetic variation and covariation in response to host plants by *Alsophila pomentaria* (Lepidoptera: Geometridae). *Evolution*, 41, 269–279.

Gaines, S. D., and Rice, W. R. (1990) Analysis of biological data when there are ordered expectations. *American Naturalist*, 135, 310–317.

Gardner, M. J., and Altman, D. G. (1986) Confidence intervals rather than P values; estimation rather than hypothesis testing. *British Medical Journal*, 292, 746–750.

Geisser, S., and Greenhouse, S. W. (1958) An extension of Box's results on the use of the F distribution in multivariate analysis. *Annals of Mathematical Statistics*, 29, 885–891.

Gibson, A. R., Baker, A. J., and Moeed, A. (1984) Morphometric variation in introduced populations of the Common Myna (*Acridotheres tristis*): an application of the jackknife to principal components analysis. *Systematic Zoology*, 33, 408–421.

Gilliam, J. F., and Fraser, D. F. (1987) Habitat selection under predation hazard: A test of a model with foraging minnows. *Ecology*, 68, 1856–1862.

Gittins, R. (1985) *Canonical Analysis*. Springer-Verlag, Berlin.

Glantz, S. A., and Slinker, B. K. (1990) *Primer of Applied Regression and Analysis of Variance*. McGraw-Hill, New York.

Glass, G. V., McGraw, B., and Smith, M. L. (1981) *Meta-analysis in Social Research*. Sage Publications, Beverly Hills, CA.

Gnanadesikan, R. (1977) *Methods for Statistical Data Analysis of Multivariate Observations*. Wiley, New York.

Goldberg, D. E. (1987) Neighborhood competition in an old-field plant commmunity. *Ecology*, 68, 1211–1223.

Goldberg, D. E., and Barton, A. M. (1992) Patterns and consequences of interspecific competition in natural communities: A review of field experiments with plants. *American Naturalist*, 139, 771–801.

Goldberg, D. E., and Landa, K. (1991) Competitive effect and response: hierarchies and correlated traits in the early stages of competition. *Journal of Ecology*, 79, 1013–1030.

Goldberg, D. E., and Werner, P. A. (1983) The equivalence of competitors in plant communities: A null hypothesis and a field experimental approach. *American Journal of Botany*, 70, 1098–1104.

Gonzalez, M. (1992) Effects of experimental acidification on zooplankton populations: A multiple scale approach. PhD thesis, University of Wisconsin, Madison.

Gonzalez, M., Frost, T. M., and Montz, P. (1990) Effects of experimental acidification on rotifer population dynamics in Little Rock Lake, Wisconsin, USA. *Verhandlungen der Internationale Vereinigung fur Theoretische und Angewandte Limnologie*, 24, 449–456.

Goodall, D. W. (1978) Sample similarity and species correlation. In *Ordination of Plant Communities* (ed. R. H. Whittaker). Junk, The Hague, pp. 99–149.

Goodman, D. (1987) The demography of chance extinction. In *Viable Populations for Conservation* (ed. M. E. Soule). Cambridge University Press, New York, pp. 11–34.

Gould, S. J. (1986) Evolution and the triumph of homology, or why history matters. *American Scientist*, 74, 60–69.

Gower, J. C., and Legendre, P. (1986) Metric and Euclidean properties of dissimilarity coefficients. *Journal of Classification*, 3, 5–48.

Grant, P. R., and Grant, B. R. (1992) Demography and the genetically effective size of two populations of Darwin's finches. *Ecology*, 73, 766–784.

Gray, R. D. (1987) Faith and foraging: A critique of the "paradigm argument for design". In *Foraging Behavior* (eds A. C. Kamil, J. R. Krebs, and H. R. Pulliam),.Plenum Press, New York, pp. 69–138.

Green, P. J., Jennison, C., and Seheult, A. H. (1985) Analysis of field experiments by least squares smoothing. *Journal of the Royal Statistical Society, Series B*, 47, 299–315.

Green, R. F. (1990) Putting ecology back into optimal foraging theory. *Comments Theoretical Biology*, 1, 387–410.

Greenhouse, S. W., and Geisser, S. (1959) On methods in the analysis of profile data. *Psychometrika*, 24, 95–112.

Grime, J. P. (1977) Evidence for the existence of three primary strategies in plants and its relevance to ecological and evolutionary theory. *American Naturalist*, 111, 1169–1194.

Grondona, M. O., and Cressie, N. (1991) Using spatial considerations in the analysis of experiments. *Technometrics*, 33, 381–392.

Grondona, M. O., and Cressie, N. (1993) Efficiency of block designs under stationary second-order autoregressive errors. *Sankhya A*, 55 (in press).

Gross, K. L., and Werner, P. A. (1982) Colonizing abilities of "biennial" plant species in relation to ground cover: Implications for their distributions in a successional sere. *Ecology*, 63, 921–931.

Gurevitch, J. (1986) Competition and the local distribution of the grass *Stipa neomexicana*. *Ecology*, 67, 46–57.

Gurevitch, J., and Chester, S. T. J. (1986) Analysis of repeated experiments. *Ecology*, 67, 251–254.

Gurevitch, J., Morrow, L. L., Wallace, A., and Walsh, J. S. (1992) A meta-analysis of field experiments on competition. *American Naturalist*, 140, 539–572.

Haber, R. N., and Wilkinson, L. (1982) Perceptual components of computer displays. *IEEE Computer Graphics and Applications*, 2(3), 23–35.

Hairston, N. G., Sr. (1980) The experimental test of an analysis of field distributions: Competition in terrestrial salamanders. *Ecology*, 61, 817–8226.

Hairston, N. G., Sr. (1989) *Ecological Experiments: Purpose, Design, and Execution*. Cambridge University Press, Cambridge.

Hairston, N. G., Sr., Smith, F. E., and Slobodkin, L. B. (1960) Community structure, population control, and competition. *American Naturalist*, 94, 421–425.

Hall, P., and Wilson, S. R. (1991) Two guidelines for bootstrap hypothesis testing. *Biometrics*, 47, 757–762.

Hand, D. J., and Taylor, C. C. (1987) *Multivariate Analysis of Variance and Repeated Measures*. Chapman and Hall, London.

Harper, J. L. (1977) *Population Biology of Plants*. Academic Press, New York.

Harris, R. J. (1985) *A Primer of Multivariate Statistics*. Academic Press, New York.

Hartley, H. O., and Searle, S. R. (1969) A discontinuity in mixed model analyses. *Biometrics*, 25, 573–576.

Harvey, L. E., Davis, F. W., and Gale, N. (1988) The analysis of class dispersion patterns using matrix comparisons. *Ecology*, 69, 537–542.

Hassell, M. P. (1978) *The Dynamics of Arthropod Predator-Prey Systems*. Princeton University Press, Princeton, NJ.

Hassell, M. P., Lawton, J. H., and Beddington, J. R. (1977) Sigmoid functional responses by invertebrate predators and parasitoids. *Journal of Animal Ecology*, 46, 249–262.

Hayduk, L. A. (1987) *Structural Equation Modeling with LISREL. Essentials and Advances*. Johns Hopkins University Press, Baltimore.

Hedges, L. V. (1983) A random effects model for effect sizes. *Psychological Bulletin*, 93, 388–395.

Hedges, L. V., and Olkin, I. (1985) *Statistical Methods for Meta-analysis*. Academic Press, New York.

Heltshe, J. F. (1988) Jackknife estimate of the matching coefficient of similarity. *Biometrics*, 44, 447–460.

Heltshe, J. F., and Forrester, N. E. (1985) Statistical evaluation of the jackknife estimate of diversity when using quadrat samples. *Ecology*, 66, 107–111.

Herr, D. G. (1986) On the history of ANOVA in unbalanced, factorial design: The first 30 years. *American Statistician*, 40, 265–270.

Heyde, C. C., and Cohen, J. E. (1985) Confidence intervals for demographic projections based on products of random matrices. *Theoretical Population Biology*, 27, 120–153.

Heywood, J. S. (1991) Spatial analysis of genetic variation in plant populations. *Annual Reviews of Ecology and Systematics*, 22, 335–355.

Hicks, C. R. (1982) *Fundamental Concepts in the Design of Experiments*, 3rd ed. Holt, Rinehart & Winston, New York.

Hocking, R. R. (1985) *The Analysis of Linear Models*. Brooks/Cole, Belmont, CA.

Holling, C. S. (1966) The functional response of invertebrate predators to prey density. *Memoirs of the Entomological Society of Canada*, 48, 1–87.

Holm, S. (1979) A simple sequentially rejective multiple test procedure. *Scandinavian Journal of Statistics*, 6, 65–70.

Holt, R. D. (1977) Predation, apparent competition, and the structure of prey communities. *Theoretical Population Biology*, 12, 197–229.

Holt, R. D. (1985) Population dynamics in two-patch environments: Some anomalous consequences of an optimal habitat distribution. *Theoretical Population Biology*, 28, 181–208.

Holt, R. D. (1987) Population dynamics and evolutionary processes: The manifold roles of habitat selection. *Evolutionary Ecology*, 1, 331–347.

Holt, R. D., and Kotler, B. P. (1987) Short-term apparent competition. *American Naturalist*, 130, 412–430.

Horn, J. S., Shugart, H. H., and Urban, D. L. (1989) Simulators as models of forest dynamics. In *Perspectives in Ecology Theory* (eds J. Roughgarden, R. M. May, and S. A. Levin). Princeton University Press, Princeton, NJ, pp. 256–267.

Horton, D. R., Chapman, P. L., and Capinera, J. L. (1991) Detecting local adaptation in phytophagus insects using repeated measures designs. *Environmental Entomology*, 20, 410–418.

Horvitz, C. C., and Schemske, D. W. (1986) Seed dispersal and environmental heterogenetiy in a neotropical herb: A model of population and patch dynamics. In *Frugivores and seed dispersal* (eds A. Estrada and T. H. Fleming). Dr. W. Junk Publishers, Dordrecht, pp. 169–186.

Houck, M. A., and Strauss, R. E. (1985) The comparative study of functional responses: Experimental design and statistical interpretation. *Canadian Entomologist*, 117, 617–629.

Howe, H. F., and Westley, L. C. (1988) *Ecological Relationships of Plants and Animals*. Oxford University Press, New York.

Hubert, L. J. (1987) *Assignment Methods in Combinatorial Data Analysis*. Marcel Dekker, New York.

Huberty, C. J., and Morris, J. D. (1989) Multivariate analysis versus multiple univariate analyses. *Psycological Bulletin*, 105, 302–308.

Hunt, R. L. (1976) A long-term evaluation of trout habitat development and its relation to improving management-oriented research. *Transactions of the American Fisheries Society*, 105, 361–364.

Hunter, J. E., Schmidt, F. L., and Jackson, G. B. (1982) *Meta-analysis: Cumulating Research Findings Across Studies*. Sage Publications, Beverly Hills, CA.

Hurlbert, S. H. (1984) Pseudoreplication and the design of ecological field experiments. *Ecological Monographs*, 54, 187–211.

Hutchings, M. J., Booth, K. D., and Waite, S. (1991) Comparison of survivorship by the logrank test: Criticisms and alternatives. *Ecology*, 72, 2290–2293.

Huynh, H., and Feldt, L. S. (1970) Conditions under which mean square ratios in repeated measurement designs have exact F-distributions. *Journal of the American Statistical Association*, 65, 1582–1589.

Huynh, H., and Feldt, L. S. (1976) Estimation of the Box correction for degrees of freedom from sample data in the randomized block and split-plot designs. *Journal of Educational Statistics*, 1, 69–82.

Huynh, H., and Mandeville, G. K. (1979) Validity conditions in repeated measures designs. *Psychological Bulletin*, 86, 964–973.

Hyde, J. S., and Linn, M. C. (1986) *The Psychology of Gender: Advances Through Meta-analysis*. Johns Hopkins University Press, Baltimore.

Ivlev, V. S. (1961) *The Experimental Ecology of the Feeding of Fishes*. Yale University Press, New Haven, CT.

Iwasa, Y. M., Higashi, M., and Yamamura, N. (1981) Prey distribution as a factor determining the choice of optimal foraging strategy. *American Naturalist*, 117, 710–723.

Jaccard, J., Turrisi, R., and Wan, C. K. (1990) *Interaction Effects in Multiple Regression*. Sage Publications, Newbury Park.

Jaccard, P. (1901) Distribution de la flore alpine dans le Bassin des Dranes et dans quelques regions voisines. *Bulletin Societe Vaudoise des Sciences Naturelles*, 37, 241–272.

Jackson, J. B. C. (1981) Interspecific competition and species' distributions: The ghosts of theories and data past. *American Zoologist*, 21, 889–901.

James, F. C., and McCulloch, C. E. (1990) Multivariate analysis in ecology and systematics: Panacea or Pandora's box? *Annual Review of Ecology and Systematics*, 21, 129–166.

Jassby, A. D., and Powell, T. M. (1990) Detecting changes in ecological time series. *Ecology*, 71, 2044–2052.

Jenkins, G. M. (1979) *Practical Experiences with Modelling and Forecasting Time Series*. St. Helier: Gwylim Jenkins and Partners (Overseas), Ltd.

Jennrich, R. I., and Schluchter, M. D. (1986) Unbalanced repeated-measures models with structured covariance matrices. *Biometrics*, 42, 805–820.

Johnson, C. R., and Mann, K. H. (1988) Diversity, patterns of adaptation, and stability of Nova Scotian kelp beds. *Ecological Monographs*, 58, 129–154.

Johnson, M. L., Huggins, D. G., and DeNoyelles, F. J. (1991) Ecosystem modeling with LISREL: a new approach for measuring direct and indirect effects. *Ecological Applications*, 1, 383–398.

Johnson, R. A., and Wichern, D. W. (1988) *Applied Multivariate Statistical Analysis*. 2nd ed. Prentice Hall, Englewood Cliffs, NJ.

Jones, D. (1984) Use, misuse, and role of multiple-comparison procedures in ecological and agricultural entomology. *Environmental Entomology*, 13, 635–649.

Jones, D., and Matloff, N. (1986) Statistical hypothesis testing in biology: A contradiction in terms. *Journal of Economic Entomology*, 79, 1156–1160.

Jones, R. H. (1980) Maximum likelihood fitting of ARMA models to time series with missing observations. *Technometrics*, 22, 389–395.

Jöreskog, K. G., and Sörbom, D. (1988) *LISREL 7. A Guide to the Program and Applications*, Scientific Software, Mooresville, IL.

Journel, A. G., and Hiujbregts, C. J. (1978) *Mining Geostatistics*. Academic Press, London.

Juliano, S. A. (1988) Chrysomelid beetles on water lily leaves: Herbivore density, leaf survival, and herbivore maturation. *Ecology*, 69, 1294–1298.

Juliano, S. A., and Lawton, J. H. (1990) The relationship between competition and morphology II. Experiments on co-occurring dytiscid beetles. *Journal of Animal Ecology*, 59, 831–848.

Juliano, S. A., and Williams, F. M. (1985) On the evolution of handling time. *Evolution*, 39, 212–215.

Juliano, S. A., and Williams, F. M. (1987) A comparison of methods for estimating the functional response parameters of the random predator equation. *Journal of Animal Ecology*, 56, 641–653.

Kalbfleisch, J. D., and Prentice, R. L. (1980) *The Statistical Analysis of Failure Time Data*. Wiley, New York.

Kalisz, S. (1991) Experimental determination of seed bank age structure in the winter annual *Collinsia verna*. *Ecology*, 72, 575–585.

Kalisz, S., and McPeek, M. A. (1992) Demography of an age-structured annual: Resampled projection matrices, elasticity analyses and seed bank effects. *Ecology*, 73, 1082–1093.

Karlin, S., Cameron, E. C., and Chakraborty, R. (1983) Path analysis in genetic epidemiology: A critique. *American Journal of Human Genetics*, 35, 695–732.

Kempton, R. A., and Howes, C. W. (1981) The use of neighboring plot values in the analysis of variety trials. *Applied Statistics*, 30, 59–70.

Kenny, D. A., and Judd, C. M. (1986) Consequences of violating the independence assumption in analysis of variance. *Psychological Bulletin*, 99, 422–431.

Keppel, G. (1991) *Design and Analysis: A Researcher's Handbook*. Prentice-Hall, Englewood Cliffs, NJ.

Kiefer, J. (1975) Construction and optimality of generalized Youden designs. In *A Survey of Statistical Design and Linear Models* (ed. J. N. Srivastava). North-Holland, Amsterdam, pp. 333–353.

Kiefer, J., and Wynn, H. P. (1981) Optimum balanced block and latin square designs for correlated observations. *Annals of Statistics*, 9, 737–757.

King, R. B., and King, B. (1991) Sexual differences in color and color change in wood frogs. *Canadian Journal of Zoology*, 69, 1963–1968.

Kingsolver, J. G., and Schemske, D. G. (1991) Analyzing selection with path analysis. *Trends in Ecology and Evolution*, 6, 276–280.

Koch, G. G., Amara, I. A., Stokes, M. E., and Gillings, D. B. (1980) Some views on parametric and non-parametric analyses for repeated measurements and selected bibliography. *International Statistical Review*, 48, 249–265.

Kohn, J. R., and Waser, N. M. (1986) The effect of *Delphinium nelsonii* pollen on seed set in *Ipomopsis aggregata*, a competitor for hummingbird pollination. *American Journal of Botany*, 72, 1144–1148.

Kokoska, S. M. and Johnson, L. B. (1987) A comparison of statistical techniques for analysis of growth curves. *Growth*, 51, 261–269.

Krannitz, P. G., Aarssen, L. W., and Dow, J. M. (1991) The effect of genetically based differences on seedling survival in *Arabidopsis thaliana*. *American Journal of Botany*, 78, 446–449.

Kucera, C. L. and Martin, S. C. (1957) Vegetation and soil relationships in the glade region of the southwestern Missouri Ozarks. *Ecology*, 38, 285–291.

Laird, N. M., and Ware, J. H. (1982) Random effects models for longitudinal data. *Biometrics*, 38, 963–974.

Lakatos, I. (1974) Popper on demarcation and induction. In *The Philosophy of Karl Popper* (ed. P. A. Schilpp). Open Court Publishing Company, La Salle, IL. pp. 241–273.

Lakatos, I. (1978) *The Methodology of Scientific Research Programmes.* Cambridge University Press, New York.

Lande, R. (1982) A quantitative genetic theory of life history evolution. *Ecology*, 63, 607–615.

Lande, R. (1987) Extinction thresholds in demographic models of territorial populations. *American Naturalist*, 130, 624–635.

Lande, R. (1988) Demographic models of the northern spotted owl (*Strix occidentalis caurina*). *Oecologia*, 75, 601–607.

Lande, R., and Arnold, S. J. (1983) The measurement of selection on correlated characters. *Evolution*, 37, 1210–1226.

Lanyon, S. M. (1987) Jackknifing and bootstrapping: Important "new" statistical techniques for ornithologists. *The Auk*, 104, 144–146.

Lauter, J. (1978) Sample size requirements for the T^2 test of MANOVA (Tables for one-way classification). *Biometrical Journal*, 20, 389–406.

Law, R. (1983) A model for the dynamics of a plant population containing individuals classified by age and size. *Ecology*, 64, 224–230.

Law, R., and Watkinson, A. R. (1987) Response-surface analysis of two-species competition: An experiment on *Phleum arenarium* and *Vulpia fasciculata*. *Journal of Ecology*, 75, 871–886.

Lawless, J. F. (1982) *Statistical Models and Methods for Lifetime Data.* Wiley, New York.

Lee, C.-S., and Rawlings, J. O. (1982) Design of experiments in growth chambers. Uniformity trials in the North Carolina State University phytotron. *Crop Science*, 22, 551–558.

Lee, E. T. (1980) *Statistical Methods for Survival Data Analysis.* Lifetime Learning, Belmont, CA.

Lefkovitch, L. P. (1965) The study of population growth in organisms grouped by stages. *Biometrics*, 21, 1–18.

Legendre, L., and Legendre, P. (1983) *Numerical Ecology. Developments in Environmental Modelling, 3.* Elsevier, Amsterdam.

Legendre, P. (1993) Spatial autocorrelation: Trouble or new paradigm? *Ecology*, 74 (in press).

Legendre, P., and Fortin, M.-J. (1989) Spatial pattern and ecological analysis. *Vegetatio*, 80, 107–138.

Legendre, P., Sokal, R. R., Oden, N. L., Vaudor, A., and Kim, J. (1990) Analysis of variance with spatial autocorrelation in both the variable and the classification criterion. *Journal of Classification*, 7, 53–75.

Legendre, P., and Troussellier, M. (1988) Aquatic heterotrophic bacteria: Modeling in the presence of spatial autocorrelation. *Limnology and Oceanography*, 33, 1055–1067.

Legendre, P., and Vaudor, A. (1991) *The R Package: Multidimensional Analysis, Spatial Analysis*. Departement de sciences biologiques, Université de Montreal, Quebec.

Lenski, R. E., and Service, P. M. (1982) The statistical analysis of population growth rates calculated from schedules for survivorship and fecundity. *Ecology*, 63, 655–662.

Leslie, P. H. (1945) On the use of matrices in population mathematics. *Biometrika*, 33, 183–212.

Leslie, P. H. (1948) Some further notes on the use of matrices in population mathematics. *Biometrika*, 35, 213–245.

Levins, R. (1969) Some demographic and genetic consequences of environmental heterogeneity for biological control. *Bulletin of the Entomological Society of America*, 15, 237–240.

Lewontin, R. C. (1965) Selection for colonizing ability. In *The Genetics of Colonizing Species* (eds H. G. Baker and G. L. Stebbins). Academic Press, New York, pp. 77–94.

Li, C. C. (1975) *Path Analysis. A Primer*. Boxwood Press, Pacific Grove, CA.

Light, R. J., and Pillemer, D. B. (1984) *Summing Up: The Science of Reviewing Research*. Harvard University Press, Cambridge, MA.

Livdahl, T. P. (1979) Evolution of handling time: The functional response of a predator to the density of sympatric and allopatric strains of prey. *Evolution*, 33, 765–768.

Livdahl, T. P., and Stiven, A. E. (1983) Statistical difficulties in the analysis of functional response data. *Canadian Entomologist*, 115, 1365–1370.

Ljung, G. M., and Box, G. E. P. (1978) On a measure of lack of fit in time series models. *Biometrika*, 65, 297.

Loehlin, J. C. (1987) *Latent Variable Models*. Lawrence Erlbaum, Hillsdale, NJ.

Long, J. S. (1983) *Covariance Structure Models. An Introduction to LISREL*. Sage Press, Beverly Hills, CA.

Ludwig, D., and Rowe, L. (1990) Life-history strategies for energy gain and predator avoidance under time constraints. *American Naturalist*, 135, 686–707.

Ludwig, J. A., and Reynolds, J. F. (1988) *Statistical Ecology*. Wiley, New York.

MacCullum, R. (1986) Specification searches in covariance structure modeling. *Psychological Bulletin*, 100, 107–120.

Maddox, G. D., and Antonovics, J. (1983) Experimental ecological genetics in *Plantago*: A structural equation approach to fitness components in *P. aristata* and *P. patagonica*. *Ecology*, 64, 1092–1099.

Madenjian, D. P., Jude, D. J., and Tesar, F. J. (1986) Intervention analysis of power plant impact on fish populations. *Canadian Journal of Fisheries and Aquatic Sciences*, 43, 819–829.

MacIsaac, H. J., Hutchinson, T. C., and Keller, W. (1987) Analysis of planktonic rotifer assemblages form Sudbury, Ontario, area lakes of varying chemical composition. *Canadian Journal of Fisheries and Aquatic Sciences*, 44, 1692–1701.

Magurran, A. E. (1988) *Ecological Diversity and Its Measurement*. Princeton University Press, Princeton, NJ.

Mangel, M. (1991) Adaptive walks on behavioural landscapes and the evolution of optimal behaviour by natural selection. *Evolutionary Ecology*, 5, 30–39.

Manly, B. F. J. (1986a) Randomization and regression methods for testing for associations with geographical, environmental, and biological distances between populations. *Researches in Population Ecology*, 28, 201–218.

Manly, B. F. J. (1986b) *Multivariate Statistical Methods: A Primer*. Chapman and Hall, London.

Manly, B. F. J. (1990a) On the statistical analysis of niche data. *Canadian Journal of Zoology*, 68, 1420–1422.

Manly, B. F. J. (1990b) *Stage-Structured Populations*. Chapman and Hall, London.

Manly, B. F. J. (1991) *Randomization and Monte Carlo Methods in Biology*. Chapman and Hall, London.

Mantel, N. (1967) The detection of disease clustering and a generalized regression approach. *Cancer Research*, 27, 209–220.

Mantel, N., and Haenszel, W. (1959) Statistical analysis of the analysis of data from retrospective studies of disease. *Journal of the National Cancer Institute*, 22, 719–748.

Matheron, G. (1963) Principles of geostatistics. *Economic Geology*, 58, 1246–1266.

Matloff, N. S. (1991) Statistical hypothesis testing: problems and alternatives. *Environmental Entomology*, 20, 1246–1250.

Matson, P. A., and Carpenter, S. R. (1990) Statistical analysis of ecological responses to large-scale perturbations. *Ecology*, 71, 2037.

Maxwell, S. E., and Delaney, H. D. (1990) *Designing Experiments and Analyzing Data: A Model Comparison Perspective*. Wadsworth, Belmont, CA.

McCleary, R., and Hay, R. A. (1980) *Applied Time Series Analysis for the Social Sciences*. Sage Publications, Beverly Hills, CA.

McCreary, N. J., Carpenter, S. R., and Chaney, J. E. (1983) Coexistence and interference in two submerged freshwater perennial plants. *Oecologia*, 59, 393–396.

McGill, R., Tukey, J. W., and Larsen, W. A. (1978) Variations of box plots. *American Statistician*, 32, 12–16.

McGraw, J. B., and Chapin, F. S., III (1989) Competitive ability and adaptation to fertile and infertile soils in two *Eriophorum* species. *Ecology*, 70, 736–749.

McIntosh, R. P. (1985) *The Background of Ecology*. Cambridge University Press, Cambridge.

McLean, R. A., Sanders, W. L., and Stroup, W. W. (1991) A unified approach to mixed linear models. *American Statistician*, 45, 54–64.

McNamara, J. M., and Houston, A. I. (1987) Foraging in patches: There's more to life than the marginal value theorem. In *Quantitative Analyses of Behavior, Volume 6: Foraging* (eds M. L. Commons, A. Kacelnik, and S. J. Shettleworth). Lawrence Erlbaum, Hillsdale, NJ, pp. 23–39.

McPeek, M. A., and Holt, R. D. (1992) The evolution of dispersal in spatially and temporally varying environments. *American Naturalist*, 140, 1010–1027.

Mead, R. (1988) *The Design of Experiments: Statistical Principles for Practical Application*. Cambridge University Press, Cambridge.

Mead, R., and Curnow, R. N. (1983) *Statistical Methods in Agricultural and Experimental Biology*. Chapman and Hall, London.

Meagher, T. R., and Antonovics, J. (1982) The population biology of *Chamaelirium luteum*, a dioeceous member of the lily family: Life history studies. *Ecology*, 63, 1690–1700.

Menges, E. (1986) Predicting the future of rare plant populations: Demographic monitoring and modeling. *Natural Areas Journal*, 6, 13–25.

Meyer, J. S., Ingersoll, C. G., McDonald, L. L., and Boyce, M. S. (1986) Estimating uncertainty in population growth rates: Jackknife vs. Bootstrap techniques. *Ecology*, 67, 1156–1166.

Mielke, P. W., Berry, K. J., Brockwell, P. J., and Williams, J. S. (1981) A class of nonparametric tests based on multiresponse permutation procedures. *Biometrika*, 68, 720–724.

Miller, R. G. (1974) The jackknife—a review. *Biometrika*, 61, 1–17.

Miller, T. E., and Kerfoot, W. C. (1987) Redefining indirect effects. In *Predation: Direct and Indirect Effects on Aquatic Communities* (eds W. C. Kerfoot and A. Sih). University Press of New England, Hanover, pp. 33–37.

Miller, T. E., and Werner, P. A. (1987) Competitive effects and responses between plant species in a first-year old-field community. *Ecology*, 68, 1201–1210.

Milliken, G. A., and Johnson, D. E. (1984) *Analaysis of Messy Data. Volume 1: Designed Experiments*. Lifetime Learning, Belmont, CA.

Mitchell, R. J. (1992) Testing evolutionary and ecological hypotheses using path analysis and structural equation modeling. *Functional Ecology*, 6, 123–129.

Mitchell, R. J. (1993) Adaptive significance of *Ipomopsis aggregata* nectar production rate: observational and experimental studies in the field. *Evolution* 47, 25–35.

Mitchell, R. J., and Waser, N. M. (1992) Adaptive significance of *Ipomopsis aggregata* nectar production rate: Pollination success of single flowers. *Ecology*, 73, 633–638.

Mitchell, W. A. (1990) An optimal control theory of diet selection: the effects of resource depletion and exploitative competition. *Oikos*, 58, 16–24.

Mitchell, W. A., and Brown, J. S. (1990) Density-dependent harvest rates by optimal foragers. *Oikos*, 57, 180–190.

Mitchell, W. A., and Valone T, J. (1990) The optimization research program: Studying adaptations by their function. *Quarterly Review of Biology*, 65, 43–52.

Mitchell-Olds, T. (1987) Analysis of local variation in plant size. *Ecology*, 68, 82–87.

Mitchell-Olds, T., and Shaw, R. G. (1987) Regression analysis of natural selection: Statistical inference and biological interpretation. *Evolution*, 41, 1149–1161.

Moloney, K. A. (1986) A generalized algorithm for determing category size. *Oecologia*, 69, 176–180.

Morrison, D. F. (1990) *Multivariate Statistical Methods*, 3rd ed. McGraw-Hill, New York.

Mueller, L. D. (1979) A comparison of two methods for making statistical inferences on Nei's measure of genetic distance. *Biometrics*, 35, 757–763.

Muenchow, G. (1986) Ecological use of failure time analysis. *Ecology*, 67, 246–250.

Muir, W. M. (1986) Estimation of response to selection and utilization of control populations for additional information and accuracy. *Biometrics*, 42, 381–391.

Muthen, B., and Kaplan, D. (1985) A comparison of some methodologies for the factor analysis of non-normal Likert variables. *British Journal of Mathematical and Statistical Psychology*, 38, 171–193.

Neter, J., and Wasserman, W. (1974) *Applied Linear Statistical Models*. Richard S Irwin, Homewood.

Neter, J., Wasserman, W., and Kutner, M. H. (1985) *Applied Linear Statistical Models*. Richard S Irwin, Homewood.

Noakes, D. (1986) Quantifying changes in British Columbia Dungeness crab landings using intervention analysis. *Canadian Journal of Fisheries and Aquatic Sciences*, 43, 634–639.

Norusis, M. J. (1990) *SPSS Advanced Statistics User's Guide*. SPSS Inc., Chicago.

O'Brien, R. G., and Kaiser, M. K. (1985) MANOVA method for analyzing repeated measures designs: An extensive primer. *Psychological Bulletin*, 97, 316–333.

Oden, N. L., and Sokal, R. R. (1992) An investigation of three-matrix permutation tests. *Journal of Classification*, 9, 275–280.

Pacala, S. W., and Silander, J. A. (1990) Field tests of neighborhood population dynamic models of two annual weed species. *Ecological Monographs*, 60, 113–134.

Pallesen, L., Berthouex, P. M., and Booman, K. (1985) Environmental intervention analysis: Wisconsin's ban on phosphate detergents. *Water Research*, 19, 353–362.

Parker, G. A., and Maynard Smith, J. (1990) Optimality theory in evolutionary biology. *Nature* (London), 348, 27–33.

Pedhazur, E. J. (1982) *Multiple Regression in Behavioral Research*, 2nd ed. Holt, Rinehart & Winston, New York.

Peters, R. H. (1991) *A Critique for Ecology*. Cambridge University Press, Cambridge.

Peto, R., and Peto, J. (1972) Asymptotically efficient rank invariant procedures (with discussion). *Journal of the Royal Statistical Society A*, 135, 185–206.

Peto, R., and Pike, M. C. (1973) Conservatism of the approximation $(O - E)^2/E$ in the log rank test for survival data or tumor incidence data. *Biometrics*, 29, 579–584.

Pianka, E. R. (1966) Latitudinal gradients in species diversity: A review of concepts. *American Naturalist*, 100, 65–75.

Platt, J. R. (1964) Strong inference. *Science*, 146, 347–353.

Platt, W. J., and Weis, I. M. (1985) An experimental study of competition among fugative prairie plants. *Ecology*, 66, 708–720.

Pons, T. L., and van der Toorn, J. (1988) Establishment of *Plantago lanceolata* L. and *Plantago major* L. among grass. I. Significance of light for germination. *Oecologia*, 75, 394–399.

Poole, R. W. (1978) The statistical prediction of population fluctuations. *Annual Review of Ecology and Systematics*, 9, 427–448.

Poorter, H., and Remkes, C. (1990) Leaf area ratio and net assimilation rate of 24 wild species differing in relative growth rate. *Oecologia*, 83, 553–559.

Popper, K. R. (1959) *The Logic of Scientific Discovery*. Hutchinson, London.

Porter, K. G. G. J., and Orcutt, J. D. J. (1982) The effect of food concentration on swimming patterns, feeding behaviour, ingestion, assimilation and respiration by *Daphnia*. *Limnology and Oceanography*, 27, 935–949.

Porter, W. R., and Trager, W. F. (1977) Improved nonparametric statistical methods for estimation of Michaelis-Menten enzyme kinetics parameters by the direct linear plot. *Biochemical Journal*, 161, 293–302.

Potvin, C., Lechowicz, M. J., Bell, G., and Schoen, D. (1990a) Spatial, temporal, and species-specific pattern of heterogeneity in growth chamber experiments. *Canadian Journal of Botonay*, 68, 499–504.

Potvin, C., Lechowicz, M. J., and Tardif, S. (1990b) The statistical analysis of ecophysiological response curves obtained from experiments involving repeated measures. *Ecology*, 71, 1389–1400.

Potvin, C., and Roff, D. A. (1993) Distribution-free and robust statistical methods: Viable alternatives to parametric statistics? *Ecology*, 74 (in press).

Potvin, C., and Tardif, S. (1988) Sources of variability and experimental design in growth chambers. *Functional Ecology*, 2, 123–130.

Pulliam, H. R. (1974) On the theory of optimal diets. *American Naturalist*, 108, 59–75.

Pulliam, H. R. (1975) Diet optimization with nutrient constraints. *American Naturalist*, 109, 765–768.

Pulliam, H. R. (1988) Sources, sinks, and population regulation. *American Naturalist*, 132, 652–661.

Pyke, D. A., and Thompson, J. N. (1986) Statistical analysis of survival and removal rate experiments. *Ecology*, 67, 240–245. (Errata in *Ecology*, 68, 232.)

Pyke, G. H., Pulliam, H. R., and Charnov, E. L. (1977) Optimal foraging: A selective review of theory and tests. *Quarterly Review of Biology*, 52, 137–156.

Quinn, J. F., and Dunham, A. E. (1983) On hypothesis testing in ecology and evolution. *American Naturalist*, 122, 602–617.

Rapport, D. J. (1991) Myths in the foundations of economics and ecology. *Biological Journal of the Linnean Society*, 44, 185–202.

Raudenbush, S. W. (1991) Summarizing evidence: Crusaders for simplicity. *Educational Researcher*, 20(10), 33–37.

Reed, D. C., and Foster, M. S. (1984) The effects of canopy shading on algal recruitment and growth in a giant kelp forest. *Ecology*, 65, 937–948.

Rice, W. R. (1990) A consensus combined P-value test and the family-wide significance of component tests. *Biometrics*, 46, 303–308.

Rogers, D. J. (1972) Random search and insect population models. *Journal of Animal Ecology*, 41, 369–383.

Rosenzweig, M. L. (1981) A theory of habitat selection. *Ecology*, 62, 327–335.

Roughgarden, J. (1983) Competition and theory in community ecology. *American Naturalist*, 122, 583–601.

Rubin, D. B. (1981) A Bayesian bootstrap. *Annals of Statistics*, 9, 130–134.

Russo, R. (1986) Comparison of predatory behavior in five species of *Toxorhynchites* (Diptera: Culicidae). *Annals of the Entomological Society of America*, 79, 715–722.

Sacks, H. S., Berrier, J., Reitman, D., Ancona-Berk, V. A., and Chalmer, T. C. (1987) Meta-analyses of randomized controlled trials. *New England Journal of Medicine*, 316, 450–455.

Salsburg, D. S. (1985) The religion of statistics as practiced in medical journals. *American Statistician*, 39, 220–223.

Salt, G. W. (1983) Roles: Their limits and responsibilities in ecological and evolutionary research. *American Naturalist*, 122, 697–705.

SAS Institute Inc. (1988) *SAS/ETS User's Guide, Version 6, First Edition*. SAS Institute, Cary, NC.

SAS Institute Inc. (1989a) *SAS/STAT User's Guide, Version 6, Fourth Edition, Volume 1*. SAS Institute, Cary, NC.

SAS Institute Inc. (1989b) *SAS/STAT User's Guide, Version 6, Fourth Edition, Volume 2*. SAS Institute, Cary, NC.

SAS Institute Inc. (1990) *SAS Procedures Guide, Version 6, Third Edition*. SAS Institute, Cary, NC.

SAS Institute Inc. (1992) *SAS Technical Report P–229, SAS/STAT Software: Changes and Enhancements, Release 6.07*. SAS Institute, Cary, NC.

Satterthwaite, F. E. (1946) An approximate distribution of estimates of variance components. *Biometrics Bulletin*, 2, 110–114.

Saville, D. J. (1990) Multiple comparison procedures: The practical solution. *American Statistician*, 44, 174–180.

Schemske, D. W., and Horvitz, C. C. (1988) Plant-animal interactions and fruit production in a neotropical herb: A path analysis. *Ecology*, 69, 1128–1137.

Schenker, N. (1985) Qualms about bootstrap confidence intervals. *Journal of the American Statistical Association*, 80, 360–361.

Schindler, D. W. (1987) Detecting ecosystem responses to anthropogenic stress. *Canadian Journal of Fisheries and Aquatic Science*, 44(Suppl. 1), 6–25.

Schindler, D. W., Mills, K. H., Malley, D. F., Findlay, D. L., Shearer, J. A., Davies, I. J., Turner, M. A., Linsey, G. A., and Cruikhank, D. R. (1985) Long-term ecosystem stress: The effects of years of experimental acidification on a small lake. *Science*, 228, 1395–1401.

Schoener, T. W. (1983) Field experiments on interspecific competition. *American Naturalist*, 122, 240–285.

Schoener, T. W. (1986) Mechanistic approaches to community ecology: A new reductionism. *American Zoologist*, 26, 81–106.

Scott, S. J., and Jones, R. A. (1990) Generation means analysis of right-censored response-time traits: Low temperature seed germination in tomato. *Euphytica*, 48, 239–244.

Searle, S. R. (1971) *Linear Models*. Wiley, New York.

Searle, S. R. (1987) *Linear Models for Unbalanced Data*. Wiley, New York.

Sedinger, J. S., and Flint, P. L. (1991) Growth rate is negatively correlated with hatch date in Black Brant. *Ecology*, 72, 496–502.

Shaffer, M. (1987) Minimum viable populations: Coping with uncertainty. In *Viable Populations for Conservation* (ed. M. E. Soule). Cambridge University Press, New York, pp. 69–86.

Shaw, R. G. (1987) Maximum-likelihood approaches applied to quantitative genetics of natural populations. *Evolution*, 41, 812–826.

Shaw, R. G., and Mitchell-Olds, T. (1993) Anova for unbalanced data: An overview. *Ecology*, 74 (in press).

Shrader-Frechette, K. S., and McCoy, E. D. (1992) Statistics, costs and rationality in ecological inference. *Trends in Ecology and Evolution*, 7, 96–99.

Siegel, S. (1956) *Nonparametric Statistics for the Behavioral Sciences*. McGraw-Hill, New York.

Sih, A. (1980) Optimal behavior: Can foragers balance two conflicting demands? *Science*, 210, 1041–1043.

Silvertown, J. (1987) *Introduction to Plant Population Ecology*, 2nd ed. Longman, Harlow.

Simberloff, D. (1983) Competition theory, hypothesis-testing, and other community ecology buzzwords. *American Naturalist*, 122, 626–635.

Simes, R. J. (1986) An improved Bonferroni procedure for multiple tests of significance. *Biometrika*, 73, 751–754.

Simken, D., and Hastie, R. (1987) An information-processing analysis of graph perception. *Journal of the American Statistical Association*, 82, 454–465.

Simms, E. L., and Burdick, D. S. (1988) Profile analysis of variance as a tool for analyzing correlated responses in experimental ecology. *Biometrical Journal*, 30, 229–242.

Simms, E. L., and Rausher, M. D. (1992) Uses of quantitative genetics for studying the evolution of plant resistance. In *Plant Resistance to Herbivores and Pathogens: Ecology, Evolution, and Genetics* (eds R. S. Fritz and E. L. Simms). University of Chicago Press, IL, pp. 42–68.

Simpson, E. H. (1951) The interpretation of interaction in contingency tables. *American Statistician*, 13, 238–241.

Sjoberg, S. (1980) Zooplankton feeding and queueing theory. *Ecological Modelling*, 10, 215–225.

Slansky, F. J., and Feeny, P. (1977) Stabilization of the rate of nitrogen accumulation by larvae of the cabbage butterfly on wild and cultivated food plants. *Ecological Monographs*, 47, 209–228.

Slice, D. E. (1990) *DS-DIGIT: Basic Digitizing Software*. In *Proceedings of the Michigan Morphometrics Workshop, Special Publication 2* (eds J. F. Rohlf and F. L. Bookstein). University of Michigan Museum of Zoology, Ann Arbor, pp. 5–6.

Smith, E. P. (1985) Estimating the reliability of diet overlap measures. *Environmental Biology of Fishes*, 13, 125–138.

Smith, E. P., Genter, R. B., and Cairns, J. (1986) Confidence intervals for the similarity between algal communities. *Hydrobiologia*, 139, 237–245.

Smith, W. D., Kravitz, D., and Grassle, J. F. (1979) Confidence intervals for similarity measures using the two sample jackknife. In *Multivariate Methods in Ecological Work* (eds L. Orloci, C. R. Rao, and W. M. Stiteler). International Cooperative Publishing House, Fairland, MD, pp. 253–262.

Smouse, P. E., Long, J. C., and Sokal, R. R. (1986) Multiple regression and correlation extensions of the Mantel test of matrix correspondence. *Systematic Zoology*, 35, 627–633.

Snedecor, G. W., and Cochran, W. G. (1989) *Statistical Methods*. Iowa State University Press, Ames.

Sober, E. (1984) *The Nature of Selection: Evolutionary Theory in Philosophical Focus*. MIT Press, Cambridge, MA.

Sokal, R. R. (1979) Testing statistical significance of geographic variation patterns. *Systematic Zoology*, 28, 227–231.

Sokal, R. R., Lengyel, I. A., Derish, P. A., Wooten, M. C., and Oden, N. L. (1987) Spatial autocorrelation of ABO serotypes in medieval cemeteries as an indicator of ethnic and familial structure. *Journal of Archeological Science*, 14, 615–633.

Sokal, R. R., Oden, N. L., Legendre, P., Fortin, M.-J., Kim, J., Thomson, B. A., Vaudor, A., Harding, R. M., and Barbujani, G. (1990) Genetics and language in European populations. *American Naturalist*, 135, 157–175.

Sokal, R. R., Oden, N. L., Thomson, B. A., and Kim, J. (1993) Testing for regional differences in means: Distinguishing inherent from spurious spatial autocorrelation by restricted randomization. *Geographical Analysis*, 25, 199–210.

Sokal, R. R., and Rohlf, F. J. (1981) *Biometry*, 2nd ed. Freeman, New York.

Stephens, D. W., and Krebs, J. R. (1986) *Foraging Theory*. Princeton University Press, Princeton, NJ.

Stevens, J. (1992) *Applied Multivariate Statistics for the Social Sciences*. 2nd ed. Lawrence Erlbaum, Hillsdale, NJ.

Stewart-Oaten, A., Bence, J. R., and Osenberg, C. W. (1992) Assessing effects of unreplicated perturbations: No simple solutions. *Ecology*, 73, 1396–1404.

Stewart-Oaten, A., Murdoch, W. W., and Parker, K. R. (1986) Environmental impact assessment: "Pseudoreplication" in time. *Ecology*, 67, 929–940.

Stimson, J. (1985) The effect of shading by the table coral *Acropora lyacinthus* on understory corals. *Ecology*, 66, 40–53.

Strong, D. R. J., Lawton, J. H., and Southwood, R. (1984) *Insects on Plants*. Blackwell Scientific, Oxford.

Strong, D. R. J., Szyska, L. A., and Simberloff, D. S. (1979) Tests of community-wide character displacement against null hypotheses. *Evolution*, 33, 897–913.

Suppe, F. (1977) *The Structure of Scientific Theories*, 2nd ed. University of Illinois Press, Urbana.

Tamura, R. N., Nelson, L. A., and Naderman, G. C. (1988) An investigation on the validity and usefulness of trend analysis for field plot data. *Agronomy Journal*, 80, 712–718.

Tanaka, J. S. (1987) "How big is big enough?": Sample size and goodness of fit in structural equation models with latent variables. *Child Development*, 58, 134–146.

Taylor, B. E., and Mahoney, D. L. (1988) Extinction and recolonization: Processes regulating zooplankton dynamics in a cooling reservoir. *Verhandlungen der Internationale Vereinigung fur Theoretische und Angewandte Limnologie*, 23, 1536–1541.

Taylor, R. J. (1984) *Predation*. Chapman and Hall, New York.

Tessier, A. J., and Consolatti, N. L. (1991) Resource quantity and offspring quality in *Daphnia*. *Ecology*, 72, 468–478.

Thompson, D. J. (1975) Towards a predator-prey model incorporating age structure: The effects of predator and prey size on the predation of *Daphnia magna* by *Ischnura elegans*. *Journal of Animal Ecology*, 44, 907–916.

Tilman, D. (1987) The importance of the mechanisms of interspecific competition. *American Naturalist*, 129, 769–774.

Tilman, D. (1988) *Plant Strategies and the Dynamics and Structure of Plant Communities*. Princeton University Press, Princeton, NJ.

Toft, C. A., and Shea, P. J. (1983) Detecting community-wide patterns: Estimating power strengthens statistical inference. *American Naturalist*, 122, 618–625.

Trexler, J. C., McCulloch, C. E., and Travis, J. (1988) How can the functional response best be determined? *Oecologia*, 76, 206–214.

Trexler, J. C., and Travis, J. (1993) Nontraditional regression analysis. *Ecology*, 74 (in press).

Tritton, L. M., and Hornbeck, J. W. (1982) Biomass equations for major tree species of the northeast. *USDA Forest Service PL Northeastern Forest Experiment Station General Technical Report*, NE–69.

Tufte, E. R. (1983) *The Visual Display of Quantitative Information*. Graphics Press, Cheshire.

Tufte, E. R. (1990) *Envisioning Information*. Graphics Press, Cheshire.

Tukey, J. W. (1977) *Exploratory Data Analysis*. Addison-Wesley, Reading, MA.

Tuljapurkar, S. D. (1982a) Population dynamics in variable environments: II. Correlated environments, sensitivity analysis and dynamics. *Theoretical Population Biology*, 21, 114–140.

Tuljapurkar, S. D. (1982b) Population dynamics in variable environments: III. Evolutionary dynamics of r-selection. *Theoretical Population Biology*, 21, 141–165.

Tuljapurkar, S. (1989) An uncertain life: Demography in random environments. *Theoretical Population Biology*, 35, 227–294.

Tuljapurkar, S. D., and Orzack, S. H. (1980) Population dynamics in variable environments: I. Long-run growth rates and extinction. *Theoretical Population Biology*, 18, 314–342.

Turner, P. R. (1989) *Guide to Numerical Analysis*. CRC Press, Boca Raton, FL.

Turner, T. (1985) Stability of rocky intertidal surfgrass beds: Persistence, preemption and recovery. *Ecology*, 66, 83–92.

Upton, G. J. G., and Fingleton, B. (1985) *Spatial Data Analysis by Example. Volume 1: Point Pattern and Quantitative Data*. Wiley, New York.

Valone, T. J., and Brown, J. S. (1989) Measuring patch assessment abilities of desert granivores. *Ecology*, 70, 1800–1810.

Vandermeer, J. (1978) Choosing category size in a stage projection matrix. *Oecologia*, 32, 79–84.

van Es, H. M., van Es, C. L., and Cassel, D. K. (1989) Application of regionalized variable theory to large-plot field experiments. *Soil Science Society of America Journal*, 53, 1178–1183.

van Groenendael, J., and Slim, P. (1988) The contrasting dynamics of two populations of *Plantago lanceolata* classified by age and size. *Journal of Ecology*, 76, 585–599.

van Latesteijn, H. C., and Lambeck, R. H. D. (1986) The analysis of monitoring data with the aid of time-series analysis. *Environmental Monitoring and Assessment*, 7, 287–297.

van Lenteren, J. C., and Bakker, K. (1976) Functional responses in invertebrates. *Netherlands Journal of Zoology*, 26, 567–572.

Via, S. (1991) The genetic structure of host plant adaptation in a spatial patchwork: Demographic variability among reciprocally transplanted pea aphid clones. *Evolution*, 45, 827–852.

Wade, M. J. (1991) Genetic variance for rate of population increase in natural populations of flour beetles, *Tribolium* spp. *Evolution*, 45, 1574–1584.

Ward, D. (1992) The role of satisficing in foraging theory. *Oikos*, 63, 312–317.

Watras, C. J., and Frost, T. M. (1989) Little Rock Lake (Wisconsin): Perspectives on an experimental ecosystem approach to seepage lake acidification. *Archives of Environmental Contamination and Toxicology*, 18, 157–165.

Watt, A. S. (1947) Pattern and process in the plant community. *Journal of Ecology*, 35, 1–22.

Webster, R. (1985) Quantitative spatial analysis of soil in the field. *Advances in Soil Science*, 3, 1–70.

Wei, W. W. S. (1990) *Time Series Analysis: Univariate and Multivariate Methods.* Addison-Wesley, New York.

Weiner, J., and Solbrig, O. T. (1984) The meaning and measurement of size hierarchies in plant populations. *Oecologia*, 61, 334–336.

Weis, A. E., Wolfe, C. L., and Gorman, W. L. (1989) Genotypic variation and integration in histological features of the goldenrod ball gall. *American Journal of Botany*, 76, 1541–1550.

Weller, D. E. (1987) A reevaluation of the −3/2 power rule of self-thinning. *Ecological Monographs*, 57, 23–43.

Welsh, A. H., Peterson, A. T., and Altmann, S. A. (1988) The fallacy of averages. *American Naturalist*, 132, 277–288.

Werner, E. E., and Anholt, B. R. (1993) Ecological consequences of the tradeoff between growth and mortality rates mediated by foraging activity. *American Naturalist* 142, 242–272.

Werner, P. A., and Caswell, H. (1977) Population growth rates and age versus stage-distribution models of teasel (Dipsacus sylvestris Huds.). *Ecology*, 58, 1103–1111.

Wilkinson, G. N., Eckert, S. R., Hancock, T. W., and Mayo, O. (1983) Nearest neighbor (NN) analysis of field experiments. *Journal of the Royal Statistical Society, Series B*, 45, 152–212.

Wilkinson, L. (1988) *SYSTAT: The System for Statistics.* SYSTAT, Inc., Evanston, IL.

Wilkinson, L. (1990) *SYGRAPH: The System for Graphics.* SYSTAT, Inc., Evanston, IL.

Williams, F. M. (1980) On understanding predator-prey interactions. In *Contemporary Microbial Ecology* (eds D. C. Ellwood, J. N. Hedger, M. J. Latham, J. M. Lynch, and J. H. Slater). Academic Press, London, pp. 349–376.

Williams, F. M., and Juliano, S. A. (1985) Further difficulties in the analysis of functional-response experiments, and a resolution. *Canadian Entomologist*, 117, 631–640.

Wilson, S. D., and Tilman, D. (1991) Components of plant competition along an experimental gradient of nitrogen availability. *Ecology*, 72, 1050–1065.

Winer, B. J., Brown, D. R., and Michaels, K. M. (1991) *Statistical Principles in Experimental Design.* McGraw-Hill, New York.

Wolf, L. L., and Hainsworth, F. R. (1990) Non-random foraging by hummingbirds: Patterns of movement between *Ipomopsis aggregata* (Pursch) V. Grant inflorescences. *Functional Ecology*, 4, 149–158.

Woolfenden, G. E., and Fitzpatrick, J. W. (1984) *The Florida Scrub Jay.* Princeton University Press, Princeton, NJ.

Wootton, J. T. (1990) Direct and indirect effects of bird predation and excretion on the spatial and temporal patterns of intertidal species. PhD thesis, University of Washington, Seattle.

Wright, S. (1920) The relative importance of heredity and environment in determining the piebald pattern of guinea pigs. *Proceedings of the National Academy of Science*, U.S.A., 6, 320–332.

Wright, S. (1921) Correlation and causation. *Journal of Agricultural Research*, 20, 557–585.

Wright, S. (1934) The method of path coefficients. *Annals of Mathematics and Statistics*, 5, 161–215.

Wright, S. (1960) Path coefficients and path regression: Alternative or complementary concepts? *Biometrics*, 16, 189–202.

Wright, S. (1983) On "path analysis in genetic epidemiology: A critique." *American Journal of Human Genetics*, 35, 733–756.

Wu, C. F. J. (1986) Jackknife, bootstrap, and other resampling methods in regression analysis. *Annals of Statistics*, 14, 1261–1295.

Yates, F. (1938) The comparative advantages of systematic and randomized arrangements in the design of agricultural and biological experiments. *Biometrika*, 30, 444–466.

Yoccoz, N. G. (1991) Use, overuse, and misuse of significance tests in evolutionary biology and ecology. *Bulletin of the Ecological Society of America*, 72, 106–111.

Young, L. J., and Young, J. H. (1991) Alternative view of statistical hypothesis testing. *Environmental Entomology*, 20, 1241–1245.

Zar, J. H. (1984) *Biostatistical Analysis*, 2nd ed. Prentice-Hall, Englewood Cliffs, NJ.

Zerbe, G. O. (1979) Randomization analysis of the completely randomized design extended to growth and response curves. *Journal of the American Statistical Association*, 74, 215–221.

Zimmerman, D. L., and Harville, D. A. (1991) A random field approach to the analysis of field-plot experiments and other spatial experiments. *Biometrics*, 47, 223–239.

Zimmerman, G. M., Goetz, H., and Mielke, P. W. (1985) Use of an improved statistical method for group comparisons to study effects of prairie fire. *Ecology*, 66, 606–611.

Author Index

Aarssen, L. W. 264
Abrami, P. C. 383
Abrams, P. A. 159
Agresti, A. 133
Aiken, L. S. 189
Allen, T. F. H. 138, 140–1
Alonso, J. A. 146
Alonso, J. C. 146
Altman, D. G. 293
Altmann, S. A. 212
Amara, I. A. 133–4
Ancona-Berk, V. A. 378
Anholt, B. R. 70
Antonovics, J. 16, 223, 229, 233, 240
Arnold, S. J. 205, 229
Austin, M. P. 69
Ayala, F. J. 74
Ayres, M. P. 53

Baird, D. 323–4, 335
Baker, A. J. 294
Baker, F. B. 123
Baker, H. G. 229
Baker, L. A. 142, 155
Bakker, K. 161
Ball, F. M. 232
Barbujani, G. 344, 359
Bard, Y. 161, 168
Barker, B. M. 111
Barker, H. R. 111
Barlow, R. E. 89
Barnett, V. 27
Bartell, S. M. 138, 140–1
Bartha, S. 16

Bartholomew, D. J. 89
Bartlett, M. S. 335
Barton, A. M. 69–70, 75, 79, 397
Bautista, L. M. 146
Bayes, T. 9
Becker, R. A. 15
Beddington, J. R. 161, 164, 169
Bedford, B. L. 21, 31
Bell, G. 47–8, 50, 56
Belovsky, G. E. 239, 362
Bence, J. R. 140, 154
Benditt, J. 18
Bentler, P. M. 223
Berger, J. O. 9
Berrier, J. 378
Berry, K. J. 111
Berthouex, P. M. 146
Besag, J. E. 324, 335, 359
Bierzychudek, P. 233, 239, 249
Bollen, K. A. 212, 217, 222–4, 227–9
Bookstein, F. L. 228–9
Booman, K. 146
Booth, K. D. 262, 278
Bormann, F. H. 139
Box, G. E. P. 33, 123, 143–6, 148, 154, 156
Boyce, M. S. 294, 305, 308
Bradley, J. V. 344
Bray, J. H. 105, 111
Breckler, S. J. 222, 229
Bremer, J. M. 89
Brens M. D. 33, 35, 38
Brezonik, P. L. 142, 155
Brockwell, P. J. 111
Brody, A. K. 225

Brooks, D. R. 376
Brown, D. R. 47, 51–2, 55, 59, 64, 114, 116–7, 121, 123, 129, 131, 133–4
Brown, J. H. 79
Brown, J. S. 362, 368–9, 373
Brownlee, K. A. 91, 372
Brunk, H. D. 89
Buckland, S. T. 294, 303
Bull, J. J. 241
Buonaccorsi, J. P. 294
Burdick, D. S. 111
Burgman, M. A. 344
Burton, P. J. 388, 394

Cairns, J. 295, 305–6
Cameron, E. C. 229
Campbell, B. D. 70, 77
Campbell, D. R. 212, 215–6
Capinera, J. L. 132
Caraco, T. 364
Carpenter, S. R. 138–9, 147, 154, 319, 388, 394
Cassel, D. K. 343–4, 347
Cassini, M. H. 370–1
Caswell, H. 75, 232–5, 237–8, 240, 242, 245, 247, 251–2
Cerato, R. M. 133
Chakraborty, R. 229
Chalmer, T. C. 378
Chalmers, I. 378
Chambers, J. M. 15
Chaney, J. E. 388, 394
Chapin, F. S., III 70
Chapman, P. L. 132
Charlesworth, B. 251
Charnov, E. L. 214, 362, 364
Chatterjee, S. 184, 187, 190–1, 202, 205
Cheng, C. S. 335
Chester, S. T. J. 2, 114, 129, 134
Cheverud, J. M. 348
Cleveland, W. S. 14–8, 25, 40
Cliff, A. D. 342
Cloninger, C. R. 228–9
Cochran, W. G. 11, 47, 55, 102, 114, 324
Cochran-Stafira, D. L. 105
Cock, M. J. W. 164–5, 167
Cohen, J. 228, 381, 390
Cohen, J. E. 239–40
Cohen, P. 228
Cohen, P. A. 383
Coley, P. D. 183, 194, 197–8, 204

Collier, R. O. Jr. 123
Connell, J. H. 69, 81, 397
Connolly, J. 72, 74
Connor, E. F. 7
Consolatti, N. L. 240
Cooper, H. M. 378, 383
Cox, C. R. 33
Cox, G. M. 47, 55
Crespi, B. J. 228
Cressie, N. 232–4, 328–9, 331–5
Crouse, D. T. 232–3
Crowder, L. B. 232–3
Crowder, M. J. 116, 123, 132–4
Cruikhank, D. R. 154
Cryer, J. D. 143
Cullis, B. R. 323, 335
Curnow, R. N. 56, 60
Czárán, T. 16

d'Appollonia, S. 383
Davidson, D. W. 79
Davies, I. J. 154
Davis, F. W. 352
Day, R. W. 2, 9, 87, 324, 344
DeAngelis, D. L. 138, 140–1
de Kroon, H. 233, 238
Delaney, H. D. 54, 116–8, 123, 134–5
DeNoyelles, F. J. 223
Denslow, J. S. 33, 35, 38
Derish, P. A. 344, 348, 359
Diamond, J. M. 5, 7, 8
Digby, P. G. N. 2
Diggle, P. J. 143
Dixon, P. M. 300, 308
Dorfman, J. H. 294, 308
Douglas, M. E. 344
Dow, J. M. 264
Dow, M. M. 348
Drapeau, P. 344
Draper, N. R. 152, 184, 191
Dungan, M. L. 389, 394
Dunham, A. E. 8, 367
Dunning, J. B. 371
Dutilleul, P. 55, 57, 319
Dykhuizen, D. 240

Eaton, J. 142, 155
Eberhardt, L. L. 141, 147, 319
Eckert, S. R. 335, 343
Edwards, A. W. F. 267
Edwards, W. 9

Efron, B. 21, 25, 37, 245, 247, 291, 299–301, 303–5, 308–9
Ehrenfeld, J. G. 74
Elhourne, D. 378
Ellison, A. M. 13, 21, 27, 31, 33, 35, 38, 40
Endler, J. A. 187, 205, 240, 344
Enkin, M. 378
Epling, C. 232
Estabrook, G. F. 352
Evans, J. 295

Feeny, P. 205
Feinsinger, P. 133
Feldt, L. S. 116, 123
Findlay, D. L. 154
Fingleton, B. 342, 352
Firbank, L. G. 87
Fisher, R. A. 323
Fitzpatrick, J. W. 239
Flint, P. L. 178
Forrester, N. E. 294
Fortin, M.-J. 342–4, 346, 359
Foster, M. S. 388–9, 394
Fowler, N. L. 2, 11, 16, 33, 87, 388, 394
Fox, G. A. 256, 258, 264, 266, 284
Fraser, D. F. 362
Fretwell, S. D. 241
Freund, R. J. 184, 190
Frost, T. M. 138–42, 146–7, 154–5
Fry, J. D. 84
Futuyma, D. J. 184

Gaines, S. D. 2, 89
Gale, N. 352
Gardner, M. J. 293
Garrison, P. 142, 155
Garthwaite, P. H. 294
Gates, B. 352
Geisser, S. 123
Genter, R. B. 295, 305–6
Gibson, A. R. 294
Gilliam, J. F. 362
Gillings, D. B. 133–4
Gilpin, M. E. 7, 8, 74
Gittins, R. 111
Glantz, S. A. 161, 164–6, 168, 176
Glass, G. V. 378
Gleeson, A. C. 323, 335
Gnanadesikan, R. 24
Goetz, H. 111

Goldberg, D. E. 69–70, 72, 75, 78–9, 87, 397
Gonzalez, M. 142, 146, 154–5
Goodall, D. W. 290, 294
Goodman, D. 232, 238–9
Gorman, W. L. 229
Gould, S. J. 6
Gower, J. C. 351
Grant, B. R. 232
Grant, P. R. 232
Grassle, J. F. 305–6
Gray, R. D. 363, 365, 373
Green, P. J. 335
Green, R. F. 362, 365
Greenhouse, S. W. 123
Grime, J. P. 70, 77
Grondona, M. O. 232–4, 334–5
Gross, K. L. 388, 394
Gurevitch, J. 2, 16, 70, 75, 114, 129, 134, 348–9, 359, 378, 381, 384–6, 396

Haber, R. N. 21, 23
Haenszel, W. 277
Hainsworth, F. R. 214
Hairston, N. G., Sr. 1, 5, 8, 70, 183, 212
Hall, D. J. 138, 140–1
Hall, P. 294
Hancock, T. W. 335, 343
Hand, D. J. 102, 116, 123, 129, 132–4
Harding, R. M. 344, 359
Harper, J. L. 78
Harris, R. J. 96, 99–101, 103, 108, 111, 118, 120, 134, 216
Hartley, H. O. 53
Harvey, L. E. 352
Harville, D. A. 323–4, 332, 334–5
Hassell, M. P. 161–2, 164, 169
Hastie, R. 18
Hastie, T. J. 15
Hawkins, D. M. 334
Hay, R. A. 143, 146
Hayduk, L. A. 213, 217–8, 223–4, 228–9
Hayes, T. F. 123
Hedges, L. V. 378, 382–3, 385, 387, 391, 393
Heisey, D. 139, 147, 154
Heltshe, J. F. 294, 394, 305
Herr, D. G. 52
Hetherington, J. 378
Heyde, C. C. 240
Heywood, J. S. 241

Hicks, C. R. 324
Higashi, M. 362, 368
Hiujbregts, C. J. 329
Hocking, R. R. 84
Holling, C. S. 25, 159, 162
Holm, S. 249
Holt, R. D. 81, 241, 362
Horn, J. S. 25
Hornbeck, J. W. 25
Horton, D. R. 132
Horvitz, C. C. 213, 228–9, 233
Houck, M. A. 161, 163–4, 168
Houston, A. I. 362
Howe, H. F. !83
Howes, C. W. 335
Hubert, L. J. 343, 348
Huberty, C. J. 97
Huggins, D. G. 223
Hunt, R. L. 141
Hunter, J. E. 380
Hunter, J. S. 33, 148, 154
Hunter, W. G. 33, 148, 154
Hurlbert, S. H. 2, 55, 138, 140–1, 328, 335, 343
Hutchings, M. J. 262, 278
Hutchinson, T. C. 155
Huynh, H. 116, 123
Hyde, J. S. 378

Ingersoll, C. G. 294, 305, 308
Inouye, R. S. 79
Ivlev, V. S. 164
Iwasa, Y. M. 362, 368

Jaccard, J. 189
Jaccard, P. 294
Jackson, G. B. 380
Jackson, J. B. C. 69
James, F. C. 2, 3
Jassby, A. D. 139
Jenkins, G. M. 143–4, 146, 148, 156
Jennison, C. 335
Jennrich, R. I. 156
Johnson, C. R. 389, 394
Johnson, D. E. 51, 53–4, 60, 66, 84, 102
Johnson, L. B. 133
Johnson, M. L. 223
Johnson, R. A. 111
Jones, D. 9
Jones, R. A. 263
Jones, R. H. 153

Journel, A. G. 329
Jöreskog, K. G. 223
Judd, C. M. 344
Jude, D. J. 146
Juliano, S. A. 161, 164–5, 167–9, 176–8

Kacelnik, A. 370–1
Kaiser, M. K. 116–7, 120, 131, 133–5
Kalbfleisch, J. D. 254–5, 264, 266–7, 272, 276, 278, 280
Kalisz, S. 233–4, 239–40, 242–4, 249–51
Kaplan, D. 228
Karlin, S. 229
Keirse, M. J. N. C. 378
Keller, W. 155
Kempton, R. A. 2, 324, 335, 359
Kenny, D. A. 344
Keppel, G. 120
Kerfoot, W. C. 80
Kiefer, J. 335
Kim, J. 342, 344, 359
King, B. 132
King, R. B. 132
Kingsolver, J. G. 222, 228
Kling, C. L. 294, 308
Koch, G. G. 133–4
Kohn, J. R. 215
Kokoska, S. M. 133
Kotler, B. P. 362, 373
Krannitz, P. G. 264
Kratz, T. 139, 142, 147, 154–5
Kravitz, D. 305–6
Krebs, J. R. 159, 362–3
Kucera, C. L. 320
Kutner, M. H. 184, 193

Laird, N. M. 156
Lakatos, I. 1, 6, 361, 376
Lambeck, R. H. D. 146
Landa, K. 72, 87
Lande, R. 205, 232
Lanyon, S. M. 294
Larsen, W. A. 37
Lauter, J. 108
Law, R. 74, 234
Lawless, J. F. 254–6, 262, 276, 278–80
Lawton, J. H. 161, 164, 169, 178, 183
Lechowicz, M. J. 2, 47–8, 50, 56, 113–4, 118, 124, 127, 129, 132–5
Lee, C.-S. 47, 57, 65
Lee, E. T. 262, 279, 282

Lefkovitch, L. P. 234
Legendre, L. 347, 351
Legendre, P. 319, 342–4, 346–8, 350–1, 355, 359
Lengyel, I. A. 344, 348, 359
Lenski, R. E. 245, 290
LePage, R. 305
Leslie, P. H. 234
Levins, R. 232
Lewis, H. 232
Lewontin, R. C. 232
Li, C. C. 215, 218–9, 221, 228
Liebhold, A. M. 294
Light, R. J. 383
Likens, G. E. 139
Lindman, H. 9
Linn, M. C. 378
Linsey, G. A. 154
Littell, R. C. 184, 190
Livdahl, T. P. 161, 164, 175–6
Ljung, G. M. 145
Loehlin, J. C. 223–4, 226, 228–9
Loiselle, B. A. 33, 35, 38
Long, J. C. 347–8
Long, J. S. 224
Lucas, H. L. 241
Ludwig, D. 362
Ludwig, J. A. 2, 3, 178
Lynch, E. A. 212, 215–6

MacCullum, R. 222
MacIsaac, H. J. 155
Maddox, G. D. 223, 229
Madenjian, D. P. 146
Magnuson, J. J. 142, 155
Magurran, A. E. 95, 290
Mahoney, D. L. 305
Malley, D. F. 154
Mandeville, G. K. 116, 123
Mangel, M. 362
Manly, B. F. J. 2, 111, 241, 251, 294, 305, 343, 346–8, 351
Mann, K. H. 389, 394
Mantel, N. 277, 343–4, 346–7, 353
Martin, S. C. 320
Matheron, G. 329
Matloff, N. S. 9, 292
Matson, P. A. 139
Maxwell, S. E. 54, 105, 111, 116–8, 123, 134–5
Maynard Smith, J. 363

Mayo, O. 335, 343
McCleary, R. 143, 146
McCoy, E. D. 9
McCreary, N. J. 388, 394
McCulloch, C. E. 2, 3, 159, 161, 164–6, 168–9
McDonald, L. L. 294, 305, 308
McGill, R. 37
McGraw, B. 378
McGraw, J. B. 70
McIntosh, R. P. 319
McLean, R. A. 53, 84, 109, 133
McLennan, D. A. 376
McNamara, J. M. 362
McPeek, M. A. 233–4, 239–42, 249–51
Mead, R. 56, 60, 114, 133, 135, 323–4, 335, 343
Meagher, T. R. 233
Menges, E. 238
Meyer, J. S. 294, 305, 308
Michaels, K. M. 47, 51–2, 55, 59, 64, 114, 116–7, 121, 123, 129, 131, 133–4
Mielke, P. W. 111
Miller, R. G. 303–4
Miller, T. E. 78, 80
Milliken, G. A. 51, 53–4, 60, 66, 84, 102
Mills, K. H. 154
Mitchell, R. J. 212, 215–6, 215, 223–4
Mitchell, W. A. 360–4
Mitchell-Olds, T. 54, 60, 84, 109, 190, 192, 205, 215, 222, 225, 227–9, 300, 308
Moeed, A. 294
Moloney, K. A. 233
Montz, P. 142, 146, 154
Moore, R. 241
Morin, R. 233, 238
Morris, J. D. 97
Morrison, D. F. 100, 111, 120, 128, 132
Morrow, L. L. 16, 359, 378, 384–6, 396
Morton, N. E. 228–9
Mueller, L. D. 294
Mueller-Dombois, D. 388, 394
Muenchow, G. 256, 258, 263, 266, 272–3, 276
Muir, W. M. 133
Munger, J. C. 79
Murdoch, W. W. 139–41, 147, 154, 156
Muthen, B. 228

Naderman, G. C. 343
Naiman, R. J. 233, 238

Nelson, L. A. 343
Neter, J. 102, 161, 165, 169, 176, 184, 193
Ng, D. 241
Noakes, D. 146
Norusis, M. J. 23, 33–4, 120, 134

O'Brien, R. G. 116–7, 120, 131, 133–5
Oden, N. L. 342, 344, 348, 359
Olkin, I. 380, 383, 385, 387, 391, 393
Orcutt, J. D. J. 165
Ord, J. K. 342
Orzack, S. H. 239
Osenberg, C. W. 140, 154

Pacala, S. W. 87, 169
Pallesen, L. 146
Parker, G. A. 363
Parker, K. R. 139–41, 147, 154, 156
Pedhazur, E. J. 212, 221, 228
Perry, J. 142, 155
Peters, R. H. 6, 8, 165
Peterson, A. T. 212
Peto, J. 277–8
Peto, R. 277–8
Philippi, T. E. 184
Pianka, E. R. 4
Pierce, D. A. 145
Pike, M. C. 278
Pillemer, D. B. 383
Plaisier, A. 233, 238
Platt, J. R. 5, 224, 364, 367
Platt, W. J. 388, 394
Pons, T. L. 388, 394
Poole, R. W. 148, 156
Poorter, H. 58
Popper, K. R. 6
Porter, K. G. G. J. 165
Porter, W. R. 178
Potvin, C. 2, 47–8, 50, 56, 62, 66–7, 113–4, 118, 124, 127, 129, 132–5, 294
Powell, T. M. 139
Prentice, R. L. 254–5, 264, 266–7, 272, 276, 278, 280
Price, B. 184, 187, 190–1, 202, 205
Price, M. V. 212, 215–6
Pulliam, H. R. 214, 241, 362
Pyke, D. A. 278
Pyke, G. H. 214

Quinn, G. P. 2, 9, 87, 324, 344
Quinn, J. F. 8, 367

Rao, D. C. 228–9
Rapport, D. J. 363
Raudenbush, S. W. 379
Rausher, M. D. 205
Rawlings, J. O. 47, 57, 65
Reed, D. C. 388–9, 394
Reich, T. 228–9
Reitman, D. 378
Remkes, C. 58
Reynolds, J. F. 2, 3, 178
Rice, J. 228–9
Rice, W. R. 2, 9, 89, 96, 249
Roff, D. A. 294
Rogers, D. J. 163–4
Rohlf, F. J. 11, 21, 33, 37, 221, 228, 344
Rose, W. 142, 155
Rosenzweig, M. L. 362
Roughgarden, J. 6, 7
Rowe, L. 362
Rubin, D. B. 303
Russo, R. 161, 164

Sacks, H. S. 378
Salsburg, D. S. 293
Salt, G. W. 6
Sanders, W. L. 53, 84, 109, 133
SAS Institute Inc. 13, 52–3, 84, 99, 110, 120–1, 127, 133, 150, 156, 165–6, 169, 173, 192, 216–7, 223, 228, 264, 273–4
Satterthwaite, F. E. 84
Savage, L. J. 9
Saville, D. J. 9
Schemske, D. W. 213, 222, 228–9, 233
Schenker, N. 299–300
Schindler, D. W. 138, 154
Schluchter, M. D. 156
Schmidt, F. L. 380
Schoen, D. 47–8, 50, 56
Schoener, T. W. 69–71, 78, 290, 397
Scott, S. J. 263
Searle, S. R. 50–1, 53, 82–3, 91
Sedinger, J. S. 178
Segura, E. T. 370–1
Seheult, A. H. 335
Service, P. M. 245, 290
Sexton, R. J. 294, 308
Shaffer, M. 239
Shaw, R. G. 53–4, 60, 84, 109, 190, 192, 205, 222, 228, 267
Shea, P. J. 6, 9
Shearer, J. A. 154

Sheperd, B. 142, 155
Shrader-Frechette, K. S. 9
Shugart, H. H. 25
Siegel, S. 11, 368–9
Sih, A. 364
Silander, J. A. 87, 169
Silvertown, J. 72, 74
Simberloff, D. S. 6–8
Simes, R. J. 9
Simken, D. 18
Simms, E. L. 111, 205
Simpson, E. H. 95
Sjoberg, S. 164
Slansky, F. J. 205
Slice, D. E. 384
Slim, P. 233
Slinker, B. K. 161, 164–6, 168, 176
Slobodkin, L. B. 183
Smith, E. P. 294–5, 305–6
Smith, F. E. 183
Smith, H. 152, 184, 191
Smith, M. L. 378
Smith, R. J. 373
Smith, W. D. 305–6
Smouse, P. E. 347–8
Snedecor, G. W. 11, 102, 114, 324
Sober, E. 363
Sokal, R. R. 11, 21, 33, 37, 221, 228, 342–4, 353, 347–8, 359
Solbrig, O. T. 290, 294, 296
Southwood, R. 183
Sörbom, D. 223
Stephens, D. W. 159, 362–3
Stevens, J. 96, 105, 108, 111, 116–8, 134–6, 216
Stewart-Oaten, A. 139–41, 147, 154, 156
Stimson, J. 388, 394
Stiven, A. E. 164
Stokes, M. E. 133–4
Strauss, R. E. 161, 163–4, 168
Strong, D. R. J. 7, 183
Stroup, W. W. 53, 84, 109, 133
Suppe, F. 4, 6
Swenson, W. 142, 155
Szyska, L. A. 7

Tamura, R. N. 343
Tanaka, J. S. 228
Tardif, S. 2, 62, 66–7, 113–4, 118, 124, 127, 129, 132–5
Taylor, B. E. 305

Taylor, C. C. 102, 129, 134
Taylor, R. J. 159, 164
Tesar, F. J. 146
Tessier, A. J. 240
Thomas, D. L. 53
Thomas, J. M. 147, 319
Thompson, C. 241
Thompson, D. J. 161
Thompson, J. N. 278
Thomson, B. A. 342, 344, 359
Tiao, G. C. 145
Tibshirani, R. 25, 247, 291, 308–9
Tiebout, H. M. I. 133
Tilman, D. 70–1
Toft, C. A. 6, 9
Trager, W. F. 178
Travis, J. 159, 161, 164–6, 168–9
Trexler, J. C. 159, 161, 164–6, 168–9
Tritton, L. M. 25
Troussellier, M. 344, 348, 359
Tufte, E. R. 14–5, 18
Tukey, J. W. 14, 16, 19, 21, 25, 27, 31, 37
Tuljapurkar, S. D. 239–40, 251
Turner, M. A. 154
Turner, P. R. 167
Turner, T. 339, 394
Turrisi, R. 189

Upton, G. J. G. 342, 352
Urban, D. L. 25

Valone, T. J. 360–1, 363–4, 368
Vandermeer, J. 233
van der Toorn, J. 388, 394
van Es, C. L. 343–4, 347
van Es, H. M. 343–4, 347
van Groenendael, J. 233, 238
van Latesteijn, H. C. 146
van Lenteren, J. C. 161
Vaudor, A. 342, 344, 350–1, 355, 359
Via, S. 240, 251

Wade, M. J. 240, 251
Waite, S. 262, 278
Wallace, A. 16, 359, 378, 384–6, 396
Walsh, J. S. 16, 359, 378, 384–6, 396
Wan, C. K. 189
Ward, D. 364, 376
Ware, J. H. 156
Waser, N. M. 212, 215–6

Wasserman, W. 102, 161, 165, 169, 176, 184, 193
Watkinson, A. R. 74, 87
Watras, C. J. 142, 155
Watt, A. S. 334
Webster, K. 142, 155
Webster, R. 332
Wei, W. W. S. 143, 153
Weiner, J. 290, 294, 296, 300, 308
Weis, A. E. 229
Weis, I. M. 388, 394
Weller, D. E. 25
Welsh, A. H. 212
Werner, E. E. 70
Werner, P. A. 72, 78, 240, 388, 394
West, S. G. 189
Westley, L. C. 183
Wichern, D. W. 111
Wilkinson, G. N. 335, 343
Wilkinson, L. 13, 15–6, 21, 23, 100
Wilks, A. R. 15
Williams, F. M. 161, 164–5, 167–9, 176–8
Williams, J. S. 111
Wilson, S. D. 70
Wilson, S. R. 294

Winer, B. J. 47, 51–2, 55, 59, 64, 114, 116–7, 121, 123, 129, 131, 133–4
Wirtz, W. O. 373
Wolf, L. L. 214
Wolfe, C. L. 229
Woodley, R. 300, 308
Woolfenden, G. E. 239
Wooten, M. C. 344, 348, 359
Wootton, J. T. 229
Wright, S. 212–3, 215, 222, 228–9
Wu, C. F. J. 192
Wynn, H. P. 335

Yamamura, N. 362, 368
Yates, F. 323
Yoccoz, N. G. 2
Young, J. H. 9, 292
Young, L. J. 9, 292

Zar, J. H. 11, 21, 33, 37, 84–5, 91, 126, 292, 301
Zerbe, G. O. 133
Zimmerman, D. L. 323–4, 332, 334–5
Zimmerman, G. M. 111

Subject Index

Accelerated failure-time models (*see* Failure-time analysis)
Aedes triseriatus 175–8
Ailanthus altissima 19–20, 22–4, 295
Analysis of covariance (ANCOVA)
  analogy for Mantel test 358
  competition experiments 84–7
  general linear model 184
  relation to multiple regression 184
  spatial pattern 343
Analysis of variance (ANOVA) (*see also* Assumptions, Fixed effects, Interactions, Mixed models, Multivariate analysis of variance, Random effects)
  assumptions 114, 206, 333–4, 342–4
  collinearity 189–90
  compared with GLS-variogram 326–8
  compared with Mantel test 356–7
  distinguished from multivariate analysis of variance (MANOVA) 94, 96–7, 99
  field competition experiments 81–4
  fixed effects models 52–3
  general linear model 184
  multiple regresion example 199
  ordinary vs repeated measures analysis 114
  principles of 50–1
  protected 96
  random vs fixed factors 52–3
  relation to multiple regression 184
  repeated measures 115–7, 121–4, 134–5
    (*see also* Experimental design, Multivariate analysis of variance, Split plot)
  unmeasured factors 206
Anthesis

failure-time analysis 268–72
life table analysis 264
Apparency, plant 197
Aristida glauca 349, 356–8
Asellus aquaticus 25–6, 169–72
Assumptions
  analysis of variance (ANOVA) 114, 342–4
  autoregressive integrated moving average (ARIMA) models 148
  best linear unbiased estimator (BLUE), in regression models 185
  circularity 115–6, 133
  compound symmetry 116
  concept of 3–5
  demographic analysis 238, 244–5
  fixed effects models, meta-analysis 382
  foraging theory 361
  functional response models 163
  hard core 361, 363, 375–6
  homogeneity of variances 87, 343–4
  hypothesis testing 363–4
  mixed models, meta analysis 382, 393
  independence
    among samples (individuals) 114, 275, 370, 385
    among experiments 384–5
    bootstrapping 309
    spatial autocorrelation 322–3, 342
    randomization 139–40, 344
  linearity, path analysis 228
  multiple regresion 185–6, 192–3
  normal distribution
    convex hulls 27, 30
    failure times 257, 259

meta-analysis 385
multivariate 110
path analysis 216
meta-analysis 383–5, 397
multivariate analysis of variance
    (MANOVA) 97, 109–11
path analysis 216, 227–8
randomized block designs 115–6
repeated measures (MANOVA) 118
reproduction, plant, separate experiments
    212
sphericity 116, 123
split-plot designs 115–6
survival analysis 254–6
unmeasured factors 206
violation
    general consequences 5, 364, 375–6
    functional response analysis 168–9
    revealed by exploratory data analysis
        (EDA) 16
    sphericity 123
Autocorrelation
function 144
genetic 343
spatial
    causes 322, 342
    sample size 334, 346–7
temporal 140, 343
time series (ARMA) processes 143
type I error, resulting in 342
Autoregressive integrated moving average
    (ARIMA) models
ARMA models 143
autoregressive (AR) models, described 143
moving average (MA) models, described
    143
parameter number 144
strengths 143
techniques, time series 139

Bar charts 35
alternatives 35
disadvantages illustrated 38
Bayesian statistics 9
Bias
accelerated correction 301–3
bootstrapping 247, 298, 301
demographic data 233, 241, 251
due to small sample size 387
estimating missing data 54

Greenhouse-Geisser correction (repeated
    measures) 123–4
incomplete principal components (IPC) re-
    gression 201–2
jackknifing 304
measuring, multiple regression 188
modeling functional responses (predation)
    164, 174–5
multiple regression
    coefficients 187, 190, 192
    precision 187
    principal components 191–2
    principal components, omitted (multiple re-
        gression) 191–2
    publication 383
    ridge regression 191, 201–2
Blocking
applicability 66–7
benefits and costs 56–7
demographic data 241
incomplete 66
repeated-measures designs 115
spatial patterns 224–5
Bodo 105–6
Bonferonni correction 9
bootstrapping 249
multiple univariate analyses of variance 96
pairwise comparisons in multivariate analy-
    sis of variance 101–2
profile analysis 128
Bootstrapping
accelerated 301–3
bias corrected 298–9, 301
Bonferonni correction 249
complex designs 242, 245–8
definition of 292
demographic data 233
multiple regression 193
sampling 304
Box-and-wisker plot
alternative to histograms 21
illustrated 23, 39
notched 37
obscuring bimodality 21

Canonical analysis 103–5
canonical correlations and canonical coeffi-
    cients 104–5
related to other procedures 111
Carbon dioxide (CO_2)
atmospheric 379

greenhouses 52, 113
latin square design 58, 62–5
photosynthetic rate 116–8
χ^2 distribution (*see* Goodness of fit)
Centroid
   data set 102
   defined 99
Chaetophractus vellerosus (armadillo) 370–1
Circularity (*see* Assumptions)
Clematis lingusticifolia 273–4
Coefficient of determination (r^2) 185
Coefficient of multiple determination (R^2) 186
Cohort, estimating demographic parameters 242–3
Collinearity (*see also* Correlation)
   analysis of variance (ANOVA) 190
   example of 198, 205
   experimental design 193–4
   problems presented by 185, 189–92
   strong, detecting (multiple regression) 188
   warnings regarding 229
   weak, selection of models 188
Collinsia verna 242
Colpoda 105–8
Competition
   among plants, review of 386, 390, 392, 395–6
   Aristida glauca, field experiment 356–7
   ecological questions 69–71
   effects on plant growth 381
   exploitation 81
   functional responses of predators 161
   general effects, plants in field experiments 396–7
   mechanisms 70–1, 161
   productive vs unproductive habitats (plants) 70, 379
   resource, and fruit set 215
   Stipa neomexicana, field experiment 349, 356–8
Competitive effect
   defined 72
   per-unit measures 72, 78–9, 85, 87
Competitive response
   complex 90
   defined 72
Condition number, multiple regression 190, 198
Confidence interval
   bootstrapped 299–303
   definition of 292

demographic parameters 233, 247–9
Contour plot 35–6
Contrasts
   alternative to repeated-measures analysis of variance 135
   analysis of variance (ANOVA) 324–6
   multivariate analysis of variance (MANOVA) 100–3, 108
   meta-analysis 292–3
   profile analysis 124–9
Convex hulls
   correlations, relationship 27
   illustrated 30
   nonnormality of data 27
   peeled 27, 31
Coreopsis lanceolata 97, 108
Correlation (*see also* Collinearity)
   canonical (*see* Canonical analysis)
   coefficient 185
   direct and indirect effects 218–20
   matrix 215–6, 218, 221
   path analysis 215
   plant traits, between 212, 214–6
   response variables, among 95–6
Cox proportional hazards models 184, 272
Cross validation, in multiple regression 193
Crossover experiments (*see* Experimental design)
Curve fitting, repeated measures designs 132–3
Cyclidium 105–8

Data
   censored 254–5
   long term 139
   heteroscedastic (*see* Transformations, data)
   missing
      analysis of variance (ANOVA) 54, 84
      bias 54
      listwise deletion 217
      multivariate analysis of variance (MANOVA) 109–10
      pairwise deletion 217
      path analysis 216–7
      randomized blocks 60
   smoothing
      techniques 25, 27
      lowess (locally weighted regression) 25–6, 167, 172
      vs conventional regression 27
Deduction, building and testing theory 3–4
Defenses, plant (*see* Herbivory)

Degrees of freedom
  interaction term (multiple regression) 189
  path analysis 223
Demography (*see also* Survivorship)
  analysis, goals 232–3
  "classical" 239
  models 234
  parameters
    calculating 235–6, 244–5
    confidence limits 233
  sampling
    biases 233, 241
    types 234
  stochasticity (*see* Stochasticity, demo-
    graphic)
  uses 232–3
Density plot 19, 21–2
Designs (*see* Experimental design, Unbalanced
  designs)
Differencing (*see* Transformations)
Digitizing, obtaining data from graphs 16, 384
Dipodomys merriami (kangaroo rat) 368–9
Direct effects (*see* Effects, direct and indirect)
Disc equation, Holling 162
Distance matrices
  described, Mantel test 345–6
  constructing 351–3
Dit plots
  alternative to histograms 21
  displaying bimodality 21
  illustrated 22
Diversity (*see* Index, diversity)

Ecology (*see also* Experiments, Unbalanced designs)
  applied 378–9
  behavioral 159, 361–2
  experiments 4–5, 397–8
  intervention analysis 146
Effect size
  estimated average, fixed effects model 386–
    7
  grand mean 390–1
  magnitude, defined 380–2
  path analysis, pollination experiments 212,
    222
Effects, direct and indirect
  competition experiments 80–2
  separating, path analysis 213–5, 217–9,
    221–2

Eigenvalue (*see also* Multivariate analysis of
  variance)
  condition number, multiple regression 191–2
  definition of 100
  demographic analysis 235, 237–40, 251
  dominant 235
  principal components, example 198
Eigenvector 98, 100
Elasticity analysis 233, 237–8
Emergence, seedling
  failure-time analysis 270–1
  life table analysis 260, 263–4
Empirical semivariogram (*see also* GLS-variogram) 331
Envelope effect 87–9
Eriogonum albertianum 256, 276
Error (*see also* Residuals, Type I error, Type
  II error)
  bars 37, 40
  distribution 233
  population sampling 238–41
Euclidean distance, Mantel test 352
EXCEL spreadsheet (Microsoft), in meta-analysis 384, 388–9, 394
Experimental design
  before-after-control-impact (BACI) 141–2
  censored data 255–6
  classical 55
  completely randomized 55–6, 322–3
  crossover experiments 133
  latin square 56, 58, 61–2
  paired 141
  partially hierarchical 373–4
  randomized block 56–60, 114–5
  split plot 62–6, 114–5, 120–3
  unbalanced (*see* Unbalanced designs)
Experiments
  design (*see* Experimental design)
  disadvantages 184, 211–2
  enclosure 138
  field, manipulating density 72–4, 178
  intervention 141–2, 145–6, 155–6
  manipulative and observational 4–6, 212
  natural 4–5, 141
  observational study, after 226
  pilot 222
  pollination 211
  unreplicated 140
  strong inference 5, 212, 364, 367
Exploratory data analysis (EDA)
  definition of 14

model building, multiple regression 193,
197
violation of assumptions, revealing 16
Exponential distribution 281–2
Exponential semivariogram 332 (*see also*
GLS-variogram)
Extinction, time to 239
Extrapolation
experimental results 238–9
population dynamics 240
research review 383

F-statistic
analysis of variance (ANOVA) 51, 53
fixed, random, and mixed models 52–3
repeated-measures analysis of variance 123–
4
Factor analysis, limitations 212
Failure-time analysis
accelerated failure-time models 266–8
data
types 254–6
presentation of 275–6
definition of 254
life table analysis 259–62, 272
proportional hazard models 266, 272
regression analysis 264
Fixed effects models
analysis of variance (ANOVA) 52–3
meta-analysis
described 382
cumulating effect sizes 387, 390
Foraging
giving up density (GUD) 369, 373
giving up time 255
marginal value theorem 367–8, 368
optimal 363
patch use 364–5
theory 159, 214, 360–1
Fruit set, path analysis 215
Functional response (predators)
comparing 161, 176
distinguishing 165, 167
type I, description 159–60
type II
description 159–60
indicated by data analysis 177
models 162–4, 176
type III
description 159–60
estimating parameters 173–4

example 171–2
models 162–4
revealed, logistic regression 172–3
suggested by lowess 26, 167, 172
Fuzzygrams
alternative to histograms 21
described 21
displaying bimodality 21
illustrated 23

Gamma distribution
definition of 281, 284–5
of residuals 87, 169
General linear models (*see* Models, analysis of
variance)
Gini coefficient
bootstrap analysis of 295–9
definition of 294
jackknife analysis of 303–5
Global change 381
GLS-variogram
assumptions 333–4
calculated 331–3
compared with analysis of variance
(ANOVA) 326–8
compared with Mantel test 334
definition of 328
illustrated 332
mathematical model 336–7
method 328–33
Goodness of fit
meta-analysis 391–2, 396
path analysis 223–8
Greenhouse-Geisser correction 123–4, 135–6
Growth curves
curve fitting 132
multivariate analysis 127–9
nonlinear least squares 133, 178
profile analysis 124–7

Hazard function 261, 272
Helianthus annuus (sunflower) 33, 35–6, 61–2
Herbivory
analysis of 184, 197–202
interpretation of experimental results 203–5
missing values, causing 216
plant defenses and loss rates 186, 197
structural defenses 183, 197, 203–4
interaction with nutrient levels 201, 204–
5
variation in losses 183

Heterogeneity
    controlled environments 46–50
    field experiments, spatial 319, 342–3
Heteroscedastic data (*see* Transformations,
        data)
Histograms
    alternatives (*see* Box and wisker plots, Dit
        plots, Fuzzygrams, Jitter plots, Stem-
        and-leaf plots)
    bar charts vs 19
    bimodality, displaying 21
    illustrated 20
    problems 19, 21
Homogeneity of variances (*see* Assumptions)
Homogeneity statistic (*Q*) 391–2, 396
Hotelling-Lawley trace (*see* Multivariate analy-
        sis of variance)
Hotellling's T^2 (*see* Multivariate analysis of
        variance)
Huynh-Feldt adjustment 122–4, 136
Hypotheses
    a priori vs a posteriori 15, 101
    falsification 4
    functional responses, predators 161, 176
    null
        functional response 165
        meta-analysis, fixed effects model 391
        multiple regression 192–3
        null model, vs 7
        Mantel test 344, 347
    path analysis
        alternative, multiple 213
        causal 212–4
        first, pollination study 211
        testing 222
        working 224
    process of testing 4–7, 202–3, 363–4
    profile analysis 120, 124–6
    reformulation 364, 375
    testing specific 378

Independence (of observations) (*see* Assump-
        tions)
Index
    diversity 95, 290
    similarity 290, 294–5, 305–7
Indirect effects, competition experiments 80–2
Induction, building and testing theory 3–4, 6–
        7
Inference, strong 5, 212, 364, 367
Influence plots

    correlation 27
    illustrated 28–9
    outliers 27
Interaction, statistical
    blocks 59–60
    competition experiments 70–1
    detecting, herbivory experiments 184
    displaying 33, 35–6
    multivariate analysis of variance
        (MANOVA) 105–8
    mean squares (MS) for 51–3, 59–60
    multiple regression 186, 188–9, 193, 200–1
    plant reproduction 212
    problems presented 186
    profile analysis 119, 124–8
    repeated measures
        analysis of variance (ANOVA) 212–2
        factorial experiments 130–1
    testing, using Mantel test 353
    time 66–7
    treatment × habitat 82–3
Ipomopsis aggregata
    clumping 214
    correlation matrix 216
    fruit set, causes 213–5
    path analysis example 220–1
    pollination 211
Irradiance, growth chambers 47–8

Jaccard index (*see* Index, similarity)
Jackknifing (*see also* Randomization, tests)
    bias 304
    definition of 303–4
    independence of samples 305–6
    path coefficients 227
Jitter plot
    alternative to histograms 19
    bimodality, obscured 21
    described 19, 21
    illustrated 22

Keratella taurocephala 142, 146–7, 149, 154–
        5

Lag
    distance 331
    time series (ARIMA models) 142–5
    moving average (MA) and autoregressive
        (AR) models 143
Lagrange multiplier test 268–9
Lakes

acidification 138–42, 150
Keratella abundance 149, 153
Latin square design 61–2
  displaying results 33, 35
  illustrated 58
  randomized block design, vs 61
  spatial pattern 343
Leaves (*see also* Herbivory)
  damage to 197
  toughness 198–9
Life cycle graph 234, 236
Life expectancy 255
Life history stages, responses measured 75
Life table analysis (*see* Failure-time analysis)
Linear models, mixed 52–3, 132–3
Linearity (*see* Assumptions)
Log-logistic distribution 281, 284, 286–7
Lognormal distribution
  definition of 281, 284, 286
  species abundances, estimating parameters
    178
Log-rank test
  life table analysis 259, 261–3
  N samples 278–9
  two samples 277–8
Lotka-Volterra models, competition 77
Lowess (*see* Data smoothing)

Mantel test
  compared with GLS-variogram 334
  description 344–7
  null hypothesis 346
  partial (3 matrix)
    uses 347–8
    calculations 357
  statistical significance, determining 345–6,
    355–6
  uses 344
Matrix
  algebra 235
  correlation 216–8
  inversion, and missing values 217
  Lefkovitch 234
  Leslie 234
  nonsymmetrical 251
  projection analysis 234–40
Maximum likelihood estimation
  analysis of variance (ANOVA) 51, 53–4, 84
  failure-time analysis 267, 269, 279–80
  functional response analysis 166

multivariate analysis of variance
    (MANOVA) 109–10
  mixed models, restricted (REML) 53–4, 134
  repeated measures 134
  unbalanced univariate designs 54, 84
Mean squares (MS)
  analysis of variance (ANOVA) 51
  erroneously reduced 65–6
  error, experimental design 91
  interaction 51–3, 59–60
  repeated-measures analysis of variance 122
Meta-analysis
  data
    criteria for 382–6
    obtaining from graphs 16
  definition of 378
  fixed effects models
    described 382, 386–7
    cumulative effect sizes 387, 390
  mixed effects models 382, 393, 395–6
  multiple comparisons 390, 392, 395–6
  standard deviation required 40
Michaelis-Menten kinetics, functional response
    models 164
Missing values (*see* Data, missing)
Mixed models
  analysis of variance (ANOVA) 52–3, 84
  controversies, analysis of 53
  described, meta-analysis 382
Model (*see also* Mixed models, Fixed effects
    models, Null model)
  building
    auxiliary assumptions 361–2
    correcting misspecification 186–8
    example 197–202
  data
    agreement with, path analysis 222
    fitting 169–75
  general linear 183–4
  identification, path analysis 223–4
  Markovian, population dynamics 239
  nested, path analysis 225–7
  parsimony 187
  refutation 364, 375–6
  selection 161, 363–4
Mortality rate, unconditional 261
Multicollinearity (*see* Collinearity)
Multiple comparisons 279 (*see also* Bonfero-
    nni correction, Meta-analysis)
Multivariate analysis of variance (MANOVA)
  assumptions 110–1

distinguished from analysis of variance (ANOVA) 94
greatest characteristic root (*see* Roy's greatest root)
Hotelling-Lawley trace 100
Hotelling's T^2 102, 124, 126
missing data 109–10
Pillai's trace 100, 110
post-hoc tests 100–5
power 108–9, 132
repeated-measures designs
  described 117–20
  power 118
  profile analysis 124, 126–9
  recommendations concerning 134–6
  sample size 118
Roy's greatest root 100–3
Wilk's lambda 100, 103

Nearest neighbor
  designs 343
  distance 214, 224–5
Nespera aquatica 35, 38–9
Newton's method (root of implicit equation) 167–8, 174
Nonparametric statistics (*see also* Randomization, Mantel test, Wilcoxon test)
  confidence intervals, bootstrapped 302–3
  functional responses 169, 178
  sign test 369
Normality (*see* Assumptions)
Notonecta glauca 25–6, 169, 172
Null hypothesis (*see* Hypothesis, null)
Null model, use 7–8

Optimal foraging theory (*see* Foraging)
Orthogonal polynomials 129–30
Orthonormal variables 116
Outliers, searching for 27–9
Overidentification (*see* Path analysis)

Panama, Barro Colorado Island, trees 197
Path analysis
  assumptions (*see* Assumptions, path analysis)
  complicated 219–20
  description 121–3
  features, useful and beautiful 219–20
  interpretation of 221–2
  multiple regression 212
  overidentification 224

path coefficients 218, 220–1
path diagram 213–5
statistical control 212
Perognathus amplus (pocket mouse) 373–4
Permutation tests, statistical significance in
  Mantel test 346–7, 355–6
Phenols 197, 200 (*see also* Herbivory)
Pie charts 35
  alternatives to 35, 37
  disadvantages illustrated 38
Pillai's trace (*see* Multivariate analysis of variance)
Pilot experiments (*see* Experiments, pilot)
plant-insect interactions
  ecological importance 183
  repeated-measures experiments 132
Plantago major 33–5, 58–9
Polar category plots 37, 39
Pollination
  approach rate 214, 217–9, 224
  complications in studying 211–2
  crossover experiment 133
  effects of gender 273
  timing, analysis of 273
  visitation rate 216
Population
  comparison (*see* Failure-time analysis)
  dynamics 239–40
  growth rate
    descriptor 239
    predicting 232–3
  projection matrix 234, 236
  sampling 233, 238, 241
Post-hoc tests, repeated measures 117 (*see also* Multivariate analysis of variance)
Power, statistical
  analysis of variance (ANOVA) 91
  herbivory, experimental design 184
  multivariate analysis of variance (MANOVA) 108–9, 132–3
  meta-analysis 380
  multiple regression 193–4, 197
  repeated-measures analysis of variance 118, 135
  type II error 8–9
Precision
  bias, multiple regression 187
  definition of 292
  IPC regression 191, 205
  multiple regression coefficients, estimating 187

Predation, density dependence 159 (*see also* Foraging theory, Functional response, Herbivory)
Prey density
  Asellus aquaticus 169
  constant, in models 163
  functional response, predator 159
Principal components
  collinearity, eliminating in multiple regression 191–2
  multiple regression example 199–201
Probability level
  basis for 5
  choice of 8–9
  statistical power 8–9
Probability plots
  normal 21
  illustrated 24
  profile analysis
    repeated-measures designs 120, 124–9
    relation to other approaches 111
Proportional hazard model (*see* Failure-time anlaysis)
Pseudoreplication 2, 55, 140
Publication bias (*see* Bias)

Quadrats
  bootstrapping data 246
  permanent, demography 244

Rana sylvatica (wood frog) 132
Raphanus sativus (radish) 257
Random effects
  analysis of variance (ANOVA) 51–2
  models 52–3
Randomization (*see also* Bootstrapping, Jackknifing)
  experimental design 55, 138, 320, 335
  tests, multivariate designs 111
  time series analysis 154
Randomization intervention analysis (RIA) 154
Randomized block designs
  analysis of 59–60
  definition of 57–9
  spatial pattern 334–5, 343
Regression
  best linear unbiased estimator (BLUE) 185
  envelope effect 87
  failure times 264, 266–8, 272
  incomplete principal components (IPC) 191–2, 201–2, 205

isotonic 89
linear vs nonlinear 86–7, 164–5
logistic 161, 169, 178, 184
multiple
  backward selection 188
  description 184–6
  disadvantage 88, 212
  forward selection 188
  path analysis and 212, 218
  statistical control 88, 212, 218
  stepwise 188
nonlinear least squares 161
nonparametric 169
ordinary least squares, described 185
ridge 191, 202
univariate, problems 206
Replication
  competition experiments, multiple densities 79
  experimental design 60, 156
  experimental error 139–40
  experiments 378
  growth chambers 62
  predation experiments 167
  repeated-measures experiments 118, 135
  sites 81
  species 79–80
  time 66–7
  treatments, hindered by cost and scale 62, 138
Residuals (error) (*see also* Gamma distribution)
  ARMA (time series) models 143
  establishing linearity 185
  fitting 150
  misspecification errors, diagnosing 186
  normality, path analysis 216, 227
  plotting 169, 174
Review
  extrapolation from data 383
  literature, narrative 379–80
Ridge regression (*see* Regression, ridge)
Rotifers 138, 141
Roy's greatest root (*see* Multivariate analysis of variance)

Sample size (*see also* Power)
  bootstrapping 247
  failure time 258
  multiple regression 194

multivariate analysis of variance
(MANOVA) 108–9, 118
path analysis 216
reporting 11, 37, 40, 382
spatial autocorrelation 334, 346
statistical significance 380
Sampling
design 241, 251
life history 233–4
horizontal (cross-sectional), demographic
234
population 238
vertical (longitudinal), demographic 234
Scale, experimental 138–9
Scatterplot
bivariate data, exploring and presenting 25,
27, 31
illustrated 26, 28–9, 32
symmetrical matrix 31
three dimensional 36
Seed bank
bootstrapping data 245–8
Collinsia verna 242–5
experimental 244
genotypes 250
life cycle with 242–3
Seedling emergence (*see* Emergence, seedling)
Seed production
Ipomopsis aggregata 213–5
pollination experiments 212
Selasphorus platycercus 211
Selasphorus rufus 211
Selection, natural
demography 240
estimating 205
Sensitivity analysis, described 233, 237
Signal-to-noise ratio, in data 16
signficance, statistical (*see also* Type I error)
"data dredging" 222
Mantel test 346–7, 355–7
sample size 366, 380
tests, compromise of 213
Similarity index (*see* Index, similarity)
Simpson's paradox 95
Size hierarchies (*see* Gini coefficient)
Spatial pattern (*see* Assumptions, Blocking,
GLS-variogram, Mantel test)
Spermophilus tereticaudus (ground squirrel)
368–9
Sphericity (*see* Assumptions)

Split plots (*see* Experimental design)
Spread vs level plot 33–4
Squirrels, diet 113–4
Stable age distribution, calculating 232
Stable stage distribution
achieving 239
calculating 232
defined 235
Standard deviation
bivariate ellipse 27, 29
calculation, from published data 383–4
effect size 381
error bars 37–40
Standard error (of the mean)
bivariate ellipse 27, 29
definition of 37, 292
error bars 37–40
Stationary process (time series) 143–4, 148,
150, 239, 241
Statistical
control 212
power (*see* Power, statistical)
interactions (*see* Interactions, statistical)
Stem-and-leaf plot
alternative to histograms 19, 21
disadvantages 19
displaying bimodality 21
illustrated 22
Stipa neomexicana 349–50, 356–8
Stochasticity, demographic
causes 238–41
variation in vital rates 248–9
Structural equation modeling 223, 227, 229
Sums of squares (SS)
analysis of variance (ANOVA) 51–2
Types I - IV, SAS 54, 84, 190
Survivorship (*see also* Failure-time analysis)
competition, affected by 379
curves
comparing using logistic regression 178
determining shape 267
function 261
mechanisms affecting 232–3

t-test 102, 126, 140, 154, 185, 371
Tannins 197, 199–200 (*see also* Herbivory)
Theories, scientific
building 3–4, 361
confirmation 6–7
falsification 6, 363–4

nature of 3, 361–2
testing 4–5, 363–4
Timing (*see* Failure-time analysis)
Toxorhynchites rutilus 175, 177
Transformations, data
  contrasts, profile analysis 126, 128
  differencing 144
  heteroscedastic data 33, 87, 372
  log
    displaying outliers 27, 29
    heteroscedastic data 33, 35, 372
    ratio series 147–8
  multiple regression 186, 198
  negative exponential 33–5
  orthogonal 161
  path analysis 216, 227
  time series analysis 148
Trend analysis 129, 343
Trends
  deterministic 143
  random 143
  seasonal and long term 144
Type I error (*see also* Bonferonni correction,
    Significance, statistical)
  autocorrelation, spatial 342
  bootstrapping tests 248–9
  experimental design 7
  explained 8–9, 371
  inflation 9–10
    multiple univariate analyses of variance
      95
    repeated-measures analysis of variance
      135–6
  split-plot designs 65–6
Type I functional response (*see* Functional re-
    sponse)
Type II error (*see also* Power, statistical)
  experimental design 7
  explained 8–10
Type II functional response (*see* Functional re-
    sponse)
Type III functional response (*see* Functional re-
    sponse)

Unbalanced designs
  analysis 52–4
  ecological experiments 83–4
  multivariate analysis of variance (MA-
    NOVA) 109–10
  restricted maximum likelihood estimation
    (REML) 53

Variables
  continuous, relationships 21, 25
  dependent
    influences 220–1
    relationships 212
    and independent, path analysis 217–8
    unexplained influences 221
  inclusion, path analysis 214
  omitted, path analysis 225–7
  sample size and path analysis 216
Variance
  conditional and unconditional, meta-analysis
    387, 390, 396
  inflation factor, multiple regression 190
Variation
  demographic parameters 233
  vital rates
    genetic, spatial 240–1, 250
    temporal 239, 248–9
Variogram 328 (*see also* GLS-variogram)
Vote counting, meta-analysis 380

Weibull distribution 281–3
Wilcoxon test
  life table analysis 262
  N samples 278–9
  two samples 278
Wilk's lambda (*see* Multivariate analysis of
    variance)

zooplankton (*see also Keratella taurocephala*,
    Rotifers)
  abundance 142
  community similarity 305–7